Investing in Climate, Investing in Growth

OECD
BETTER POLICIES FOR BETTER LIVES

Please cite this publication as:
OECD (2017), *Investing in Climate, Investing in Growth*, OECD Publishing, Paris.
http://dx.doi.org/10.1787/9789264273528-en

ISBN 978-92-64-27351-1 (print)
ISBN 978-92-64-27352-8 (PDF)

Revised version, June 2019
Details of revisions available at:
http://www.oecd.org/about/publishing/Corrigendum-Investing-in-Climate-Investing-in-Growth.pdf

Photo credits: Cover © Baseline Arts

Corrigenda to OECD publications may be found on line at: *www.oecd.org/about/publishing/corrigenda.htm*.

Foreword

Governments around the world are striving to re-ignite growth in their economies while reducing widening inequalities. At the same time, they are working hard to implement the climate goals agreed by the global community under the Paris Agreement. These challenges are not mutually exclusive. We have a unique window of opportunity to bring the climate and economic growth agendas together and to generate inclusive economic growth in the short term, while ensuring that we meet the climate challenge in the longer term.

Investing in Climate, Investing in Growth *lays out the case for governments to pursue an integrated policy approach that combines climate action with fiscal initiatives and structural reforms. It is clear from the report's analysis that countries can achieve strong and inclusive economic growth while reorienting their economies towards development pathways with low greenhouse gas emissions and high resilience to the effects of climate change. The report sees potential to increase long-run output by up to 2.8% on average across G20 countries in 2050, with a net effect of nearly 5% if mitigated climate impacts are taken into account. Importantly, growth impacts are positive in the near-term too: the report sees potential for a net GDP effect of around 1% for G20 economies by 2021.*

However, it is also increasingly clear that meeting the Paris Agreement's goals will require countries to step up ambition, enhance co-operation across borders and strengthen domestic policies and implementation on the ground as a matter of urgency. Moreover, there is a need for governments to take immediate action. The decisions that we take now on key issues such as infrastructure and the structure of our economies will be critical in ensuring a longer term future that enhances rather than diminishes well-being. Proactive, forward-looking policies to facilitate a just transition for affected businesses and households will also be vital to ensure that reform is inclusive, progressive and good for business, particularly in vulnerable regions and communities.

This report has been produced in the context of the German G20 Presidency, which has placed climate squarely on the G20 agenda in recognition of the fact that the growth and climate agendas are mutually supportive. Indeed, adopting an inclusive, low-emission and climate-resilient growth agenda is an opportunity to reorient G20 growth objectives as the group's 2014 Brisbane commitment to 2% growth nears its end in 2018.

The OECD is supporting countries, developed and developing, to create more effective policy approaches to address the growth, inclusiveness and climate challenges in a holistic way. For OECD member countries, it will not be easy to achieve the transition from carbon-intensive to low-emissions economies while seeking to re-ignite growth. For partner economies, the challenge is to avoid locking in emissions-intensive development paths while pursuing growth and development opportunities. Our report shows that there are significant benefits to an integrated approach to the climate and growth challenge.

Angel Gurría
OECD Secretary-General

Acknowledgements

This report was prepared under the supervision of Anthony Cox (OECD Environment Directorate) and Andreas Schaal (OECD Sherpa Office and Global Relations Secretariat), supported by a core team of Andrew Prag, Justine Garrett, Richard Baron, Galina Alova, Sarah Finke, Elizabeth Del Bourgo and Makeda Yohannes (OECD Environment Directorate) and Roger Dubach, Nicolas Pinaud and Nejla Saula (Sherpa Office). Gabriela Ramos (OECD Sherpa and Chief of Staff), Catherine Mann (OECD Chief Economist) and Simon Upton (OECD Environment Director) provided strategic guidance to the project.

The OECD would like to thank the German Federal Ministry for the Environment, Nature Conservation, Building and Nuclear Safety (BMUB) for providing financial support to the project. A number of staff from the BMUB provided guidance and comments on the report including Norbert Gorißen, Carolin Zerger and Claudia Keller.

The report was authored by teams drawn from a number of OECD directorates: Environment Directorate (ENV); Economics Department (ECO); Directorate for Financial and Enterprise Affairs (DAF); Development Cooperation Directorate (DCD); Trade and Agriculture Directorate (TAD); Centre for Tax Policy and Administration (CTP) and the International Transport Forum (ITF). The Directorate for Public Governance also contributed. Simon Buckle (ENV) supervised the preparation of Chapters 2 and 3, Andrés Fuentes-Hutfilter and Christian Kastrop (ECO) supervised the preparation of Chapter 4, and Ana Novik (DAF) supervised the preparation of Chapters 5, 6 and 7.

The authors for the individual chapters were:

Chapter 1	Lead author was Andrew Prag with contributions from Galina Alova, Richard Baron, Anthony Cox, Roger Dubach, Justine Garrett, Andreas Schaal, Simon Upton and the other chapter authors below.
Chapter 2	Lead authors were Simon Buckle and Michael Mullan, with contributions from Jo Cadilhon, Jennifer Calder, Robert Dellink, Guillaume Gruère, Marit Hjort, Frank Jesus, Virginie Marchal, Mariana Mirabile, William Wright, Ziga Zarnic, and staff from the Met Office Hadley Centre for Climate Science and Services.
Chapter 3	Lead authors were Virginie Marchal and Mariana Mirabile, with contributions from Richard Baron, Simon Buckle, Jo Cadilhon, Jenny Calder, Naeeda Crishna Morgado, Marit Hjort, Bérénice Lasfargues, Ziga Zarnic, and staff from Grantham Institute, Imperial College London.
Chapter 4	Lead authors were Andrés Fuentes-Hutfilter and Annabelle Mourougane, with contributions from Shardul Agrawala, Jarmila Botev, Rob Dellink, Dennis Dlugosch, Jean-Marc Fournier, Olivier Durand-Lasserve, Christian Kastrop and Tomasz Kozluk.
Chapter 5	Lead authors were Andrew Prag and Enrico Botta, with contributions from Geraldine Ang, Johanna Arlinghaus, Richard Baron, Jo Cadilhon, Kurt Van Dender, Luisa Dressler, Florens Flues, Franck Jesus, Michael Mullan, Ana Novik, Jehan Sauvage, Ronald Steenblik, Cristina Tebar-Less and Nicolas Wagner.
Chapter 6	Lead author was Richard Baron with contributions from Gianpiero Mattera, Filipe Silva, Ronald Steenblik, Ziga Zarnic, staff from E3G and the Just Transition Centre, and Genevieve McInnes.
Chapter 7	Lead authors were Joel Paula, Naeeda Crishna Morgado, Raffaele della Croce and Berenice Lasfargues, with contributions from Abderrahim Assab, Wiebke Bartz, Flore-Anne Messy, Ana Novik, Haje Schutte, and Guillaume Simon.

Particular thanks are expressed to the members of the project's Advisory Council for their very helpful review and comments:

- Isher Ahluwalia (Chairperson, Board of Governors, the Indian Council for Research on International Economic Relations (ICRIER))
- Luis Alberto Moreno (President, IDB) and his team (Amal Lee-Amin)
- Kate Brandt (Lead for Sustainability, Google) and Francois Sterin (Director, Global Infrastructure, Google)
- Sharan Burrow (General Secretary, ITUC)
- Ottmar Edenhofer (Deputy Director and Chief Economist, PIK)
- Justin Lin (Director, Center for New Structural Economics; Professor and Honorary Dean National School of Development, Peking University)
- Ajay Mathur (Director General, TERI)
- Axel Ockenfels (Professor, University of Cologne)
- Ngozi Okonjo-Iweala (Former Managing Director, World Bank; Chair of the Board, Global Alliance for Vaccines and Immunization (GAVI) and African Risk Capacity (ARC))
- Paddy Padmanathan (CEO and President, ACWA Power) and David Ungar (Executive Director, ACWA Power)
- Robert N. Stavins (Albert Pratt Professor of Business & Government, Harvard Kennedy School)
- Nicholas Stern (Chair, Grantham Research Institute on Climate Change and the Environment)
- Jean-Pascal Tricoire (CEO, Schneider Electric)
- Laurence Tubiana (Former French Ambassador for Climate Change; Professor, Sciences-Po Paris; CEO of the European Climate Foundation).

The OECD would also like to thank the International Energy Agency for its helpful input and comments on the report, in particular Laura Cozzi and Timur Guel, along with Pierpaolo Cazzola, Christophe McGlade, Jacob Teter, Michael Waldron and Brent Wanner. The International Renewable Energy Agency also provided helpful comments.

In addition, the OECD thanks the project's Knowledge Partners for their contributions to and comments on the report:

- Climate Disclosure Standards Board (CDSB)
- COPPE at Universidade Federal do Rio de Janeiro (UFRJ) (Brazil)
- Development Research Center of the State Council of China (DRC, China)
- E3G
- European Development Finance Institutions Association (EDFI)
- Fiel (Argentina)
- Fondazione ENI Enrico Mattei (FEEM)
- Imperial College London
- Institute for Sustainable Development and International Relations (IDDRI) Deep and the Decarbonization Pathways Project (DDPP)
- International Development Finance Club (IDFC) Secretariat, KfW
- Just Transition Centre of the International Trade Union Confederation
- King Abdullah Petroleum Studies and Research Center (KAPSARC, Saudi Arabia)
- Met Office, Hadley Centre for Climate Science and Services (United Kingdom)
- Observer Research Foundation (ORF, India)

Thanks must also go to the following individuals for their input and comments: Amar Battacharya, Brookings Institute; Emmanuel Guérin, European Climate Foundation, Joeri Rogelj, International Institute for Applied Systems Analysis (IIASA); Ted Nace, Global Coal Tracker; Guineng Chen and Jari Kauppila, ITF; Dale Rothman, Pardee Centre; and David Blandford, Penn State College of Agriculture Science. Thanks also to the many OECD colleagues who provided helpful comments including Dorothée Allain-Dupré, Ruben Bibas, Timothy Bishop, Jonathan Brooks, Jean Chateau, Emilio Chiofalo, Jan Corfee-Morlot, Jane Ellis, Luis-Mario Hernandez, Kaori Miyamoto, Paul O'Brien, Giuseppe Nicoletti, Jens Sedemund and David Turner.

The many comments and suggestions provided by the following committees are gratefully acknowledged:

Committee for Agriculture, Committee for Scientific and Technological Policy, Committee on Financial Markets, Committee on Fiscal Affairs, Committee on Industry, Innovation and Entrepreneurship, DAC Network on Environment and Development Co-Operation, Development Assistance Committee, Employment, Labour and Social Affairs Committee, Economic Policy Committee, Environment Policy Committee, Insurance and Private Pensions Committee, Investment Committee, Joint Working Party on Agriculture and Environment, Joint Working Party on Trade and Environment, Joint Meeting of Tax and Environment Experts, Senior Officials from Centres of Government, Trade Committee, Working Party No. 1 on Macro-Economic and Structural Policy Analysis, Working Party on Climate, Investment and Development, Working Party on Integrating Environmental and Economic Policies.

The OECD is also grateful to the participants of the different workshops:

- OECD Workshop on "*Financing green infrastructure*", Paris, 3 November 2016
- OECD Roundtable on "*Scaling up climate investment through National Development Banks (NDBs), Multilateral Development Banks (MDBs), and Development Finance Institutions (DFIs)*", Casablanca, Morocco, 5 November 2016
- OECD Workshop on "*Make the economic case for the Paris Agreement: Modelling the macroeconomic consequences of climate change action*", Paris, 24-25 November 2017
- OECD Workshop on "*Low-emission and climate-resilient development pathways*", Paris, 11 January 2017
- Roundtable discussion on "*The role of national development banks, multilateral development banks and bilateral development finance institutions in scaling up investment in low-carbon and climate-resilient infrastructure*", Paris, 22 February 2017
- Workshop co-organised by ORF and the OECD on "*Growth, Investment and the Low-Carbon Transition in India*", Delhi, 7 March 2017
- Workshop co-organised by DRC and the OECD on "*Investment and the Low-carbon transition in China and the world*", Beijing, 12 April 2017

The report was prepared for publication under the direction of Elizabeth Del Bourgo from the OECD Environment Directorate, with editorial work being undertaken by Andrew Johnston (Words for Change), layout by Pierre-Yves Perez-Queyroi (Pykha) and cover design by Andrew Esson (Baseline Arts Ltd). Thank you to OECD Public Affairs and Communications Directorate colleagues Janine Treves for editorial and content support, and Catherine Bremer for media relations support. Production support was provided by Stéphanie Simonin-Edwards of the OECD Environment Directorate.

Table of contents

Tables

Figures

Executive summary

Achieving a growth path that is resilient, inclusive and sustainable is one of the top policy priorities of our time. Governments around the world are facing the triple imperatives of re-invigorating growth while improving livelihoods and urgently tackling climate change, in line with the goals of the Paris Agreement. This report argues that boosting economic growth, improving productivity and reducing inequalities need not come at the expense of locking the world into a high-emissions future. It is the *quality* of growth that matters.

With the right policies and incentives in place – notably strong fiscal and structural reform combined with coherent climate policy – governments can generate growth that will significantly reduce the risks of climate change, while also providing near-term economic, employment and health benefits. Such a climate-compatible policy package can increase long-run GDP by up to 2.8% on average across the G20 in 2050 relative to a continuation of current policies. If the positive impacts of avoiding climate damage are also taken into account, the net effect on GDP in 2050 rises to nearly 5% across developed and emerging economies of the G20.

Investment in modern, smart and clean infrastructure in the next decade is a critical factor for sustainable economic growth, especially as infrastructure generally has suffered from chronic underinvestment since before the financial crisis. The report estimates that USD 6.3 trillion of investment in infrastructure is required annually on average between 2016 and 2030 to meet development needs globally. An additional USD 0.6 trillion a year over the same period will make these investments climate compatible, a relatively small increase considering the short and long-term gains in terms of growth, productivity and well-being. The additional investment cost is likely to be offset over time by fuel savings resulting from low-emission technologies and infrastructure.

Furthermore, the current fiscal environment provides a window of opportunity to take action now. Low interest rates have increased fiscal space in many countries and, where there is less fiscal space, opportunities exist to optimise the tax and spending mix to align stronger economic growth with inclusive, low-emission, resilient development. Well-aligned climate, fiscal and investment policies will further maximise the impact of public spending to leverage private investment.

Finance will be a key factor: capital must be mobilised from both public and private sources, supported by a variety of financial instruments tuned for low-emission, climate-resilient infrastructure. Public financial institutions need to be geared for the transition, while the financial system itself should take greater steps to correctly value and incorporate climate-related risks. Development banks and finance institutions – multilateral, bilateral and national – all have a critical role to play here too, not only using their balance sheets to amplify available resources, but also developing green finance in partner countries, including through policy and capacity building support.

Getting the fundamental climate policies right is essential to aligning incentives. There is a need to accelerate the reform of inefficient fossil-fuel subsidies and broaden the carbon pricing base, focusing on tracking the impact and sharing policy experiences. Making greater use of public procurement to invest in low-emission infrastructure can trigger industrial and business model innovation through the creation of lead markets.

At the same time, we must recognise that sustainable growth also means inclusive growth. Coherent climate and investment policies, effective fiscal and structural policy settings and reforms must work together to facilitate the transition of exposed businesses and households, particularly in vulnerable regions and communities. Early planning for the transition is essential if societies are to avoid stranded assets in fossil-fuel-intensive industries and stranded communities alongside them.

Looking beyond energy production and use, developments in agriculture, forestry and other land-use sectors will enable scaling up the pace of the transformation needed elsewhere in the economy. Current stocks of carbon in tropical forests and other ecosystems need to be protected and their ability to act as carbon sinks enhanced wherever possible. Research and development needs to be significantly strengthened and followed by rapid demonstration and diffusion of technological breakthroughs that will reduce and eliminate greenhouse gas emissions from energy, industry and transport, and improve agricultural yields and crop resilience. In addition, the feasibility to deploy "negative emissions" at scale remains a major uncertainty, despite being an important feature of most scenarios consistent with the Paris Agreement's goals.

Finally, international co-operation remains fundamental to managing climate risks. Countries' current contributions to emissions reduction beyond 2020 are not consistent with the Paris temperature goal, and need to be scaled up rapidly. Support for action in developing countries will be important, not just for mitigation but also to improve the resilience and adaptive capacity of countries facing the greatest climate challenges. Climate impacts will grow, even if we achieve the Paris temperature goal. We need flexible and forward-looking decision-making to increase resilience in the face of these risks. Managing the interdependences between climate, food security and biodiversity goals will be critical to achieving the Sustainable Development Goals and long-term robust growth.

Chapter 1

A decisive transition for growth and climate

Governments around the world are facing the triple imperatives of re-invigorating growth while improving livelihoods and urgently tackling climate change. This chapter contains an extended synthesis of the report, showing how acting on climate change can also be good for growth, provided the right policies and structural reforms are put in place. After setting the scene for combined action on climate and growth, the synthesis presents results on the macro-economic implications of a "decisive transition" to a low-emission, high-growth and resilient future. The synthesis then lays out development pathways compatible with the Paris Agreement and how they vary across country types, as well as the need to scale and shift infrastructure investment. Turning to policy, the synthesis also presents the mix of structural and targeted climate policies required, the implications of the transition for exposed businesses and workers and how governments can address them, and changes needed to the financial system. It concludes with the main policy messages arising from the report.

Creating the conditions for sustainable growth

The global economy is not generating the level or quality of growth to which the citizens of G20 countries aspire. Productivity growth, the key factor that increases income per capita, has been declining for years in many countries. Widening inequalities, often related to the slowdown in productivity growth, are forcing a rethink about how the benefits of growth are shared. Many advanced countries face concerns about persistent unemployment and how to meet expectations about pensions, health and education. For some economies, this is exacerbated by ageing societies. Developing and some emerging economies have the benefit of a more dynamic demography, though many have concerns about the quality of investment and regulation. In their 2016 communiqué, G20 leaders recognised that "the use of all policy tools – monetary, fiscal and structural – individually and collectively" is needed both to support aggregate demand in the short term and to build the foundations for resilient, longer-term growth prospects.

The top priority for many G20 countries is to reinvigorate their economies, but the quality of that growth is vital. To improve lives and well-being in the short-term, growth needs to be inclusive, with benefits felt by the whole population. Economic growth over the last two centuries has led to staggering increases in wealth and well-being for much – but not all – of the world's population. To continue to improve well-being over a longer time horizon, the sources of growth need to be sustainable economically, socially and environmentally. To date, growth has exploited natural capital to meet the demands of rising populations, using technology largely based on abundant fossil fuels. Those fuels have been cheap because little account has been taken of their social and environmental costs.

Climate change: a systemic risk for growth

The impact of the current growth model on the natural environment now threatens the foundations of continued growth. While local pollution is increasingly driving momentum for reform, environmental pressures, including climate change, are no longer just local or regional; they pose profound challenges to global development. The scale of potential damage from climate change poses a major systemic risk to our future well-being and the ecosystems on which we depend, in particular for societies in less-developed, less-resilient countries. The pace and scale of the required economic transformation is unprecedented, if the worst of these risks are to be avoided; planning and investment in adaptation and resilience are also essential to reduce vulnerability to climate change.

Governments acknowledged the intrinsic importance of climate change for sustainable development and poverty alleviation in both the Paris Agreement and the 2030 Agenda for Sustainable Development. In Paris, countries collectively agreed to strengthen the global response to climate change including by limiting the global average surface temperature increase to well below 2°C and to pursue efforts to limit it to 1.5°C above pre-industrial levels, while increasing the ability to adapt to adverse impacts of climate change.

Most countries have proposed national action plans under the Paris Agreement, but collectively these are insufficient to achieve the long-term objective of the agreement. The Nationally Determined Contributions to 2030 are a positive step, but even if they were fully implemented, warming would reach around 3°C, leading to severe disruption and economic damage. Reasons for insufficient ambition vary, but commonly include perceived high economic and social costs of climate policies, and concerns about competitive disadvantage if stringent climate policies are not mirrored elsewhere. These concerns persist despite the "enhanced transparency framework" of the Paris Agreement. In addition, political and investment horizons have pitted the long-term benefits of low-emission development against the short-term (but ultimately unsustainable) benefits of cheaper high-carbon options. The threat of future damage from climate change has been too distant to drive sufficient early action, and short-term gain has tended to come first. But the threat of climatic disruption is not a conventional risk management issue, either temporally or spatially. While short-term costs are often local, a failure to address them will put future local and global benefits beyond reach.

Inclusive and climate-compatible growth

This report shows how action on climate change can generate inclusive economic growth in the short term, in addition to securing longer-term growth and well-being for all citizens. Governments can not only build strong growth but also avoid future economic damage from climate change if they collectively act for a "decisive transition" towards low-carbon economies. This requires combining climate-consistent, growth-enhancing policies with well-aligned policy packages for mobilising investment in low-emission infrastructure and technologies.

Investment in modern infrastructure is an important basis for economic growth, but underinvestment has been prevalent since the financial crisis. Energy, water supply, sanitation and waste management, mobility services and communications are foundations for economic activity and also essential for achieving the Sustainable Development Goals. Many advanced economies have suffered from a deficit of public infrastructure investment, hurting growth. Most emerging economies need massive investment to provide a growing population with universal access to modern services.

Countries are now facing a fundamental choice: the type of infrastructure investments they make will either support or seriously undermine future global well-being. As well as being a source of growth, infrastructure investment is a key determinant of future GHG emissions and resource efficiency, both directly (for example, through the type of power plants installed) and indirectly, by influencing behaviours (for example, through transport systems and urban planning). The window for making the right choices is uncomfortably narrow. The lifespans of much infrastructure and related physical investment means that future GHG emissions are going to be locked in by investment choices in the next decade, as infrastructure needs expand with the world economy. While investing in new and improved infrastructure is an important part of getting growth going now, investing in the right kind of infrastructure will deliver growth that can last. To manage climate risks and deliver long-term sustainable growth, infrastructure investment needs to be low-emission, energy-efficient and climate-resilient.

A unique opportunity

Current economic conditions – including low real interest rates in most countries – afford many governments the opportunity to invest in the right infrastructure now, to reignite growth while also paving the way to achieving the Paris Agreement goals. Governments need to bring together structural policy reforms, effective climate policies and the progressive alignment of regulatory frameworks to ensure effective action. A combined agenda for climate and growth offers numerous economic opportunities, including enhanced markets for low-emission infrastructure, technologies and services; increased market confidence spurred by greater climate policy clarity; and enhanced incentives for innovation and efficiency. These and other opportunities are relevant as the G20 prepares to revisit its Brisbane "2% upside growth" commitment and strengthen performance on growth; up to now, G20 countries have reached less than half of the 2% goal. The timing and mix of the policy interventions required will very much depend on countries' different developmental imperatives and exposure to climate risks.

The transition will not succeed unless the low-carbon economy is inclusive. To make pro-climate growth policies politically feasible, their implications for both households and businesses need to be taken into account. Beyond a well-functioning tax and welfare system, targeted measures can compensate for any potentially regressive impacts of climate policy on poor households. Past experience of industrial transitions shows that workers and communities relying on GHG-intensive activities should be actively engaged early in planning the transition. Where restructuring or plant closures are likely, authorities should aim for transparency and work with relevant companies, sectors and communities to develop economically sustainable alternatives and gain political and social support for policy measures. Clear policy signals are also essential to guide the transformation of technologies and business models for a low-GHG economy.

Acting together for better growth

The benefits of combined action on growth and climate increase as more countries act in a concerted way. Simultaneous action by countries generates economies of scale in climate solutions, magnifies the gains from learning and hastens a decline in technology costs, increasing the penetration of new technologies. Simultaneous action can also reduce the concerns of firms that competitors in countries not facing carbon pricing or regulation would be at an advantage.

Recognising their different economic structures and level of development, members of the G20 are well positioned to take the lead in uniting climate and growth efforts. The G20 countries not only account for 85% of global GDP and 80% of CO_2 emissions, they have far-reaching influence on the rest of the world through innovation, trade and development finance. They are also, collectively, leading the transition: G20 countries are home to 98% of global installed capacity of wind power, 97% of solar photovoltaic (PV) power and 93% of electric vehicles (IEA, 2017). While efforts to reduce emissions and sequence policies will vary from country to country, the G20 could spearhead the transition to low-carbon growth, generating technology cost reductions and best practices that will further accelerate the transition globally. Solar PV costs have declined about 80% in leading markets since 2010, for example. If G20 countries do not take the lead, it is hard to see how the transition can be effected.

A "decisive transition" for climate and growth

The current global macroeconomic environment – including low interest rates – provides an opportunity to take swift action to address climate change while boosting economic growth. Spurring investment in smart, modern, clean and resilient infrastructure, if combined with stronger fiscal and structural policies in a synergistic way, can boost growth in the short term and underwrite robust long-term growth, in both advanced and emerging economies. Low interest rates have increased fiscal space, giving governments more flexibility over spending choices without compromising their future financial position. Even in countries where there is less fiscal space, there are opportunities to optimise the tax and spending mix to align stronger economic growth with inclusive and low-carbon development.

Many policies aimed at strengthening growth can also support the transition to low-emission pathways; by the same token, measures aimed at stimulating investment in low-emission infrastructure can be good for growth. Economic growth and the low-carbon transition both depend on the development and diffusion of new technologies and efficient reallocation of resources towards both low-carbon and high-productivity economic activity. Policies that stimulate technological diffusion and facilitate resource reallocation thus work for both objectives and can ensure a cost-effective low-carbon transition. Such measures can be disruptive, but effects can be offset by spreading the benefits of growth widely, and through policies that improve access to new economic opportunities (education, vocational training) and provide an adequate social safety net to workers.

A decisive transition to spur growth while limiting climate change

New OECD modelling work presented in this report builds on IEA (2017) to show how combining economic reforms with ambitious climate policies in an integrated, synergistic manner can spur economic growth while also mobilising the investment needed to achieve longer-term climate objectives. Results suggest that such a collective "decisive transition" can boost long-run output by 2.8% on average across the G20, when comparing a current policies trajectory to a pathway set to hold warming below 2°C with a probability of 50% (Figure 1.1, right-hand panel). Importantly, the net effect on growth is also positive in the short term (left-hand panel).

The modelled growth effect is driven by a combination of investment in low-emission, climate-resilient infrastructure; an additional fiscal initiative to fund climate-consistent non-energy infrastructure; pro-growth reform policies to improve resource allocation; technology deployment; and green innovation. The benefits of combined growth and climate policies more than offset the impact of higher energy prices, tighter regulatory settings, and high-carbon assets that may become economically stranded before the end of their economic life. Carbon-tax revenues are assumed to be used to lower public debt in most countries. The overall macroeconomic benefits of the modelled policy package therefore also include substantial reductions in most countries' public debt-to-GDP ratios.

Avoided climate damages bring additional economic gains

If estimates for the positive impacts of avoiding damage from climate change are also accounted for, the net effect for 2050 rises to 4.7% higher than it would be if governments take no further action. While some economic damages are already captured in the modelling baseline, damages from climate change could pose a much greater threat to economic growth and well-being through mechanisms difficult to capture in economic modelling. The impact of these severe non-linear and unpredictable economic damages, such as flooding of coastal regions and increased frequency and strength of extreme weather events, could be very significant. Complementing model results with fuller damage estimates is important to give a more realistic picture of the long-term benefits of climate-friendly growth now. In addition, in the absence of action to reduce emissions, significant further damage can be expected between 2050 and 2100, outside of the timeframe of this exercise. Upper estimates of GDP costs without climate action range between 10 and 12% annually on a global scale by 2100.

The implications of a decisive transition will vary depending on a country's economic structure, but even fossil fuel exporters can offset losses and boost economic growth if policies are well chosen. This is a significant finding as climate action is usually expected to impose costs on fossil fuel exporters, including lower output and less employment in fossil fuel export activities. However, in a decisive transition these costs can be mitigated if carbon-tax revenues are judiciously recycled, in parallel with well-managed pro-growth reforms and proactive fiscal policies. The resulting positive effect on growth can more than outweigh the impact of stranded assets and higher energy prices. Results suggest the GDP boost would vary from 2% to 3% by 2050 in different G20 economies, not including avoided damages.

Figure 1.1. **Positive growth effects for the G20 by combining climate action with economic reforms in a decisive transition (50% probability of achieving 2°C)**

Average across G20, GDP difference to baseline, %

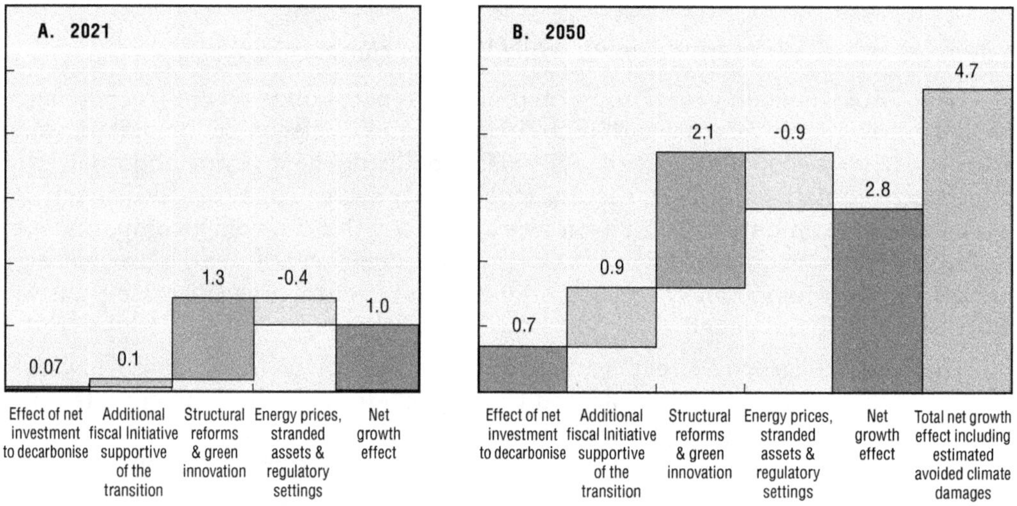

Note: See note under Figure 1.2.

Figure 1.2. **Positive growth effects in 2050 for the G20 by combining climate action with economic reforms in a more ambitious scenario (66% probability of achieving 2°C)**

Average across G20, GDP difference to baseline, %

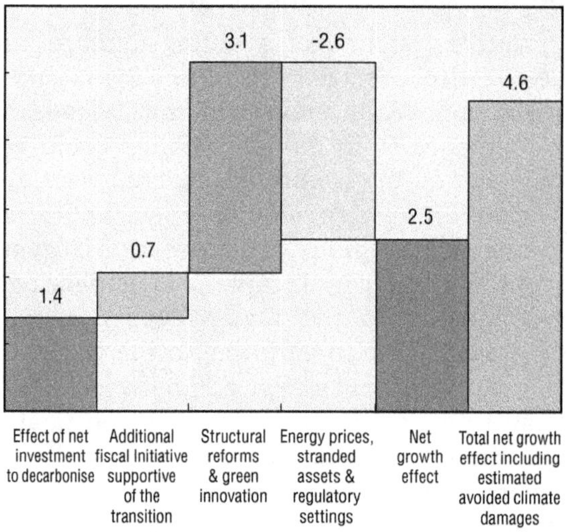

Note: The average G20 is a weighted average of selected G20 economies, which represents 88% of the G20 countries (excluding the European Union). "Net investment to decarbonise" comprises the effects of specific investment needed to achieve a 2°C climate objective. "Fiscal initiative" includes additional investment in climate-friendly non-energy infrastructure and soft infrastructure (e.g. education and research). Total investment corresponds to an increase in public investment in all countries of 0.5% of GDP. Countries that experienced disinvestment as a result of mitigation policies are assumed to compensate for this disinvestment. The structural reform modelled here includes a package of measures to improve economic flexibility and resource allocation, calculated using the OECD Product Market Regulation index. Innovation captures the increase in R&D spending necessary to reach a 2°C scenario (50% scenario) and equivalent to 0.1% GDP (66% scenario). Stranded assets are consistent with IEA estimates. Regulatory setting captures the reduced costs of the transition in a more flexible regulatory environment. For damages, simulations presented here include only a subset of potential damages, excluding for instance damages from extreme climate events, due to difficulties in projecting their frequency, severity and location. The exercise models global damages associated with temperature increases, using the Nordhaus (2016) damage function.

Pursuing a more ambitious climate scenario

Limiting warming to 2°C is not enough to satisfy the objectives of the Paris Agreement. While it is difficult to precisely define what "well below 2°C" and "efforts to limit to 1.5°C" mean, a step towards a more ambitious scenario can be described in which more stringent action raises the probability of holding warming below 2°C from 50% to 66%. Such a scenario is set out in a parallel report to the German G20 Presidency which this analysis draws upon (IEA, 2017). New OECD simulations suggest that this more stringent mitigation scenario can also be a strong basis for economic growth, with a GDP increase of around 2.5% for the G20 on average in 2050, further increased to about 4.6% if avoided climate damages are accounted for. Ambitious pro-growth reforms coupled with innovation, and in some countries the recycling of carbon-tax revenues, can outweigh losses resulting from potential energy price increases and stranded assets (Figure 1.2). However, this result requires caution. The macroeconomic effects of this scenario are hard to model because the speed and depth of the necessary economic changes are profound and difficult to anticipate. These changes include the stranding of some fossil-fuel-intensive energy activities, massive investments in the global stock of buildings and radical changes to transport systems. The extent of important developments cannot yet be known, such as a more resource-efficient circular economy, new business models and technological breakthroughs that could change the economics of the transition.

Costs of delaying action

There are also significant costs involved in delaying action to reduce emissions. Countries may be tempted to delay decarbonisation for several reasons, including the long-term nature of the climate threat and political resistance based on perceived short-term risk of economic, distributional or competitiveness impacts of climate policies. Such a delay would simply increase the transition costs and require a more abrupt adjustment when action does finally start. If more stringent policies were introduced later they would affect a larger stock of high-carbon infrastructure built in the intervening years, leading to higher levels of stranded assets across the economy. In a delayed action scenario where action on climate change accelerates only after 2025, GDP losses are estimated to be 2% on average across the G20 after 10 years, relative to the decisive transition, and would be higher for net fossil fuel exporting countries. The losses could materialise as soon as the delayed transition starts and could be aggravated by financial market instability. The main uncertainty concerns how many assets might be stranded. Further research is warranted on how those assets should be measured.

Decisive action by leading countries

Even if action is not fully co-ordinated internationally, pro-active countries could still see benefits of combining climate and growth policies through a leadership alliance, demonstrating the benefits to other countries over time. The competitive advantage for such leadership economies is not likely to suffer in aggregate, due to the growth benefits of action described above and because their policies would drive demand for low-GHG products and spur innovation. They would also gain from short-term co-benefits of action, such as improved human health due to lower pollution. However, the pro-active countries may need to plan for significant structural changes in the economy, especially if some firms in carbon-intensive sectors relocate to countries with less stringent policies. This reinforces the case for accompanying structural reforms, as well as measures to ensure a proper transition of the work force. Cost-efficient decarbonisation policies, including carbon pricing with astute use of revenues, are even more important in this scenario. While countries outside of the leading group could gain some short-term competitive advantage in carbon-intensive industries, they would likely face higher stranded assets later. And the burden these countries impose on other countries, including higher climate risks, will become increasingly clear and may have broader implications for a range of international geopolitical issues.

Regardless of the international picture, the appropriate combination of pro-growth policies and action on climate change will vary from country to country, depending on governance, economic and social structures. The following sections show how country characteristics will shape emissions pathways and infrastructure choices, before exploring how different combinations of structural reform and climate policies can trigger growth in various country contexts.

Pathways and priorities for a decisive transition

The long-term temperature goals of the Paris Agreement can be translated into a fixed quantity of long-lived GHGs to be released to the global atmosphere over time. This global "carbon budget" is best presented as a range, reflecting uncertainties on how the temperature target is interpreted, how the climate responds to GHG concentrations (climate sensitivity), and the role of non-CO_2 GHG emissions. The level of gross GHG emissions consistent with a given (net) carbon budget will also depend on assumptions about technologies for "negative emissions", which would allow for a temporary overshoot before emissions are removed from the atmosphere to maintain net emissions within the overall budget. The global carbon budget compatible with a 66% likelihood of remaining below 2°C is estimated to be 590-1 240 $GtCO_2$ from 2015 to the time of peak warming – roughly 15 to 30 years of fossil fuel-related CO_2 emissions at current rates.[1]

To remain within the carbon budget compatible with the Paris goals, the global emissions pathway created by a decisive transition requires three main features:

- an early peak in global emissions, as soon as possible;
- a subsequent rapid fall in GHG emissions;
- net GHG emissions near zero or net negative in the second half of the century.

The later the peak in global emissions, the greater the rate of emissions reduction required subsequently to stay within the carbon budget. Options for achieving ambitious mitigation goals may be lost if emissions peak too high or too late, and delayed action would lead to higher costs as described above. Further, failure to reach a global emissions peak before 2030 may make it impossible to limit global average surface temperature increase to well below 2°C, let alone 1.5°C. This is particularly important because although total global CO_2 emissions from energy have been flat for the past three years, the CO_2 intensity of primary energy across the G20 remains high. As growth picks up, global CO_2 emissions could therefore start to increase again unless governments take further action.

Low-emission pathways

The mitigation objective in the Paris Agreement is extremely stringent. A deep transformation of the energy sector is needed to decarbonise electricity supply, improve energy efficiency, deploy smart grids and storage to better manage electricity demand and supply, and electrify other energy end-uses such as transport and buildings. However, the energy sector is only part of the low-carbon transition story. Agriculture, forestry and other land use contribute around a quarter of total GHG emissions, around half of which is from agriculture. The land sectors act as both sources of GHGs (including methane from cattle and rice, nitrous oxide from fertiliser use) and sinks of CO_2 (from forestry and carbon stocks in soils), so they have an important influence over the carbon budget remaining for energy-related emissions.

Most scenario modelling of global pathways that keep warming "well below 2°C" require not only reducing emissions of all GHGs but also "net negative" emissions later this century.[2] Land-use and forestry will have to go from being a net emitter to a net sink of GHG emissions, including through reforestation, avoided deforestation, and conservation and recovery of soils as carbon stocks. Agriculture also has the potential to become more GHG-efficient while meeting increased food demand from rising populations, though this is dependent on demographics and dietary preferences, as well as technological progress in crop yields. Energy-related CO_2 emissions can also be reduced by using bioenergy, either for advanced biofuels or in power plants fitted with carbon capture and storage (CCS). Potentially a means to create "negative emissions", the required technologies are still not yet proven at commercial scale across relevant applications. Concerns remain over competition for land and whether enough biomass can be produced sustainably, while meeting food demand, maintaining carbon stocks and protecting biodiversity.

Adaptation pathways are important planning tools

Adaptation is also at the heart of the Paris Agreement. Strong action to reduce emissions will lower the need for adaptation by reducing the intensity of climate-change impacts. Nevertheless, significant climate impacts are already locked in, so planning for and investing in adaptation and resilience is critical. Vulnerability to climate change varies greatly across sectors and within countries, shaped by geography, income, governance and development choices. Socio-economic trends and trans-boundary impacts are also relevant.

Decisions being made today will affect future vulnerability to climate change, intentionally or not. However, climate vulnerabilities are diverse and projections of local and regional change are uncertain, so it is neither possible nor desirable to address the need for adaptation comprehensively at one point in time. "Adaptation pathways" can be developed to shape near-term planning and policy decisions that reduce short-term and long-term risks. These pathways provide a means to identify path dependencies and critical decision points, creating a flexible, forward-looking approach to decision-making. National adaptation plans can strengthen the capacity of national and local decision makers to account for climate change and direct investments in resilience. Relevant tools for adaptation strategies include national risk assessment, indicator sets and in-depth evaluations of large infrastructure projects.

Pathways for different countries

Both low-emission and adaptation pathways are specific to individual countries. This is highlighted by the diversity of current CO_2 intensity of energy and energy intensity of GDP, both key determinants of CO_2 emissions. The lines in Figure 1.3 show different combinations of these two determinants resulting in the level of CO_2 emissions per unit of GDP required to be on course for the IEA 66% 2°C scenario, which this report builds on, in 2030, 2040 and 2050. The 2014 positions of G20 countries are also plotted, highlighting the different starting points and challenges facing different countries as they choose the most appropriate pathways towards the Paris objectives.

Figure 1.3. The carbon and energy intensity of G20 economies in 2014 and the path to 2050

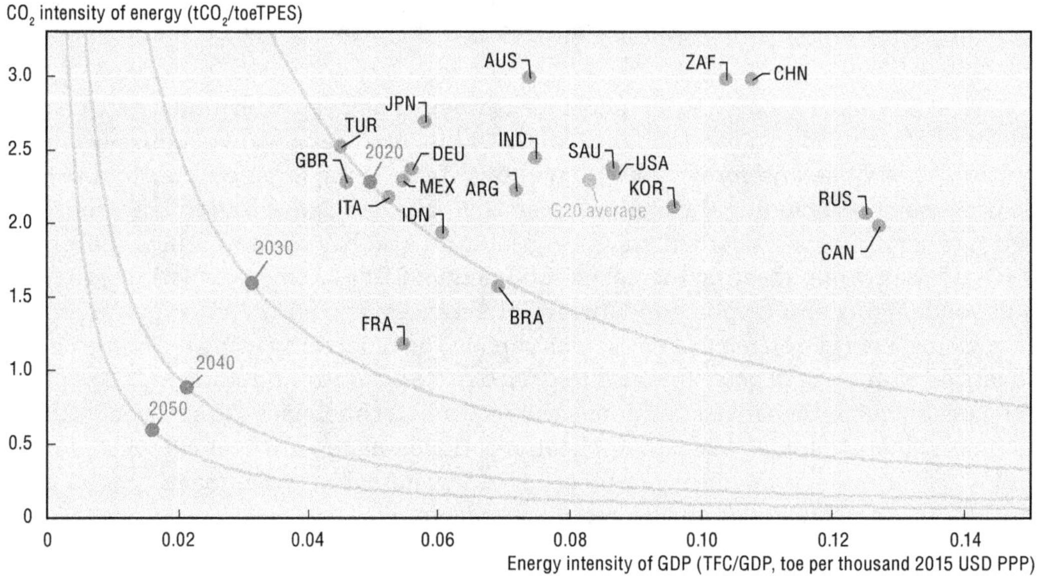

Notes: The average levels for G20 countries (excluding the European Union) refer to 2014 statistical data and the IEA 66% 2°C scenario projections for 2020, 2030, 2040 and 2050. The iso lines show other feasible combinations of CO_2 intensity and energy intensity levels. Calculations assume a constant ratio for total primary energy supply (TPES) to total final consumption (TFC). Toe = tonnes of oil equivalent.

Source: Calculations based on the IEA World Indicators and IEA 66% 2°C scenario projections.

StatLink http://dx.doi.org/10.1787/888933483964

Pathways will vary according to different country circumstances. Figure 1.4 presents a new characterisation of CO_2 emissions pathways out to 2050 under the IEA 66% 2°C scenario, showing the G20 average and also groups of advanced and emerging economies. Measured against a starting point of 2010 emissions, global CO_2 emissions fall by about 80% by 2050. Advanced economies begin rapid emissions reductions from the outset and are projected to converge at very low levels by 2050. However, pathways for emerging economies are very different. Upper middle-income countries, taken together, show a gradual decline starting from the current period, also accelerating to reach low levels by 2050. Lower middle-income countries, given their stages of economic and demographic development, show continued increases in emissions to about 2025, followed by a gradual decline back to around 2010 levels.

As well as the diversity of potential country pathways, these scenarios illustrate the importance of policies (including climate support) that can combine growth with emissions reductions, to bring forward the required peak in emissions while not harming prosperity, in particular for emerging (middle-income) market economies. Understanding the appropriateness of different policies requires understanding how low-emission pathways apply to different countries, for both energy and non-energy sectors, taking into account the relative importance of energy, industry, land-use and other sources of GHG emissions. Groups of countries that share common characteristics could gain a significant advantage from joint analysis of policy developments as they develop their plans for combined growth and climate action.

Figure 1.4. **Emissions pathways by income group**

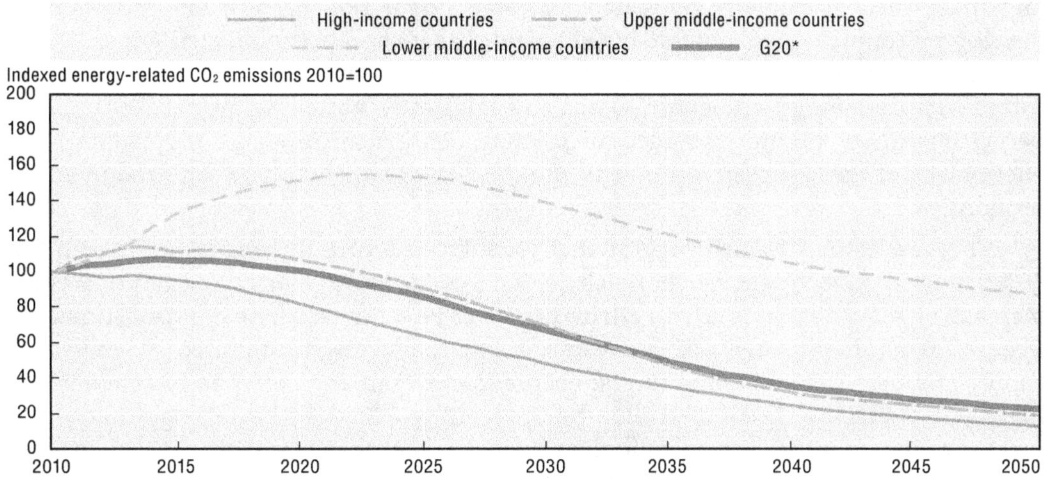

Note: Due to data limitations, G20 countries not included are Argentina, Saudi Arabia, South Africa and Turkey.
Source: IEA (2017) and OECD calculations.
StatLink http://dx.doi.org/10.1787/888933483972

Scaled-up investment in clean, resilient infrastructure

Infrastructure investment is vital to underpin economic growth as part of a decisive transition towards low-emission, climate-resilient pathways, but current levels and types of investment are inadequate. The quality of infrastructure is declining in many advanced economies, public capital stock is shrinking in some countries, and more infrastructure investment is needed in developing countries to achieve universal access to energy and basic public services. The quality of different infrastructure types, and resulting access to basic services, varies greatly across different country income groups, with implications for the quality of growth and development (Figure 1.5). For example, having nearly universal access to electricity (bottom left) does not mean that the electricity supply is of good quality (top right).

Figure 1.5. **Quality of infrastructure status and access to basic services in G20 countries, by income groups**

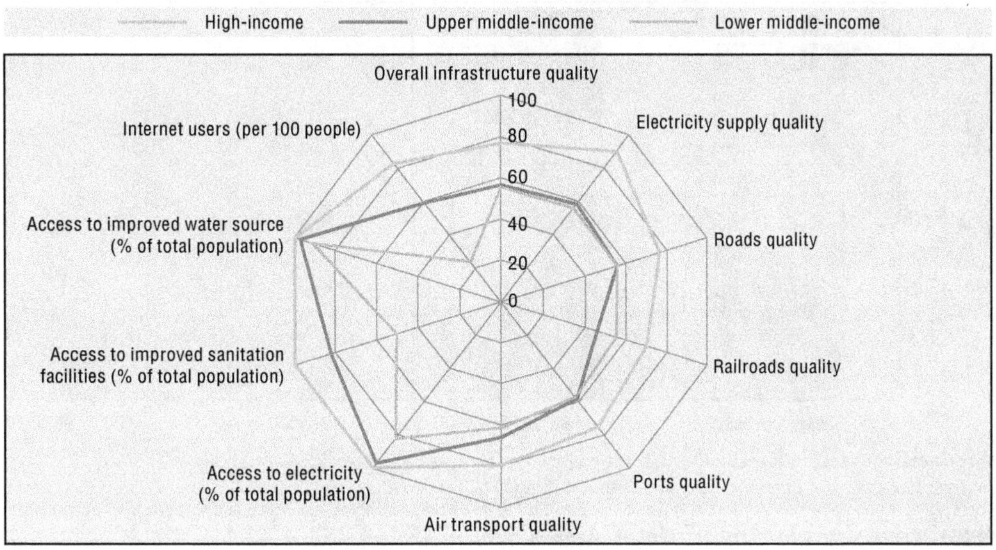

Sources: Authors, based on WEF (2015) and World Bank (n.d.) (accessed February 2017).
StatLink http://dx.doi.org/10.1787/888933483980

Unprecedented levels of infrastructure investment will be required to sustain growth and meet the basic needs generated by rapid population growth and urbanisation in developing countries, even before considering climate and pollution challenges. The OECD estimates that around USD 95 trillion of investments are needed from 2016 to 2030 in infrastructure (energy, transport, water and telecoms), equalling around USD 6.3 trillion per year without taking into account climate concerns. Transport represents 43% and energy 34% of those investment needs, 60% to 70% of which will be required by emerging economies.

The new estimates also suggest that for infrastructure to be consistent with the 2°C 66% scenario, investment needs reach USD 6.9 trillion per year in the next 15 years, an increase of about 10% in total infrastructure investment from the reference estimate above (Figure 1.6, left-hand panel). This covers transport, water and sanitation as well as energy supply and use. The additional capital cost is low overall and could be offset over time by fuel savings reaching USD 1.7 trillion per year up to 2030 (Figure 1.6, right-hand panel) – further reinforcing the case for robust low-emission economic growth.

Focusing on energy infrastructure, low-emission pathways require a deep transformation in the way energy is used and produced, requiring 29% more investment in the energy sector (Figure 1.6, top three segments). In the IEA 2°C 66% scenario, 95% of the electricity would need to be low-carbon by 2050, 70% of new cars would be electric, the entire existing building stock would have been retrofitted, and the CO_2 intensity of the industrial sector would be 80% lower than today (IEA, 2017). Achieving this would entail a major shift of energy supply investment towards low-carbon alternatives, and a significant increase in demand-side investments to make the economy more energy-efficient in the next few years.

Figure 1.6. **Annual infrastructure investment needs and fuel savings in a low-carbon future**

Global estimates (annual average for 2016-30, USD 2015 trillion)

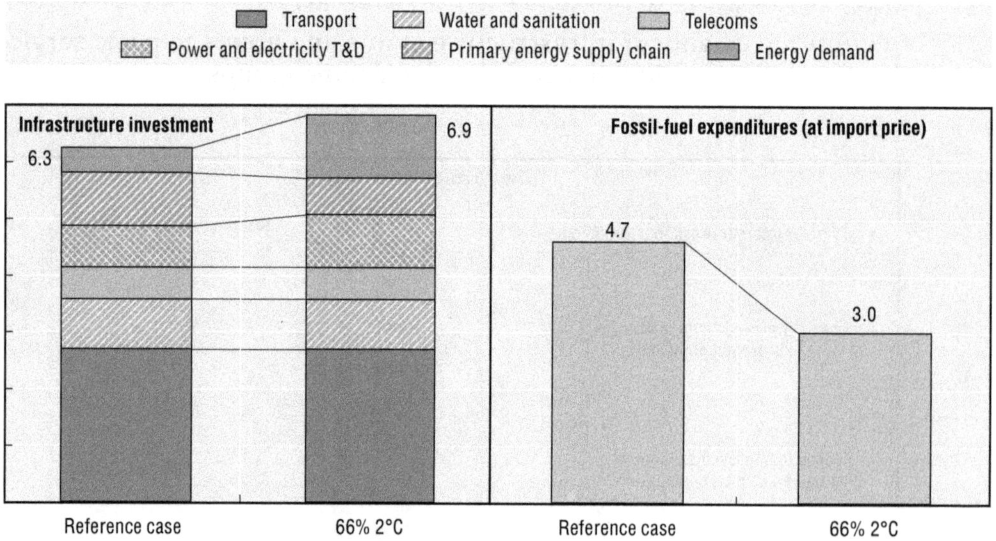

Notes: Reference case assumes no further action by governments to mitigate climate change.

Sources: IEA (2017) for energy supply and demand; IEA (2016a) for road and rail infrastructure; OECD (2012) for airports and ports; McKinsey (Woetzel et al., 2016) for telecoms. The water and sanitation estimate is an average of estimates from: Booz Allen Hamilton (2007), McKinsey (Woetzel et al., 2016) and OECD (2006).

StatLink http://dx.doi.org/10.1787/888933483994

While it is clear that a boost in investment is needed in the short term to engage on a low-emission pathway, the exact amount remains uncertain. Other modelling exercises (IEA 2016) show that in the long term (to 2050), overall investment needs could actually be lower in a low-carbon scenario than in a business-as-usual scenario. This would include savings from modal shifts to low-carbon transport, particularly at the urban level, where fewer vehicles and less parking space would be needed. In the long term, a world less reliant on fossil fuels is also likely to require less port capacity, fewer oil and gas tankers, and fewer hinterland railways to transport coal. On the other hand, digitalisation and smarter energy systems may require additional investment needs in telecommunication systems. G20 countries need to better understand the actual infrastructure investment needs associated with their low-emission development strategies.

Most existing energy and transport infrastructure was designed and built for a world in which fossil fuels were cheap and abundant. Given the long lifespan of infrastructure, failure to invest in the right type of infrastructure in the next 10 to 15 years would either lock the world into a GHG-intensive development pathway or risk stranding many assets. It would also imply serious and probably irreversible risks, not only of environmental damage, but also of financial instability that harms economic growth prospects. As explained above, the later a decisive transition begins in earnest, the more difficult and disruptive it promises to be for the energy sector and other GHG-intensive activities. Taking a low-carbon path offers an opportunity to accelerate investment in infrastructure, create a short-term boost to economic growth and development, and provide relief from persistent problems like congestion, air pollution and access to energy.

Improving the transparency of infrastructure project pipelines

While long-term planning is a vital first step for the low-carbon transition, G20 countries must also be able to transform such plans into bankable, low-emission infrastructure projects. Most countries still lack clear and transparent information on their infrastructure investment pipelines, even though G20 leaders recognised in 2014 the importance of such pipelines for tackling the global investment and infrastructure shortfall. Improving the visibility of infrastructure plans and needs is a key priority and critical to gain the confidence of private sector investors. Where current investment plans are known, they are often limited to the energy sector and generally not consistent with the commitment in the Paris Agreement to mitigate GHG emissions and support adaptation. In addition, G20 countries have a significant influence on infrastructure developments in other countries through export credits and official finance, where better alignment with the Paris Agreement should be sought.

New analysis of the current existing capacity and current pipeline of power plants in G20 countries[3] indicates that a shift towards investment in renewable energy has started and is likely to continue in the next 15 years, as two-thirds of the global capacity under construction is based on renewable energy technologies – close to what is required by the IEA 2°C 66% scenario (Figure 1.7, right-hand panel). Despite this encouraging trend, more than 20% of the projects under construction are still based on coal. This number could increase as 416 GW of coal plants are in pre-construction development, and 543 GW are on hold. Continuing this trend will put the temperature targets set out in the Paris Agreement out of reach.

Innovation will play an important role in achieving low-emission growth. While much progress can and needs to be made immediately using currently available technologies, a full low-carbon transformation will require widespread innovation and deployment of new infrastructure, technologies and business models. Beyond the need for new combinations of technologies to achieve net-negative emissions while meeting food demand sustainably, heavy industry will require technology breakthroughs to mitigate process emissions and to reduce reliance on fossil fuels. Energy sector innovation is also important, including rapid advances in energy storage to accommodate larger shares of variable renewable sources. As mentioned above, structural reforms can play an important role in facilitating this green innovation and ensuring that it is good for growth.

Figure 1.7. **Current capacity and current pipeline of power plants relative to those required in a 66% 2°C scenario**

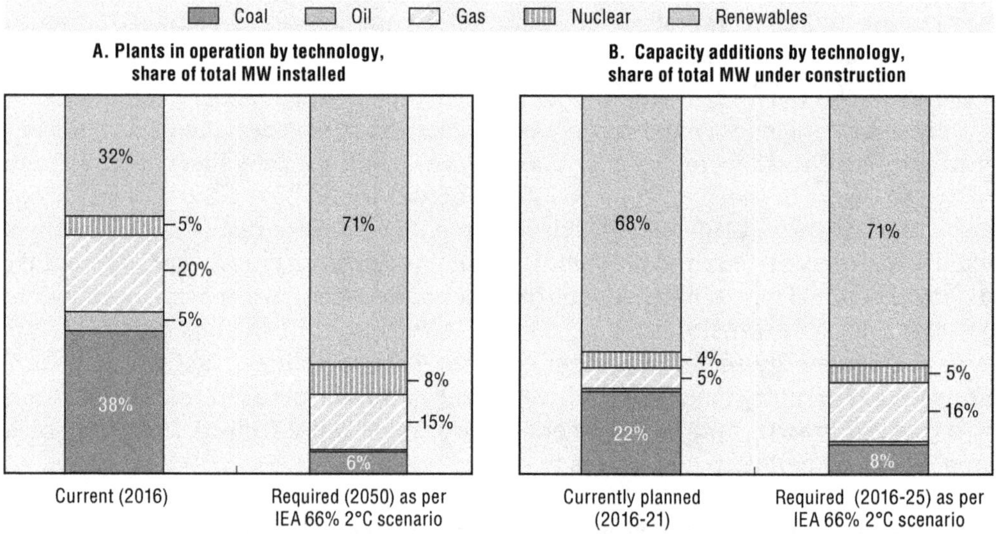

Source: Authors' analysis from *i)* Platts WEPP (2017) for oil and gas under construction in G20 countries; *ii)* the Global Coal Plant Tracker (2017) for coal under construction in G20 countries; *iii)* IAEA (2016) for nuclear under construction in G20 countries; *iv)* IEA (2016b) for renewable energy under construction in G20 countries; and v) IEA (2017) for capacity additions in the IEA 2°C 66% scenario, globally.

StatLink http://dx.doi.org/10.1787/888933484002

Combining pro-growth reforms with climate policy and well-aligned investment conditions

To mobilise the investment required for a decisive transition, governments need to support pro-growth structural reform policies with coherent climate policies and a well-aligned investment policy environment (Figure 1.8). The most effective policy combinations to mobilise investment in low-emission infrastructure vary from country to country, including the respective contributions of public and private investment.

Structural reforms that promote higher and more inclusive growth – such as measures to enhance product-market competition, facilitate access to jobs and improve skills – can be supportive of the low-carbon transition and are a key part of a decisive transition for climate and growth. The swift infrastructural, technological and industrial shifts implied by low-emission pathways to 2050 demand more rapid resource reallocation and faster technology diffusion. They can be further accompanied by improving dynamism in labour markets, provided that workers in the most affected carbon-intensive industries are supported through the transition. Pro-growth reforms that help meet these demands also generate more productive economic activity and enable new entrants to capitalise on emerging opportunities. Easier reallocation also boosts investment in R&D and other forms of knowledge-based capital, which boost adoption of new low-carbon technologies and long-term productivity growth. This requires reforms in product markets, financial markets, labour markets and housing markets. In short, policies that attempt to preserve the status quo – or at most favour an incremental transition – risk falling short from both a climate and an economic point of view.

Figure 1.8. **The three components of a well-aligned policy framework for climate and growth**

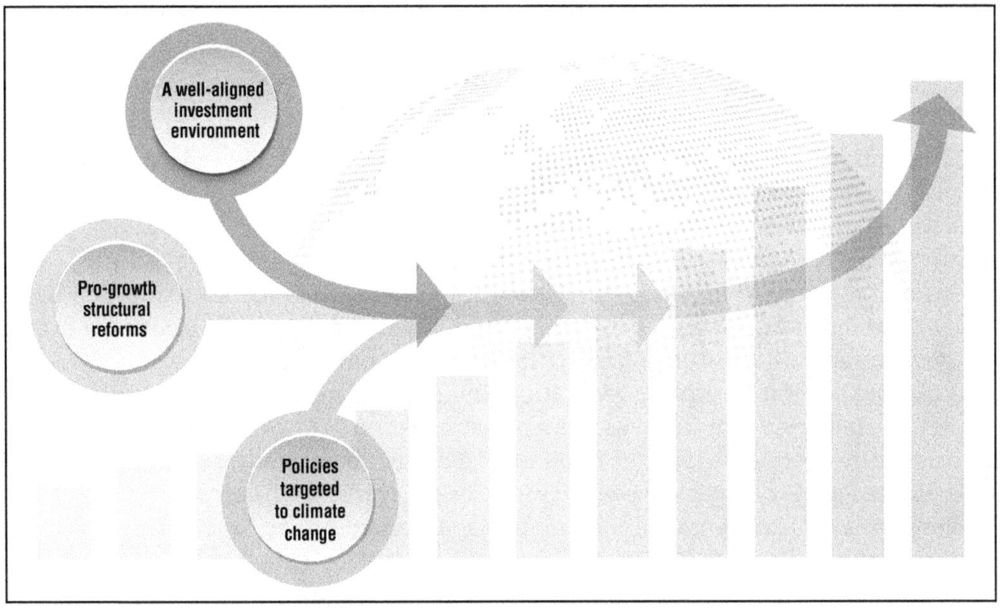

Strong and coherent climate policy as the basis for the transition

Carbon pricing can be a powerful, cost-effective tool for steering producers and households towards low-carbon and growth-oriented behaviour and investments. However, carbon prices have so far been low, especially when measured by "effective carbon rates" that incorporate the carbon price equivalent of energy taxes as well as explicit carbon prices. Currently, most CO_2 emissions within the G20 are not priced at all, and 91% are priced at less than EUR 30 per tonne of CO_2 (a conservative estimate of the lowest social costs that would result from a tonne of CO_2 emissions).

Where carbon prices exist, their impact on infrastructure investments has tended to be limited and indirect, partly because price signals have been weakened by transitional support measures or exemptions given to firms or households. Poorly targeted use of the public revenues from carbon pricing can also hinder their effectiveness and reduce the political acceptability of carbon pricing. On the other hand, intelligent use of carbon pricing revenues is an opportunity to improve fiscal space and make climate policy more inclusive and progressive, for example by reducing other taxes and lightening the burden on the poorest households.

Fossil-fuel subsidies are still widely prevalent and act as negative carbon price signals, leading to increased emissions of CO_2 and local pollutants. In 2014, G20 countries collectively provided subsidies amounting to USD 354 billion for fossil-fuel consumption, and USD 18 billion for fossil-fuel production. These subsidies translate into large fiscal costs for governments. For example, the fiscal burden of fossil-fuel subsidies reached as high as 1.4% of GDP in Mexico and 4.1% of GDP in Indonesia before both countries started reforming such subsidies; those subsidies were also regressive, benefiting mostly those on upper and middle incomes. In general, governments can make fossil fuel subsidy reform more acceptable if they precede such reform by improving energy services and introducing measures aimed at supporting the poor.

Even where carbon pricing is at the heart of countries' climate policy, local conditions and political compromises often make the design of schemes less than perfect and more susceptible to factors like information asymmetries, non-price barriers such as behavioural change, and public opposition to new taxes or tax increases. This means that carbon pricing may need to be complemented by other targeted measures such as specific investment incentives; regulations and standards; information policies; and measures aimed at low-carbon innovation. The interactions between policies need to be carefully evaluated, however.

Tuning broader investment conditions for low-emission, resilient investment

For climate policies to be more effective – and more supportive of low-emission economic growth in a decisive transition – the broader policy environment in which they operate needs to be well aligned with climate objectives. Existing policy frameworks, developed over decades to support fossil fuel-based economic growth, can inadvertently weaken the low-emission investment signal provided by carbon pricing. Potential misalignments can be identified in many policy areas, including investment, competition, trade and tax. A first priority is to ensure that pro-growth reforms are well aligned with low-carbon growth, such as ensuring a competitive level playing field for electricity generation. In addition, specific policies and regulations that weaken the business case for investment and innovation in low-emission and climate-resilient infrastructure need to be identified and fixed. For instance, poorly designed support schemes and outdated maps of domestic natural resources may hinder the attractiveness of investment in renewables. Inconsistent land-use and transport planning can lead to a locking in of carbon-intensive infrastructure and behaviour, particularly in urban areas.

Some land-use policies can also be misaligned with climate objectives. Resolving these conflicts is vital to maximise the contribution of the land-use sector to low-emission pathways while balancing land-use priorities. For example, agricultural input subsidies, price support, tariffs and subsidies on agricultural products, and in some cases subsidised crop insurance premiums, often foster more emissions-intensive practices and impede investments in adaptive technologies (though in some countries specific policy designs are aligned with sustainability objectives). Land degradation is another example, resulting from uncontrolled open access to common land. Reforming land tenure arrangements – to increase private ownership or long-term leases – or strengthening the sustainability of traditional institutions and land use rights, can foster private investment in restoring degraded landscapes or preventing land degradation, which in turn help sequester more CO_2.

Public infrastructure choices and procurement

Public procurement at central and local government levels plays a key role in the economy as a whole (averaging 13% of GDP in advanced countries, and sometimes more in emerging economies). It is particularly important for pro-growth infrastructure investment, including low-emission and resilient infrastructure. Public procurement can also create lead markets for innovative, low-GHG industrial materials and infrastructure choices. This can be done by pricing life-cycle CO_2 emissions in procurement criteria, thereby encouraging a competition to lower emissions. To unlock this potential and align procurement with Paris Agreement objectives, public procurement organisations need to be strengthened.

Efforts to improve climate resilience, in particular infrastructure resilience, need to take country and locally specific contexts into account. In general, the owners and operators of infrastructure are best placed to decide on the appropriate measures to implement. The public sector has a key role to play, however, to ensure that the current direction of infrastructure investment is aligned with the goal of increasing resilience to economic and climate-related shocks, and also catalysing private sector investment in adaptation by creating an enabling environment. A well-designed regulatory framework, information on climate risk and pricing externalities, and better aligned policies could help drive adequate investment in resilience by owners, operators and financiers.

A transition that is inclusive, progressive and good for business

Even though action on climate change can be positive for overall economic growth and welfare, most countries face political challenges in implementing ambitious policy reform. Vested interests and incumbent actors in today's high GHG-emissions societies can prevent governments from acting decisively and consistently. In a decisive transition, certain assets, especially in the fossil fuel and power sectors, will lose value and be economically stranded, with potential implications for employment opportunities. Even if the impact on overall employment is likely to be modest, jobs will shift as GHG-intensive activities change business profiles and technologies.

Most countries' economies are "entangled" with fossil fuels and other GHG-intensive sectors, reflecting the significant contribution of these activities to past economic development. Even in countries that are not fossil fuel producers, tax revenues, financial markets, pension funds and jobs depend to varying degrees on GHG-emitting activities, which can place governments in a position of significant conflict should they try to implement strong climate policies. This entanglement can render climate action ambivalent at best unless governments adopt an inter-ministry, cross-cutting approach to climate action.

Governments have previously had to learn about the modernisation and restructuring of some heavy industries, experience which may prove instructive in managing the transition to a less GHG-intensive economy, including engaging with affected firms and communities. Relevant measures used in the shipbuilding and iron and steel sectors include the creation of funds and targeted subsidies (e.g. restructuring investment aid, closure aid), special legislation and fiscal measures. Clearer decarbonisation and adaptation pathways will help governments anticipate, plan for and communicate the structural consequences of the transition away from GHG-intensive activities. This should minimise the destruction of asset values. Disruption linked to business cycles and other factors, such as the global excess capacity of iron and steel, can allow governments to prepare industry for the shift.

Creating opportunities for workers most affected by the low-carbon transition will be essential. The aggregate effect of the transition on jobs may be modest, but reallocation across sectors and activities will be necessary and in some sectors significant. Trade unions are aware of the challenges posed by the transition and advocate a role for workers in a "just transition" – a transition that includes proactive measures to plan and invest in environmentally and socially sustainable jobs, sectors and economies. Good planning to anticipate and facilitate retraining and mobility, and an active social dialogue between government, employers and workers, are vital for climate-friendly development.

The low-carbon transition will also directly affect households. Energy supply costs may increase, at least in the short term, so households could face transitional costs for new efficient equipment and infrastructure. Households could also face higher energy unit costs, for example where carbon pricing is the instrument of choice. These changes may be regressive, affecting the poorest households the most, but targeted recycling of carbon tax revenues can offset this effect. In many countries, the need to improve energy access and affordability will have a strong bearing on policy choices to facilitate the adoption of low-carbon energy practices. The reforms of fossil-fuel subsidies, initiated in some G20 countries and beyond, have shown how governments can compensate for rising energy prices and avoid regressive impacts.

The transition is unlikely to succeed, however, unless the low-carbon economy includes and provides opportunities to all actors. The transition will affect everyone, from central and local governments to the private sector, the labour force and citizens, whose divergent interests and influence will come into play. An improved understanding of aligned and divergent interests can help governments to make policy that addresses multiple needs and musters coalitions in favour of action – in business, institutions, civil society and different government portfolios. This would ensure that other pressing policy priorities, such as poverty alleviation and inclusiveness, are not compromised, making the transition more sustainable. This broad-based engagement should be an essential element in the domestic processes guiding the elaboration of low-greenhouse gas development strategies.

Overall, to improve the chances of achieving the Paris Agreement goals, it is vital to incorporate political economy considerations early in the process of elaborating domestic strategies to implement Nationally Determined Contributions. In addition, pursuing "whole-of-government" approaches to low-emission, climate-resilient growth can help governments to avoid entanglement in high-carbon sectors and activities.

Mobilising capital for a decisive transition

Coherent climate policy and a well-aligned investment framework are essential to steer the investment flows needed to pursue low-emission, resilient pathways, but in themselves are not enough. Mobilising the necessary capital also requires diverse financial instruments tuned for infrastructure financing, efficient allocation of risks and use of risk mitigation techniques, public financial institutions geared towards low-carbon investment, and a financial system that correctly values climate risk.

Private financing of infrastructure, including low-emission energy infrastructure, has undergone a major shift in the last decade. Renewable energy projects have been able to access more diversified pools of financing through project finance structures, attracting equity investors such as pension funds and sometimes project bonds. At the same time, banks are facing challenges such as non-performing loans and stricter regulation, so there is a need to open complementary sources of finance such as institutional investment and capital markets. The low-carbon transition will require substantially stronger efforts to overcome the remaining barriers to mobilising the private investment capital required for low-emission, resilient infrastructure.

New models and partnerships are scaling up financing for low-emission infrastructure, by drawing on the changing role of traditional financial actors and their respective strengths. Increased co-operation, for instance between banks and utilities, or between development finance institutions and institutional investors, has significant potential to facilitate finance for key elements of low-emission pathways, including renewables and energy efficiency in buildings and industry.

Real and perceived risks related to infrastructure financing, for example due to weak governance and regulation, currency fluctuations, and lack of domestic capital markets, continue to hamper private investment, particularly in emerging economies. There is also a need to improve the understanding of the specific risks and returns associated with investment in low-emission infrastructure. These risks often relate to infrastructure as an asset class, characterised traditionally by its long-term nature and high upfront costs, together with political, regulatory, macroeconomic and business risks and, more recently, climate change risks.

Despite the crucial role of technology and innovation, as highlighted in the pathways analysis above, new venture capital finance in clean technologies has been declining. Current investment models are not always aligned with the capital intensity and long development timelines required by clean technologies. Governments need to remove bottlenecks in clean technology finance, particularly in early stages and commercialisation, by enhancing public and private co-operation and improving business models for the financing of research and development in energy efficiency and low-emission infrastructure.

Various risk mitigation and "blended" finance approaches have been developed and need to be scaled up. Tools such as guarantees, credit enhancements, currency hedging and more diversified insurance offerings help to mitigate and better allocate risk across different actors, while instruments such as green bonds and securitised loans help to secure a reliable long-term funding basis for infrastructure projects. Blended finance models – with a focus on crowding in private finance – can de-risk and mobilise private investment in infrastructure, while optimising public investment.

Important role of development banks and finance

Development banks (national and multilateral) and development finance institutions (DFIs) have a critical role as a bridge between private and public actors, helping countries to embark on a sustainable low-carbon development path. National development banks (NDBs) are widespread in the G20, and several are initiating efforts to finance low-emission, climate-resilient infrastructure. Multilateral development banks (MDBs) and bilateral DFIs have made ambitious climate commitments and are scaling up efforts to mobilise private climate investment, while dedicating significant financing to infrastructure (Figure 1.9). MDBs are able to leverage significant capital through their shareholder governments and mobilise knowledge, expertise and innovation developed in other parts of the world. Despite this, MDBs could better align their financing for infrastructure with low-emission pathways, particularly in the transport and water sectors, by increasing the share of climate-related commitments in their portfolios, improving disclosure of portfolio-wide carbon impacts and renewing efforts to mobilise private investment. To meet their targets, MDBs and bilateral finance institutions require strong mandates. They also need to work with countries to raise awareness and build demand for low-emission, climate-resilient infrastructure, facilitated by access to concessional climate finance. Increased collaboration and joint action between MDBs, bilateral actors and NDBs will be needed to scale up financing, particularly in emerging and developing countries.

Governments also need to co-operate to guide the global financial system to more accurately value climate risk and move towards investment in low-emission and climate-resilient infrastructure. Fuller disclosure and reporting of climate impacts and risks is required to enable a broader shift in the financial system towards alignment with the Paris Agreement and the Sustainable Development Goals. Policies need to focus on the mainstreaming of climate-change risk management practices across the financial system, and the efficient pricing of assets based on disclosure of climate change risks. In spite of progress through the Financial Stability Board, public-sector finance institutions still lag behind, and individual country responses are uneven across the G20.

Figure 1.9. **Share of MDB commitments for infrastructure that are climate-related and total MDB commitments for infrastructure (USD billion) by sector, 2013-15 average**

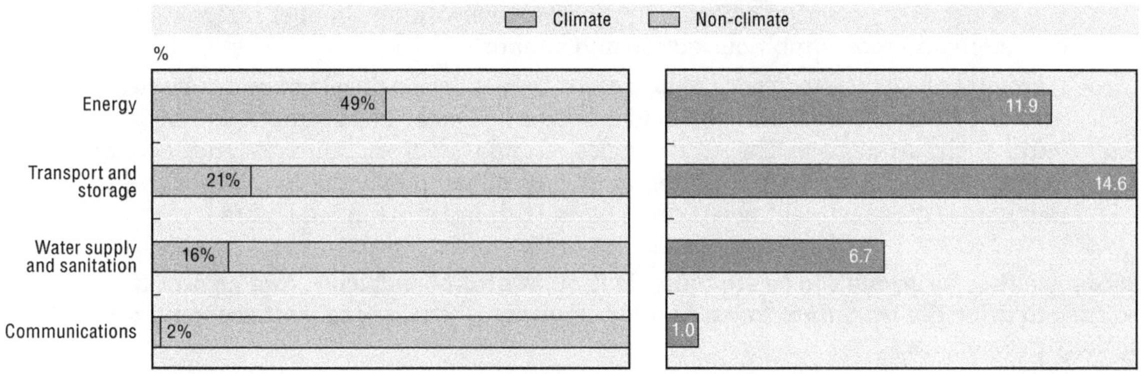

Notes: This graph is based on data reported to the OECD Development Assistance Committee by the following MDBs: the African Development Bank, the Asian Development Bank, the European Bank for Reconstruction and Development, the European Investment Bank, the Inter-American Development Bank, the Islamic Development Bank and the World Bank Group (WBG), which also includes the International Finance Corporation. Climate-related components of projects are those that target mitigation, adaptation, or both mitigation and adaptation, based on the joint MDB Climate Finance Tracking Methodology. MDB commitments include concessional and non-concessional support.

Source: OECD DAC statistical system.

StatLink http://dx.doi.org/10.1787/888933484375

Main policy messages

The analysis above points to a wide array of policy priorities that G20 countries can adopt to launch a decisive transition, creating strong, inclusive economic growth while reorienting economies towards low-emission, climate-resilient pathways:

Integrate the climate imperative into structural reform and broader national development strategies, reflecting the role of our physical environment as a fundamental pillar for strong, sustainable, balanced growth.

- Implement **structural reform policies** that boost both productivity and economic activity, as well as supporting the transition to low-emission, climate-resilient economies, through easier resource reallocation; faster technology development and diffusion; greater dynamism in labour markets; and measures to facilitate firm entry and exit.
- Reassess and optimise **national fiscal policies** to increase investment in low-emission, climate-resilient infrastructure and soft investment such as climate-focused R&D, recognising the potential of fiscal measures to revive economic growth and strengthen climate-friendly investment signals.
- Continue to develop **relevant metrics and analytical tools** to incorporate the impacts of climate change and the costs of inaction into economic policy design and implementation, to move towards a more sustainable long-term growth model.
- Pursue a **whole-of-government approach** to low-emission, climate-resilient growth and address barriers and policy misalignments with climate objectives across the investment environment, particularly in infrastructure sectors, using the OECD publication *Aligning Policies for a Low-carbon Economy* as a starting point.
- Improve understanding and management of the interdependencies between **climate change and biodiversity conservation**, in relation to **food security, poverty alleviation and human health and well-being**, which are vital to achieving the Sustainable Development Goals.

Speed up collective and national efforts towards full implementation of the Paris Agreement.

- **Jointly commit to advancing the international stocktaking and oversight mechanisms of the Paris Agreement**, including those on monitoring, reporting and review, and the robust assessment of collective progress, to encourage deeper international co-operation and more ambitious action and support.
- Develop and share experience of **long-term, low-emission development strategies**, and ensure Nationally Determined Contributions and near-term actions are consistent with such strategies. These strategies should address climate and economic development objectives in an integrated way, shaping expectations about the scale and nature of investment needs and helping minimise stranded assets.

Recognise that for growth to be sustainable it must also be inclusive, and ensure that policies to drive the transition towards a low-emission, climate-resilient economy are socially progressive.

- Integrate the **social and economic implications of the transition** more effectively into policies and planning. Support sectoral restructuring by identifying exposed labour forces, communities and regions, by assessing local capabilities, and by developing response measures, including retraining and reskilling of the exposed workforce.

Adopt flexible, forward-looking approaches to decision-making to increase climate resilience and ensure that these approaches are robust given the uncertainty surrounding climate changes effects at local and regional levels.

- Establish a **pipeline of infrastructure projects** that are consistent with long-term, low-emission development strategies, reconciling short-term action and long-term decarbonisation goals, as a means to **shift investment** to climate-resilient infrastructure
- **Bridge data gaps on infrastructure projects** and improve information on investment pipelines, for example with the support of the G20 Global Infrastructure Hub and the OECD.
- Introduce specific policies and regulations, such as spatial planning and technical standards, that promote **climate resilience of infrastructure**, including screening and factoring climate risks in public investments, including procurement procedures.

Realise GHG mitigation potential across the economy.

- Accelerate the **reform of inefficient fossil-fuel subsidies** that encourage wasteful consumption, including agreeing on a date for phasing out such subsidies. As the basis for reform, expand internationally comparable information on subsidies to more countries and types of support, for example through peer review. Share experience on successful and progressive subsidy reforms.
- Broaden the carbon pricing base, track impact and emissions reductions progress, and share policy experience of **effective carbon pricing** to inform flexible forward-looking policy decisions. Explore joint action in this area, such as minimum carbon prices, gradual increases in prices over time, and linking of emissions trading systems.
- Tap the **large mitigation potential in agriculture, forestry and other land-use sectors**. Preserve and expand existing carbon stocks in forests and other ecosystems; avoid net deforestation and forest degradation; and improve soil management, in particular of organic soils. Stimulate mitigation in the agriculture sector by increasing investment in the development and deployment of new technologies and sustainable practices. **Promote efficient and effective use of nitrogen fertilisers and limit their over-use.**
- Make greater use of **public procurement** to invest in low-emission infrastructure and to trigger industrial and business model innovation through the creation of lead markets, for example by introducing climate-related criteria to procurement decisions.
- **Implement and strengthen research, development and demonstration efforts** for breakthroughs in technologies essential for eliminating GHG emissions from industry and from road, maritime and air transport, as well as breakthroughs in energy storage and "negative emissions" technologies, including through international collaborative efforts such as Mission Innovation.

Mobilise financing for the transition.

- **Expand efforts to mobilise private investment** in low-emission, climate-resilient infrastructure by scaling up the use of diversified risk mitigation tools, improved environmental risk analysis, and diversified financial instruments and models.
- Take steps towards a **more climate-consistent global financial system** by assessing and addressing possible misalignments within financial regulations and practices, improving the ability of markets to price climate change risks, and assessing the risks climate change poses to financial stability.
- Call on all development banks and finance institutions – multilateral, bilateral and national – to put in place targets and action plans to boost support for low-emission infrastructure and climate-proofing efforts; improve disclosure of climate risks; scale up efforts to mobilise private investment; and continue to support policy and planning frameworks for climate-resilient infrastructure, especially in vulnerable countries.

Notes

1. The carbon budget from 2015 to 2100 is smaller than this for the same likelihood of remaining below 2°C, requiring negative emissions after the peak. See Chapter 2.
2. The IEA (2017) assumptions, which this report builds on, are therefore conservative in this regard.
3. The electricity sector is the only sector where enough information is available to analyse the pipeline, as surveys and commercial databases track information on capacity in operation, cancelled, announced or at pre-construction stage, as well as under construction.

References

Booz Allen Hamilton (2007), *The global infrastructure investment deficit*, Booz Allen Hamilton, McLean, Virginia.

Global Coal Plant Tracker (2017), *Global coal plant tracker*, http://endcoal.org/tracker/.

IAEA (2016), *Power Reactor Information System*, International Atomic Energy Agency, Vienna, https://www.iaea.org/pris/.

IEA (2017), Chapters 1 and 2 in *Perspectives for the Energy Transition: Investment Needs for a Low-Carbon Energy System*, International Energy Agency/OECD Publishing, Paris, www.energiewende2017.com/wp-content/uploads/2017/03/Perspectives-for-the-Energy-Transition_WEB.pdf.

IEA (2016a), *Energy Technology Perspectives 2016*, IEA/OECD Publishing, Paris, http://dx.doi.org/10.1787/energy_tech-2016-en.

IEA (2016b), *IEA Medium-Term Renewable Energy Market Report 2016: Market Analysis and Forecasts to 2021*, IEA/OECD Publishing, Paris, http://dx.doi.org/10.1787/renewmar-2016-en.

OECD (2012), *Strategic Transport Infrastructure Needs to 2030*, OECD Publishing, Paris, http://dx.doi.org/10.1787/9789264114425-en.

OECD (2006), *Infrastructure to 2030: Telecom, Land Transport, Water and Electricity*, OECD Publishing, Paris, http://dx.doi.org/10.1787/9789264023994-en.

Platts WEPP (2017), *World Electric Power Plants Database*, Platts, London, www.platts.com/products/world-electric-power-plants-database.

WEF (2015), *The Global Competitiveness Report 2015-2016*, World Economic Forum, Geneva, www3.weforum.org/docs/gcr/2015-2016/Global_Competitiveness_Report_2015-2016.pdf.

Woetzel, J. et al. (2016), *Bridging Global Infrastructure Gaps*, McKinsey Global Institute, www.mckinsey.com/industries/capital-projects-and-infrastructure/our-insights/bridgingglobal-infrastructure-gaps.

World Bank (n.d.), World Development Indicators, World Bank, Washington DC, http://data.worldbank.org/data-catalog/world-development-indicators.

Chapter 2

Pathways from Paris

Human interference with the climate system is rapidly taking us into uncharted territory, with the potential for severe and irreversible impacts and making it harder to achieve the Sustainable Development Goals (SDGs). The Paris Agreement aims to limit average global warming to well below 2°C, a political judgement based on scientific evidence. The stringency of this mitigation goal means that countries need to strengthen mitigation action without delay. After setting out the case for urgent action and the carbon budget consistent with the goal of well below 2°C, this chapter examines the characteristics of low-emission pathways and how country diversity may impact the scale, phasing and priorities for mitigation action across countries. It then summarises projected impacts, emphasising the need for flexible, forward-looking approaches to decision-making that reflect the diversity of climate vulnerabilities and confidence levels about local and regional change. Finally, the chapter looks at how countries can get to where they need to be, supported by the mechanisms of the Paris Agreement.

This chapter sets out the case for urgent action on climate change and explains in broad terms what is required to move to low-emission, climate-resilient development pathways. The first section explains why we need to act urgently. The second section assesses the carbon budget consistent with the "well below 2°C" goal in the Paris Agreement, and how this in turn depends on developments in the non-energy sector – notably in agriculture, forestry and land-use (AFOLU). The third section examines the characteristics of low-emission pathways, taking as its core a scenario consistent with a 66% likelihood of keeping the global average surface temperature increase to below 2°C throughout the century (IEA 66% 2°C scenario) from a parallel report for the German G20 Presidency on the scale and scope of energy sector investments needed to increase the chances of reaching this goal (IEA, 2017). This section also analyses the IEA 66% 2°C scenario in the context of a broader range of scenarios achieving similar outcomes. The fourth section then examines how country diversity may affect low-emission pathways and the priorities for action across countries. Even with stringent mitigation, climate change is projected to have significant negative impacts, so countries need to enhance resilience and increase their adaptive capacity. The projected changes in regional and local conditions are far less well understood than larger-scale changes in temperature, sea-level rise and ocean acidification.[1] The fifth section summarises projected impacts and emphasises the need to develop flexible, forward-looking approaches that help us to identify robust solutions. The last section of this chapter addresses the key question of how countries can get to where they need to be from where they are now, highlighting the fundamental importance of the Paris Agreement in building trust and transparency to underpin effective international action.

Climate change – why we need to act urgently

The last 60 years or so have seen unprecedented human impact on the systems that underpin life on Earth (Steffen et al., 2004). Industrial-scale agriculture and the massive use of fossil energy to drive economic growth have transformed the life chances of billions of people.[2] But they have also created an unpredictable climatic future, very different from the conditions in which humanity has thrived for the past 10 000 years. Since 1990, world GDP has more than doubled while carbon dioxide (CO_2) emissions from fossil fuel use have increased by some 60%, contributing to increasingly rapid climatic change (Figure 2.1).

Other environmental challenges have also emerged, such as ozone depletion, biodiversity loss, desertification, and local and regional pollution. Rapid progress on reducing ozone depletion has been possible, underpinned by international agreements targeting ozone depleting chemicals. Other "wicked" problems have proved more resistant to progress (Rittel and Webber, 1973). Notable among these is climate change, which both poses profound challenges to our current development paradigm and, at the same time, opens up opportunities for sustained and sustainable improvements in inclusive economic well-being.

Climate change in context

Global atmospheric concentrations of CO_2 – the major greenhouse gas (GHG)[3] – have now risen past 400 parts per million (ppm by volume) from a pre-industrial level of around 280 ppm (Figure 2.1). By 2012, the global mean surface temperature had increased by approximately 0.85°C on average from pre-industrial levels;[4] each of the last three decades has been successively warmer than any preceding decade since 1850 (IPCC, 2014a). In 2015, global mean temperatures went 1°C above pre-industrial levels for the first time, due to the combined effects of climate change and a very strong El Niño that lasted into early 2016. All but one of the 16 warmest years on record has occurred since 2001, with 2016 the hottest recorded (WMO, 2017).

Figure 2.1. **Global CO_2 emissions from fossil-fuel use and cement production, and the atmospheric concentration of CO_2**

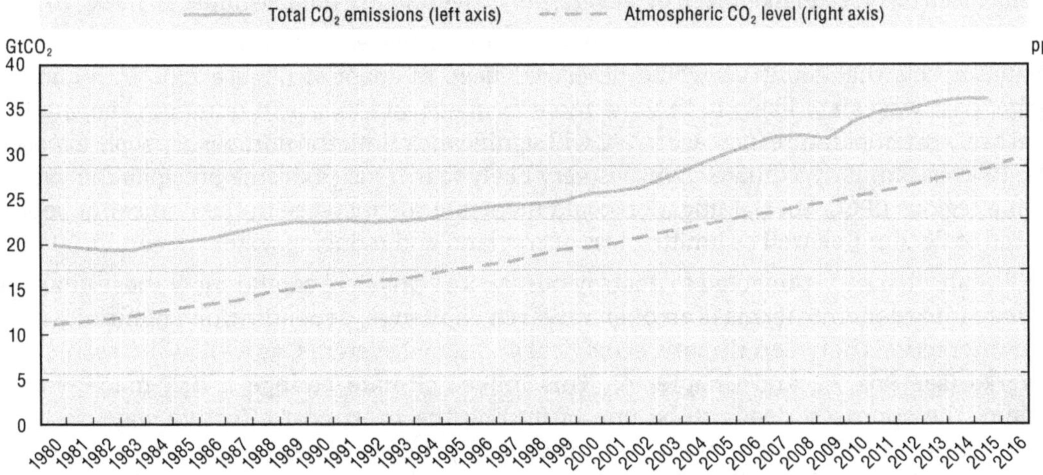

Sources: i) CO_2 emissions from Olivier et al. (2016); ii) Global atmospheric CO_2 concentrations from NOAA (2017).
StatLink http://dx.doi.org/10.1787/888933484019

So where might we be heading? Projections of climate change depend on inherently uncertain assumptions about human behaviour and future policy choices. It is also difficult to estimate the precise strength of the climate response to atmospheric GHG concentrations, due to the complexity of the climate system.[5] Scenario analysis has therefore been a vital analytical tool in helping us understand the range of plausible future outcomes and how these depend on future emissions of GHGs and atmospheric aerosols, land-use change, and many other socio-economic factors.

Table 2.1 shows end-of-century projections for global mean surface temperature relative to pre-industrial levels (1850-1900) from the most recent assessment by the Intergovernmental Panel on Climate Change (IPCC), for four Representative Concentration Pathways (RCPs) (IPCC, 2013).[6] The scenario associated with the lowest emissions, RCP2.6, is consistent with a policy target of limiting warming to below 2°C with greater than 66% likelihood, broadly in line with the IEA 66% 2°C scenario (IEA, 2017). None of the other RCPs deliver mean surface temperature changes of 2°C or lower.

Table 2.1. **Projected mean temperature changes relative to a pre-industrial (1850–1900) baseline**

Emissions scenario	Change in mean temperature (°C) by 2081-2100
Low scenario – RCP2.6	1.6
Medium scenario – RCP4.5	2.4
Medium to high scenario – RCP6.0	2.8
Very high scenario – RCP8.5	4.3

Note: The temperature changes for each RCP include an observational estimate of warming of 0.61°C between 1850-1900 and 1986-2005 and the mean warming across CMIP5 Global Climate Models between 1986-2005 and 2081-2100 for the RCP. Both the observed historical warming and GCM-derived components of the changes have uncertainties. These are not presented as methods are not generally available in the literature for combining the uncertainties in models and observations.
Source: IPCC (2013).

Climate risks and the benefits of mitigation

Climate change will lead not just to higher temperatures but also to rising sea levels, acidification of the oceans – with effects on marine ecosystems – and changing patterns of precipitation, as well as more extreme weather. Regions will be affected differently by these changes; regional (and smaller-scale) changes in weather patterns and precipitation are still highly uncertain (see for example, Shepherd, 2014). Changes could even take us beyond thresholds or "tipping points" in the climate system (Box 2.1). Greater levels of emissions will therefore lead to a greater likelihood of "severe, pervasive and irreversible impacts" (IPCC, 2014b).

Stringent mitigation action to limit temperature increases would moderate the physical climate impacts that countries would otherwise need to adapt to (Figure 2.2). With climate change, heat waves are likely to become more frequent and longer in duration; keeping the global average temperature increase to 2°C will significantly limit the number of people exposed to heatwaves. Similarly, climate change is very likely to increase extreme precipitation events in some regions (IPCC, 2013). Mitigation could moderate the increase in the number of people exposed to flooding, as well as limiting loss of cropland and reducing water stress.

Climate change is projected to destroy human and physical capital. How these changes translate into economic terms is an open research challenge, depending on potentially non-linear interactions between climate, ecological and social systems, as well as infrastructure networks (see Box 2.1 and Chapter 4). This makes climate change a risk management problem: the approach needs to be one of finding the most cost-effective ways to limit climate risks to a politically agreed level, informed by the best scientific evidence. Early and ambitious action on adaptation and mitigation can significantly reduce these risks.

Figure 2.2. **Estimates of climate change impacts avoided by 2100 through mitigation**

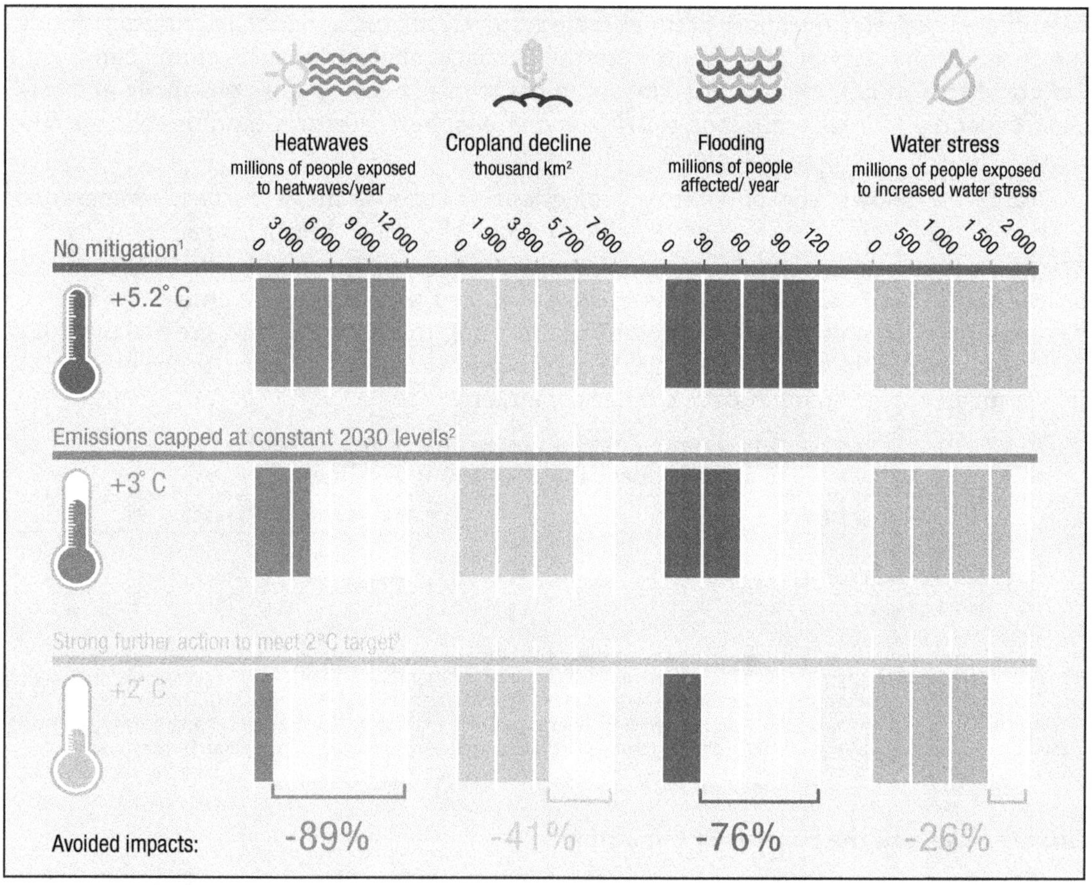

Notes: (1) Refers to RCP8.5 scenario. (2) Emissions capped 55.1 $GtCO_2e$, consistent with the NDCs, with no backtracking. (3) Strong further action for a 50% chance of meeting the 2°C target: emissions of 55.1 $GtCO_2e$ in 2030, with further large reductions in GHG emissions to meet 2°C by 2100.

Source: AVOID2 (2015).

Box 2.1. **Thresholds for abrupt and/or irreversible change**

The level of scientific understanding of thresholds in the climate systems, as well as the physical and economic implications of crossing such thresholds, is low. Such potential changes include the collapse of the Atlantic Meridional Overturning Circulation (AMOC), the disappearance of summer Arctic sea ice, ice sheet collapse, permafrost carbon release, methane release, and tropical and boreal forest dieback.

Recent research has given greater confidence to evidence that partial irreversible loss of the West Antarctic Ice Sheet has already begun. Tropical forests are being adversely affected by drought, while AMOC weakening continues. Interaction between different thresholds will be important in determining the timescales, extent and reversibility of changes throughout the climate system. For example, increased meltwater from ice sheets will further weaken the AMOC, and this may in turn alter the position of the Intertropical Convergence Zone near the equator, affecting rainfall patterns and the health of the Amazon rainforest (Lenton et al., 2008).

Figure 2.3. **Examples of thresholds for abrupt and/or irreversible climate impacts**

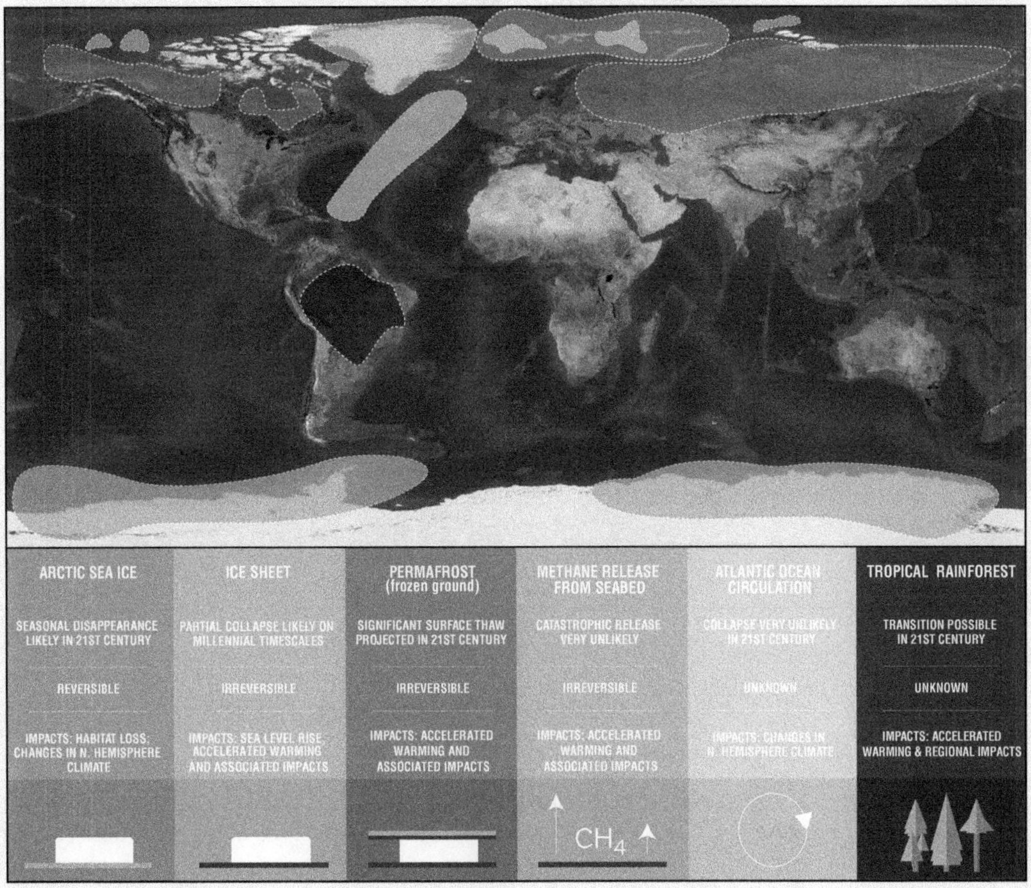

Note: There is considerable uncertainty relating to the reversibility of climate impacts. Here, impacts are considered irreversible if recovery is unlikely within 100 years after climate drops back below the relevant threshold.

Source: MOHC analysis of *i)* IPCC, 2014c and *ii)* AVOID2 WPA.5 Report.

What does the Paris Agreement mean for carbon budgets?

The interpretation of "well below 2°C"

The Paris Agreement reached at the 21st Conference of the Parties to the UNFCCC (COP21) in December 2015 aims to hold the global average surface temperature increase to "well below 2°C and to pursue efforts to limit it to 1.5°C above pre-industrial levels, recognising that this would significantly reduce the risks and impacts of climate change" (UNFCCC, 2015a). There is, however, no precise definition of what "well below 2°C" means.

It is not immediately obvious that the IEA 66% 2°C scenario used in the related IEA report (IEA, 2017) should be equated to a "well-below 2°C" goal. However, UK Meteorological Office Hadley Centre (MOHC) analysis of the many scenarios analysed as part of the IPCC's AR5[7] suggests that in general, scenarios delivering a greater than 66% likelihood would be somewhat more stringent in terms of emissions reductions than scenarios consistent with 1.75-2.0°C of median warming by 2100 (Figure 2.4). Most of these stringent IPCC mitigation scenarios (the thin coloured lines in Figure 2.4) rely on net negative CO_2 emissions, whereas the IEA 66% 2°C scenario assumes no net negative emissions. It is therefore reasonable to use the IEA 66% 2°C scenario as one representation of what a well-below 2°C scenario could look like, though of course there are other plausible pathways that could include net negative emissions.

Figure 2.4. **IPCC AR5 CO_2 emissions scenarios with a greater than 66% chance of staying below 2°C**

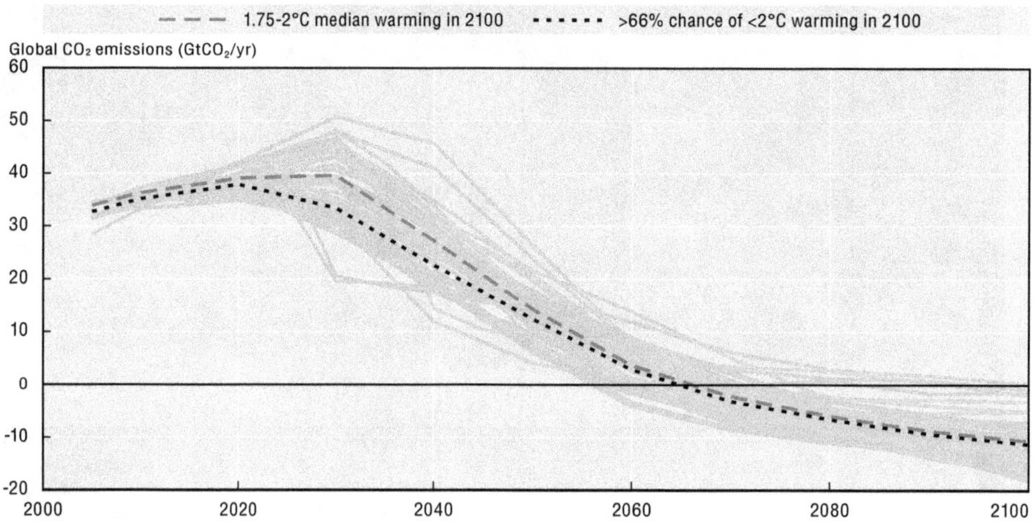

Source: IPCC AR5 Database, MOHC analysis.

StatLink http://dx.doi.org/10.1787/888933484028

Carbon budgets and temperature goals

CO_2 is the predominant GHG, but many other gases contribute to global warming (Box 2.2). For long-lived GHGs, such as CO_2, it is the cumulative level of emissions over time that determines the contribution to climate change, not just the emissions in a given year. There is a strong linear relationship between cumulative CO_2 emissions and the increase in average surface temperatures (Wigley, Richels and Edmonds, 1996; Allen et al., 2009; IPCC, 2013; Friedlingstein et al., 2014). This means that there is an upper limit on the total cumulative CO_2 emissions over time consistent with a given temperature target – the so-called "carbon budget". This budget is not a single number but a range, reflecting uncertain projections about the emissions of non-CO_2 GHGs, as well as in the climate response to GHGs in the atmosphere.[8]

Box 2.2. Greenhouse gases, aerosols and radiative forcing

Climate change is due to a net imbalance in the energy flowing into the Earth system due to human modifications of the atmosphere. CO_2 is responsible for most of the warming observed since the pre-industrial period (1.68 ± 0.035 Watts per metre squared (W/m^2) in 2011 relative to 1750), but other gases play an important role in this "radiative forcing" – tipping the balance of radiation flowing into the Earth's atmosphere.

- Atmospheric concentrations of methane (CH_4) reached 1,810 parts per billion (ppb) in 2012, 2.5 times more than in 1750. Even at these small concentrations, CH_4 has contributed about 20% of the radiative forcing of CO_2 (Ciais et al., 2013).
- Atmospheric nitrous oxide (N_2O) is another important GHG, with a radiative forcing of 0.17 ± 0.03 W/m^2 in 2011 compared with the pre-industrial period. Concentrations have risen more than 20% since pre-industrial times, mostly due to increased agricultural activity, with a lesser contribution from the burning of fossil fuels and industry (Ciais et al., 2013).
- Chlorofluorocarbons (CFCs) and hydrochlorofluorocarbons (HCFCs) contribute approximately 11% of the total radiative forcing from GHGs and also deplete stratospheric ozone (O_3). Emissions of CFCs have been drastically reduced in recent years as the Montreal Protocol has been implemented, but due to their long lifetime it will take a substantial amount of time to affect atmospheric concentrations.
- The effect of atmospheric ozone (O_3) depends on where it is situated. In the lower atmosphere, O_3 is formed when other chemical species, such as CH_4 and carbon monoxide, combine with nitrogen oxides (NOx) in sunlight, contributing to poor air quality. Stratospheric O_3 has a small cooling effect, but overall ozone has a warming effect of around 0.35 (0.15 to 0.55) W/m^2 (Myhre and Shindell, 2013).
- Aerosols are microscopic particles suspended in the atmosphere that generally cool the climate, yet some have a warming effect (e.g. black carbon). IPCC AR5 (IPCC, 2013) estimated the radiative forcing of aerosols to be -0.9 (-1.9 to -0.1) W/m^2 (Myhre and Shindell, 2013), an overall cooling effect on the climate. Aerosols and their interactions with clouds offset a substantial portion of global mean warming, but aerosols contribute the largest uncertainty to the total radiative forcing estimate.
- Land use change from human activity also affects the Earth's climate, by changing the surface albedo (how much light it reflects) and by increasing the emission of GHGs (e.g. through deforestation). Afforestation also absorbs CO_2 from the atmosphere. Land use change has significant impacts on the local water cycle and can lead to changes in rainfall in regions far away from the initial land use change (e.g. DeAngelis et al., 2010).

Carbon budgets consistent with 2°C and 1.5°C temperature targets are shown in Table 2.2, along with an indication of the likelihood of limiting warming to this level. These budgets assume non-CO_2 GHG emissions contribute the equivalent of around 420 gigatonnes of CO_2 ($GtCO_2$) (Rogelj, 2016b). The global carbon budget compatible with a greater than 66% likelihood of staying below 2°C is estimated to be 590-1 240 $GtCO_2$ from 2015 to the time of peak warming (Rogelj, 2016b). This represents roughly 15 to 30 years of fossil fuel-related CO_2 emissions at current rates – an indication of the remarkably short time remaining in which to transform the global energy system and to meet the Paris Agreement's temperature goal. Even this challenging number assumes net negative CO_2 emissions later in the century. The carbon budget to limit the temperature increase to 2°C with a 66% likelihood by 2100 is more stringent – between 470 and 1 020 $GtCO_2$.

This downwards adjustment reflects the fact that to achieve such a stringent mitigation target, modelling suggests that it would be more cost-effective to reduce emissions at a slightly lower – but still rapid – pace early on and then to compensate with "negative emissions" later in the century. Drawing CO_2 back down from the atmosphere and sequestering it safely over the long term enables such scenarios to live within their carbon budgets.[9] The most plausible options for achieving this are afforestation, bioenergy with carbon capture and storage (BECCS) and changed agricultural practices (Box 2.3).

The total carbon budget used in the IEA 66% 2°C scenario is 880 $GtCO_2$. This budget lies below the mid-point of the "peak warming" range (915 Gt CO_2) and above the mid-point of the range for the entire period 2015-2100 (745 Gt CO_2). The IEA 66% 2°C scenario assumes no net negative emissions. Out of this total budget of 880 $GtCO_2$, the IEA allocates a carbon budget of 790 $GtCO_2$ for the energy sector, and assumes that 90 $GtCO_2$ over 2015-2100 are emitted from industrial processes. Land use is assumed to generate approximately net zero cumulative emissions over the period, starting from positive emissions and becoming negative by the end of the century. Non-CO_2 GHGs are assumed to contribute around 0.5°C of warming by 2100 (IEA, 2017).

Table 2.2. **Carbon budgets from 2015 to peak warming for different temperature targets and likelihoods**

Temperature targets	>50% < 2°C	>66% < 2°C	>50% 1.5°C
Global carbon budget available from 2015 to peak warming (Gt CO_2)	990-1 240	590-1 240 [470-1 020]+	390-440

Note: Figures represent 10th-90th percentile range. The budget to peak warming may include negative emissions, but not any net negative emissions required after peak warming. +This denotes the global carbon budget over the whole period 2015-2100, taking account of net negative emissions after the peak.
Source: Adapted from Rogelj, 2016b; IPCC, 2014c.

Box 2.3. **What are negative emissions?**

Owing to the long time scales involved in the removal of carbon from the atmosphere by natural processes, recovery from an overshoot of the atmospheric CO_2 concentration may take a considerable amount of time (Lowe et al., 2009; Solomon et al., 2009). Technologies that actively remove carbon from the atmosphere – resulting in "negative emissions" – could be used to lower atmospheric CO_2 in the event of an overshoot in emissions, but could also be important in offsetting emissions from sectors where emissions reductions are more difficult (such as freight, aviation and shipping). Several options have been examined for negative emissions technologies (NETs):

- **Afforestation and reforestation (AR)** to fix atmospheric CO_2 in terrestrial biomass and soils. Potential is estimated at 4 $GtCO_2$/yr at a lower cost than BECCS and with land and nutrient requirements increasing with potential (Smith et al., 2015).
- **Changed agricultural practices (CAP)**, such as soil management practices that can improve soil quality by reducing soil erosion and increasing resilience to weather variability, while simultaneously contributing to food security objectives (OECD, 2015e). Soil carbon sequestration and biochar each have the potential to provide about 2.6 $GtCO_2$eq/yr and have fewer disadvantages than many NETs (Smith, 2016).
- **BECCS:** Farming bio-energy crops that absorb CO_2 as they grow and are then burnt for energy, with the resulting emissions captured and stored underground. Potential is estimated at around 12 $GtCO_2$/yr (Smith et al., 2015).
- **Direct air capture (DAC):** The use of chemicals to absorb CO_2 from the atmosphere before being stored in solid form or pumped into geological reservoirs. Potential is estimated at around 12 $GtCO_2$/yr but at a far greater cost and energy requirements than BECCS (Smith et al., 2015).

Box 2.3. **What are negative emissions?** *(cont.)*

- **Enhanced weathering (EW):** Natural weathering of minerals is accelerated to remove CO_2 from the atmosphere, with the products stored in soils or buried in the land surface. Potential is estimated at around 0.7 $GtCO_2$/yr (Smith et al., 2015).
- **Ocean fertilisation (OF):** Increasing the ocean's biological uptake of CO_2 by fertilising nutrient-limited areas.

These NETs each have large but varied levels of uncertainty over their social acceptability, unresolved technological issues and high costs, and variable demands for land, water, energy and fertiliser, which affect their feasibility and efficacy at scale (Smith et al., 2015). DAC is considered to have very high costs and energy requirements. EW is also a high-cost technology as well as having a limited global potential for emissions removal and significant requirements for land use. OF by contrast is seen as too risky as little is known about the ecological effect of dumping large quantities of nutrients into the sea (Schiermeier, 2007), nor does it do anything to address ocean acidification. AR and BECCS are typically the only NETs included as mitigation options in current generations of Integrated Assessment Models. The extent to which these technologies can be deployed at scale in the near- to medium-term is a key uncertainty.

Low-emission pathways

Characteristics of low-emission pathways

As can be seen from Figure 2.4 and the tight constraint on carbon budgets consistent with limiting temperature change to well below 2°C, low-emission pathways will be characterised by the following broad features:

1. A peak in global emissions as soon as possible;
2. A subsequent rapid fall in GHG emissions, particularly of CO_2 emissions;
3. Net GHG emissions approach zero or even become net negative in the second half of the century (IPCC, 2014a).

The later the peak in global CO_2 emissions, the greater the rate of emission reduction required subsequently to be consistent with the carbon budget. Options for achieving stringent mitigation goals may be lost if the peaking level is too high or too late. Delaying peaking beyond 2020 would make the Paris Agreement's goal of well below 2°C significantly more difficult to achieve, requiring even more rapid reductions of emissions and a prolonged period of net negative CO_2 emissions through major afforestation or the large-scale use of negative emissions technologies such as BECCS (Box 2.3). Action will need to come earlier and the fall-off in emissions will need to be more rapid if even more stringent targets are to be achieved (e.g. towards 1.5°C). Not reaching a global emissions peak before 2030 may preclude limiting warming to well below 2°C.

Assumptions for future non-CO_2 GHG emissions constrain the carbon budget available for the energy sector and industrial processes.[10] While CO_2 emissions will eventually need to go to zero, or below, annual emissions of short-lived GHGs such as CH_4 only need to be stabilised and can still remain positive while meeting the goal of well below 2°C (Allen et al., 2016). The higher the level at which such emissions are stabilised, however, the lower the carbon budget consistent with a given temperature goal will be (Allen et al., 2016).[11] For N_2O, a long-lived GHG, it is the cumulative level of emissions over time, not the level of emissions in a given year that matters most for maximum temperature change (Smith et al, 2012).

N_2O emissions are predominantly due to agriculture.[12] Population and economic growth are increasing demand for food, so N_2O emissions will continue for the foreseeable future to ensure food security, even if we can improve the efficiency of fertiliser use (Zhang et al., 2015). As a long-lived GHG, continued N_2O emissions would need to be offset by a reduction of other long-lived GHGs – for example, by greater negative emissions of CO_2.

The IEA pathways in context

Socio-economic developments, including economic and population growth and food demand, will influence whether future GHG emissions will be consistent with a well below 2°C target. The Shared Socio-Economic Pathways (SSPs, Riahi et al., 2017) provide a set of storylines exploring the implications of different assumptions about future economic growth, demographics and technical change. Together with the IPCC's RCPs, they provide a framework to analyse and evaluate the implications of climate policy in different socio-economic settings. In this section, the IEA 66% 2°C scenario is compared with modelling results[13] for a "middle-of-the-road" SSP scenario (SSP2), coupled with the IPCC's RCP 2.6 scenario (together, SSP2-2.6).[14]

Figure 2.5 shows the evolution of non-land-use CO_2 emissions for the IEA 66% 2°C scenario alongside the SSP2-2.6 comparison range. The IEA emissions numbers encompass both energy-related emissions and industrial process emissions:[15] the IEA non-land use CO_2 emissions pathway lies at the lower edge of the range of the SSP mitigation scenarios to 2050. The IEA's assumption of no net negative CO_2 emissions means that to meet the carbon budget constraint, emissions must peak earlier and lower than in the scenarios that do allow net negative emissions. The range of non-land-use CO_2 emissions in SSP2-2.6 becomes negative by the end of the century, due to extensive use of BECCs. The IEA 66% 2°C scenario rules out net negative CO_2 emissions and lies at the upper end or above the SSP2-2.6 range at the end of the century. Its lower CO_2 emissions early on allow the IEA scenario to still remain below 2°C with a 66% likelihood.

Figure 2.5. **Projections of non-land use CO_2 emissions**

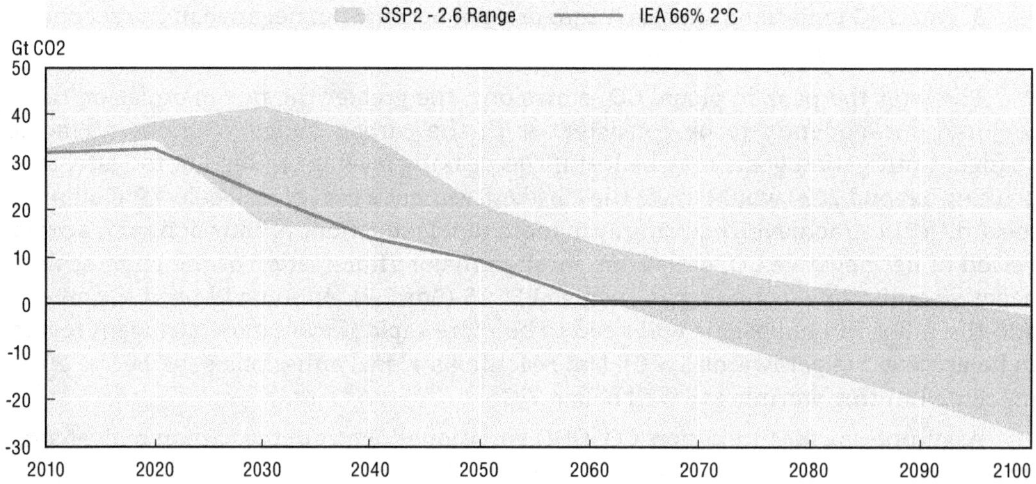

Source: IIASA (n.d.) and IEA (2017).

StatLink http://dx.doi.org/10.1787/888933484038

Figure 2.6 provides a similar comparison between the IEA and SSP scenarios for CO_2 emissions from land-use change. Land use in the IEA 66% 2°C scenario turns from a source to a small sink by 2050 and emissions lie well within the range of emissions in the SSP2-2.6 modelling results. The outcomes of one particular modelling realisation of SSP2-2.6 (the

GCAM model) display extreme changes in land-use emissions due to strong dependence on afforestation and the use of bioenergy (at different times) as mitigation options, which leads to steep projected increases in food prices towards the end of the century (Popp et al, 2017).[16]

Figure 2.6. **Projections of land-use change CO_2 emissions**

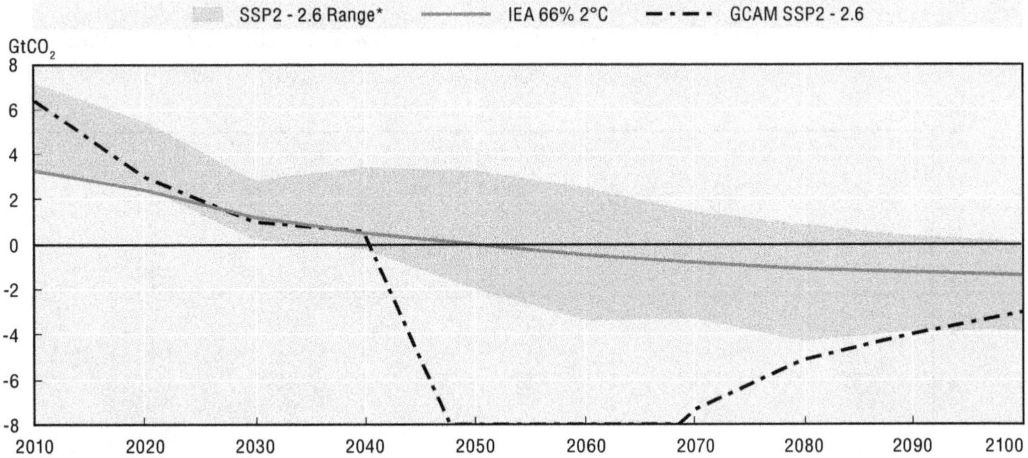

Notes: *SSP2 range excluding GCAM results.

Source: IIASA (n.d.) and IEA (2017).

StatLink http://dx.doi.org/10.1787/888933484042

Since the IEA land-use assumption aligns better with the other model realisations of SSP2-2.6, the IEA scenario would seem to be consistent with much smaller projected increases in food prices to 2100. This conclusion is further strengthened by examining projections for total bioenergy in energy demand in these different scenarios. Again, the IEA projections for total bioenergy demand align closely with the SSP2-2.6 range to 2050 as shown in Figure 2.7. In all the SSP2-2.6 scenarios, energy from traditional biomass is projected to fall sharply after 2020, while BECCS increases rapidly. The IEA assumes a modest amount of BECCS in 2050 (about 2 exajoules (EJ)/yr in the power sector), which increases the pressure on the energy system to decarbonise earlier and faster, including through the extensive use of CCS in the industrial sector (IEA, 2017)

Figure 2.7. **Bioenergy projections in the IEA 66% and SSP2-2.6 scenarios**

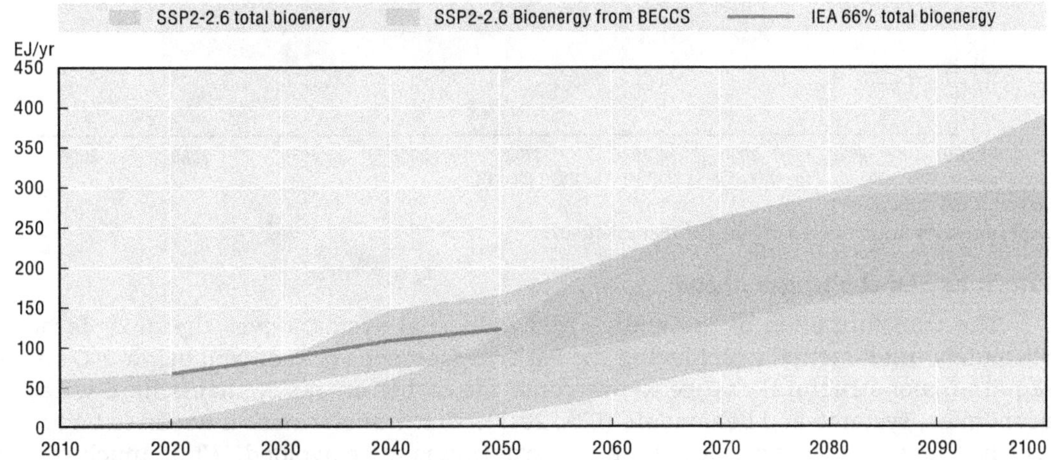

Source: IIASA (n.d.) and IEA (2017).

StatLink http://dx.doi.org/10.1787/888933484052

Turning to the main non-CO_2 GHGs, Figures 2.8 and 2.9 compare the range of CH_4 and N_2O emissions in the IEA 66% 2°C and the SSP2-2.6 scenarios. There is a wide range of projections and a much wider range still if we consider less stringent mitigation outcomes or other future socio-economic storylines. Any lack of progress in mitigating emissions to this level – particularly of N_2O – would clearly reduce the chances of staying below 2°C, or require offsetting net negative emissions through afforestation, BECCS or another approach.

Figure 2.8. **Methane emissions in the IEA 66% 2°C and SSP2-2.6 scenarios**

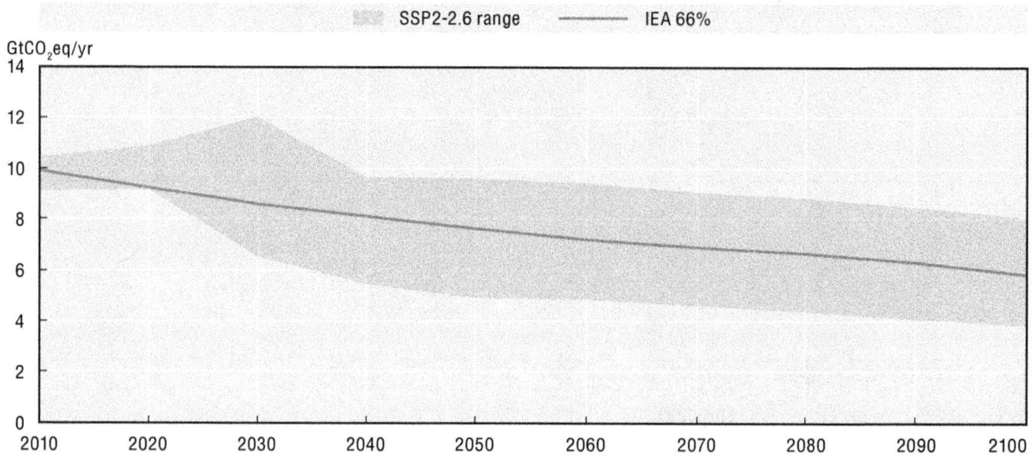

Note: uses a GWP 100 value for CH_4 of 28 (Table 8.7 of IPCC (2013)).
Source: IIASA (n.d.).
StatLink http://dx.doi.org/10.1787/888933484069

Figure 2.9. **Nitrous oxide emissions in the IEA 66% 2°C and the SSP2-2.6 scenarios**

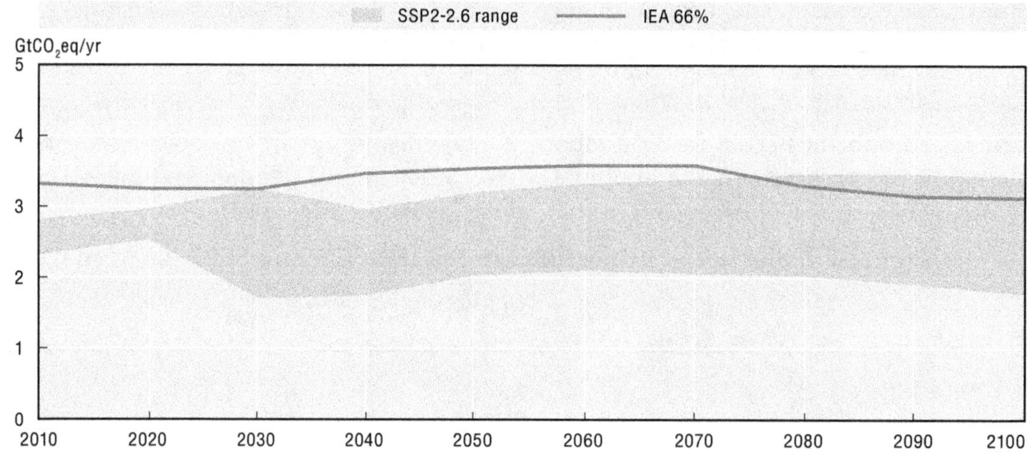

Note: uses a GWP 100 value for N_2O of 265 (Table 8.7 of IPCC (2013)).
Source: IIASA (n.d.).
StatLink http://dx.doi.org/10.1787/888933484072

Priorities and challenges ahead

The transformation of the energy and industrial systems over the next decades is absolutely fundamental to achieving the Paris Agreement's goal of well below 2°C and will require major structural change to overcome the carbon-intensity that is hard-wired into economies, systems and behaviours (IEA, 2017). That transformation needs to be effected within a few decades if serious climatic disruption is to be avoided. While much progress can and needs to be made now based on currently available technologies, we will also need to develop new technologies and infrastructure to bring us within reach of the very low or negative emissions required by the second half of the century.

Outside the energy and related end-use sectors, the extent of GHG emissions from AFOLU sectors will set the pace and nature of the transition needed in the energy sector. Additionally, mitigation options within the AFOLU sectors may be the critical determinant of whether these stringent mitigation scenarios are feasible, notably afforestation and avoided deforestation[17], bioenergy, BECCS and more GHG-efficient and productive agriculture. Availability of bioenergy is uncertain; estimates suggest it could account for 3% to 37% of the global energy share by 2050, and 23% to 50% of the global energy share by 2100 in a 2°C scenario, with models projecting more than half of modern biomass primary energy coming from non-OECD countries (Rose et al., 2014). The bioenergy share in the IEA 66% 2°C scenario falls within this range, as it does in IRENA's comparable scenario where bioenergy accounts for around 21% of total final energy consumption by 2050, growing from 13% today. Developments in AFOLU are highly uncertain, however, and depend on many factors including technical progress, demographics and demand side developments, such as dietary preferences (Box 2.4).

Box 2.4. **Competing priorities for land**

A central issue for the future of AFOLU emissions is how the demands for food production and for climate mitigation are managed. Food demand is projected to grow strongly through the century along with population and economic growth. The United Nations Food and Agriculture Organisation (FAO) estimates indicate that to meet the demand projected for 2050, global agricultural production must grow 60% above the level of 2005-07 (FAO, 2013). In parallel to increasing food production, reducing food losses and waste "from field to fork" would ease environmental pressures and climate impacts by improving efficiency along the food supply chain (OECD, 2016b).

Over the last five decades (between 1961-63 and 2007-09) agricultural production has increased by 170%. Increased agricultural demand has so far been met largely through improvements in yield (which accounted for 80% of the agricultural production increase), rather than land expansion (20% of the production increase) (OECD, 2012). But the rate of yield growth for most crops has been decelerating in the past few decades, even though it is still increasing in absolute terms (FAO, 2013). So without further yield improvements, demand for agricultural land is likely to grow, increasing the associated CH_4 and N_2O emissions. On the other hand, improving growth in agricultural Total Factor Productivity (TFP) through increased research, development and innovation has the potential to meet demand for food production while using fewer environmental resources and inputs, and emitting fewer GHGs (OECD, 2014). The AFOLU sectors could even become a net sink for CO_2 before the end of the century (IPCC, 2014a).

The demand for bioenergy for climate mitigation could grow rapidly through the century (Figure 2.7), raising questions about both the compatibility of large-scale bioenergy production with food security, and the sustainability of bioenergy in terms of life-cycle emissions and impacts on water and ecosystems, which will vary depending on the particular bioenergy technology and where and how it is applied.

Uses of bioenergy include fuels to replace fossil fuels, particularly in aviation and freight, heating for industrial processes, and as an input to negative emissions technologies (Box 2.3), such as BECCS. If deployed at sufficient scale, this sort of technology could deliver two major economic benefits: i) allow the transition to low-emission technologies to be more gradual than otherwise would be necessary; and ii) offset emissions from any sectors in which mitigation proved technically, economically or socially too difficult.

The greater the scale at which bioenergy is used and produced, however, the greater the tension with food security objectives, in the absence of demand-side measures such as dietary changes that reduced the relative demand for meat products, and reduced food waste (Smith et al., 2013).

Negative emissions technologies and other bioenergy uses will clearly affect other aspects of the Sustainable Development Goals, such as food production, water availability and biodiversity (Smith et al., 2013). The feasibility and acceptability of BECCS is uncertain, in terms of deployment of CCS technologies (see Chapter 6), as is the availability of arable land to meet the simultaneous demand for food production and for biomass for energy (Box 2.4). The IPCC AR5 mitigation scenarios consistent with a less than 2°C target require 210 $GtCO_2$ of BECCS annually by 2050 – which is of the same order of magnitude as the natural terrestrial and ocean carbon sinks – with cumulative global negative emissions typically up to 1,000 $GtCO_2$ over the century (Fuss et al., 2014). The sustainability of bioenergy feedstock is also a significant concern, in particular to guarantee a net zero carbon footprint. There is some degree of consensus among experts that the technical potential for sustainable bioenergy – the potential that is theoretically available before cost considerations are taken into account – is around 100 EJ per year (Creutzig et al., 2015).

In terms of energy use, the priority is to achieve rapid and transformational improvements in:

- *energy efficiency*, from the use of more efficient equipment, such as improved motors or internal combustion engines, from energy-efficient buildings and power plants, and from greater resource efficiency across the life-cycle of products (Box 2.5);
- *emissions intensity of energy*, by replacing emissions-intensive generation capacity and fuels with low-emission generation sources such as wind or solar, and the use of biofuels where they have a low life-cycle of emissions.

Box 2.5. **The importance of resource efficiency for climate goals**

Since 1990, the global use of material resources has grown broadly in line with global GDP, though slightly less rapidly. Global material resource consumption is projected to double by 2050 (OECD, 2016a). GHG emissions from the waste sector typically account for a few percent of total GHG emissions in OECD member countries, but this only represents direct emissions primarily from landfill methane emissions and incinerators. Resource efficiency improvements through an approach of "reduce, reuse and recycle" can support climate mitigation objectives and contribute to achievement of some of the SDGs.

The energy requirements and GHG emissions associated with the production, consumption and end-of-life management of materials can only be assessed by taking a systems view of the production of goods and fuels, transportation of goods, crop and food production and storage, and disposal of food and waste. The life-cycle GHG emissions arising from material management activities were estimated to account for 55% to 65% of national emissions for four OECD member countries, suggesting significant potential to reduce emissions through material resource efficiency measures (OECD, 2012). Substituting secondary, recycled materials for primary materials can significantly reduce GHG emissions (Table 2.3).

Table 2.3. Relative energy and carbon intensity of primary and secondary metal production

Material	Primary Energy TJ/100,000t	Secondary Energy TJ/100,000t	Primary CO_2 ktCO2/100,000t	Secondary CO_2 ktCO2/100,000t
Aluminium	4 700	240	383	29
Copper	1 690	630	125	44
Ferrous	1 400	1 170	167	70
Lead	1 000	13	163	2
Nickel	2 064	186	212	22
Tin	1 820	20	218	3
Zinc	2 400	1 800	236	56

Source: International Bureau of Recycling, 2008

Economic and population growth and increased fossil fuel use have been the main drivers behind the approximately 60% increase in global CO_2 emissions since the early 1990s. Global CO_2 emissions from energy use have increased less rapidly than GDP and energy use per unit of GDP globally has fallen by around 31%. However, at the same time, the CO_2 intensity of energy actually increased by 3%. Figures 2.10 and 2.11 show the historical performance of G20 countries on these two key measures compared with the levels projected in the IEA's 66% 2°C scenario.

The IEA estimates that the energy intensity of G20 economies would need to fall by more than 60% between 2014 and 2050 (IEA, 2017), a rate of around 3% a year from 2020 to 2050. Daunting as this sounds, it is broadly in line with historic achievements by the G20 countries. More challenging is the more than 75% reduction in CO_2 intensity of energy that is simultaneously required, an average rate of 4.4% a year from 2020 to 2050. Here historic trends are far less encouraging: achieving this scale of change will require an unparalleled increase in the deployment of low-carbon technologies (IEA, 2017).

Figure 2.10. **Energy intensity of GDP for G20 countries**

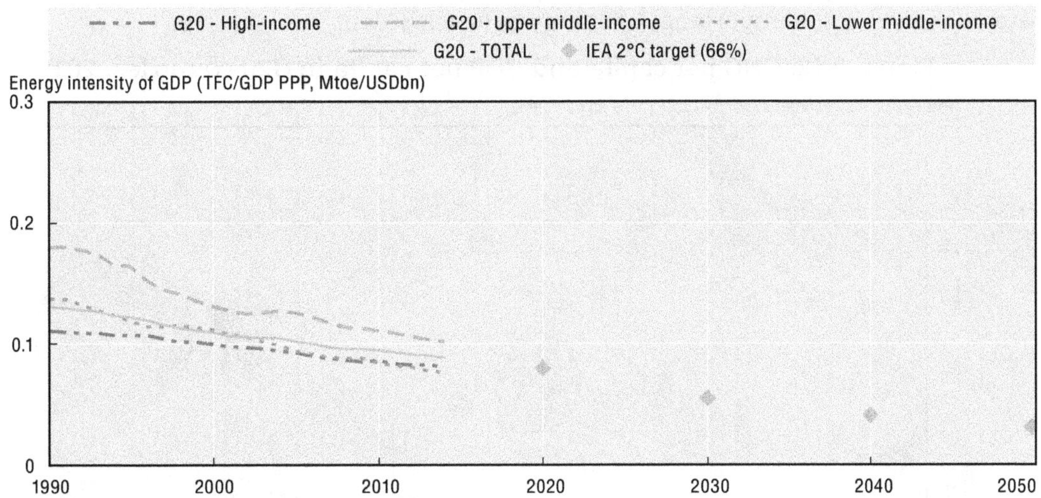

Source: World Bank (n.d.a.) and IEA (2017).

StatLink http://dx.doi.org/10.1787/888933484083

Figure 2.11. **CO_2 intensity of energy for G20 countries**

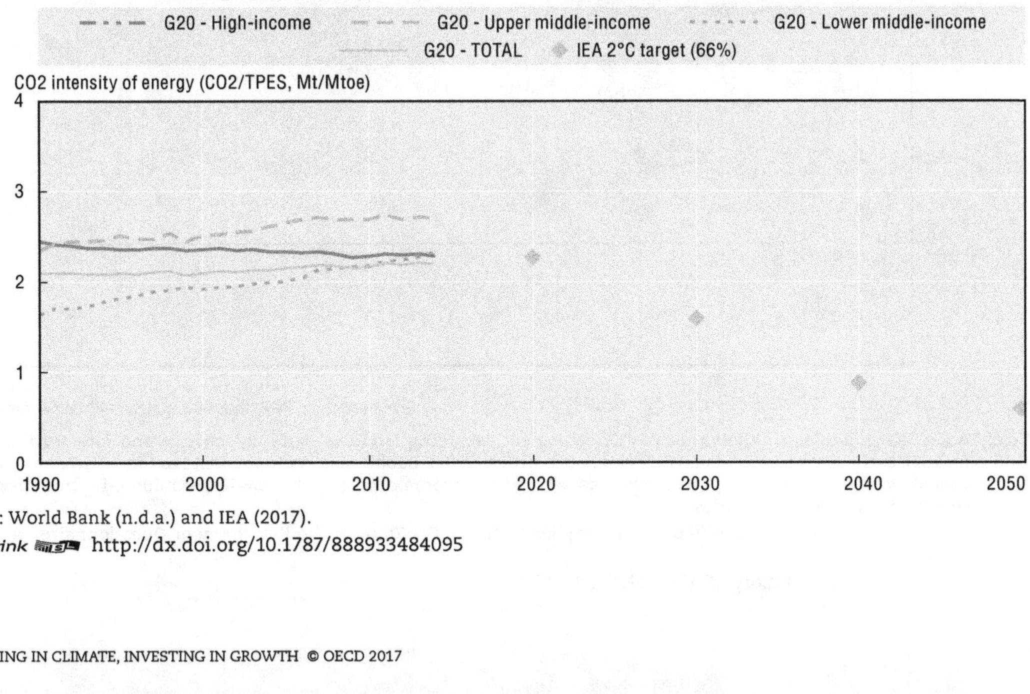

Source: World Bank (n.d.a.) and IEA (2017).

StatLink http://dx.doi.org/10.1787/888933484095

Country diversity and mitigation action

Absolute emissions reflect not just per capita income but also the size of the economy, its energy intensity and the CO_2 intensity of its primary energy supply (see above). Countries also have different income and population growth rates. These drive energy demand and future GHG emissions, as well as influence development patterns, climate resilience and adaptation capacities. Emissions from different sectors also have varying levels of importance from country to country. Finally, governance is an important factor in formulating and implementing low-emission, climate-resilient development pathways. This section analyses some of these key dimensions of country diversity.

Income levels, emissions per capita and governance

The capacity of each country to develop low-emission pathways depends on two key dimensions of country diversity: income level (GDP per capita) and average GHG emissions per person. In Figure 2.12, the size of each bubble represents the absolute level of emissions for the G20 countries (in orange), and the average emissions per G20 country included in each income group (in grey).[18] Emissions per capita are strongly correlated with GDP per person, reflecting the importance of energy to development.

Figure 2.12. **GHG per capita and GDP per capita in G20 countries, 2012**

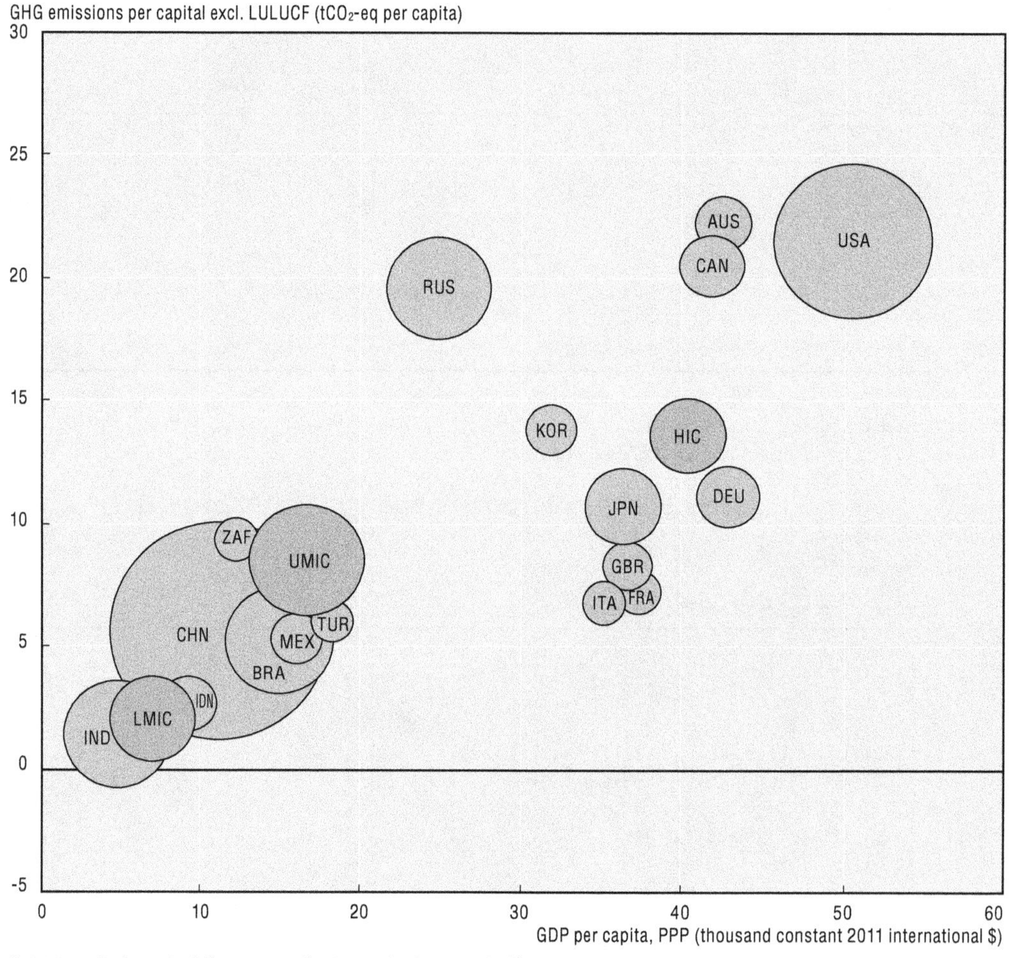

Note: Total GHG emissions in kilotonnes of CO_2 equivalent excluding land-use, land-use change and forestry (LULUCF). Values for 2012 except for Saudi Arabia (2011) and South Africa (2007). Bubble size is proportional to total GHG emissions for countries and average emissions for income groups. HIC= High-income countries, UMIC= Upper middle-income countries; LMIC= Lower middle-income countries.

Source: UNFCCC (2016), World Bank (n.d.a.), and replies to the OECD State of the Environment Questionnaire (accessed through OECD-STAT).

StatLink http://dx.doi.org/10.1787/888933484104

Income captures many dimensions of country capacity to mitigate and to adapt to climate change. More developed economies have higher levels of accumulated physical and human capital, financial and technological resources, and institutional capacity. Higher income levels are also highly correlated with standards of governance, as illustrated in Figure 2.13, which shows the results of a cluster analysis using six governance indicators and GDP per capita, and displays the results against just one of these, government effectiveness. Governance is a key factor underpinning effective and equitable adaptation across multiple actors and sectors in a context of uncertainty and complexity (Huitema et al., 2016). High income is also associated with greater levels of resilience, through mechanisms such as social safety nets, widespread insurance and infrastructure.

Figure 2.13. **Government effectiveness and GDP per capita**

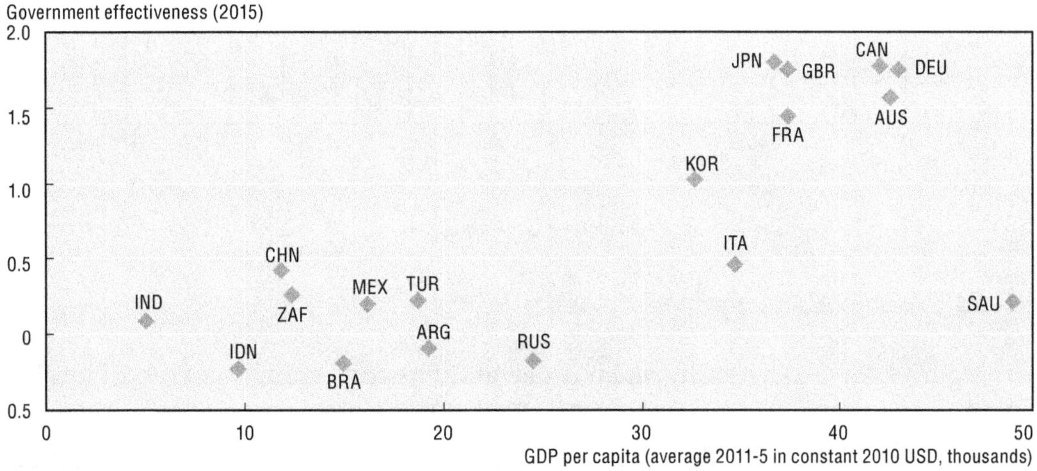

Note: Government effectiveness is an index based on World Bank data and OECD calculations.
Source: World Bank (n.d.b.) and OECD calculations.
StatLink http://dx.doi.org/10.1787/888933484116

Structure of GHG emissions across the G20

Energy emissions represent the bulk of GHG emissions in G20 countries. However, emissions from other sectors make a significant contribution to overall GHG emissions, notably in Argentina, Indonesia and Brazil (Figure 2.14). Agricultural emissions are a significant proportion of emissions in Argentina, Australia, Brazil, France, India and Indonesia, and are important in several others. Hence countries will face choices over the phasing and timing of mitigation action in different sectors and on different GHGs, with early action on long-lived GHGs essential to avoid cumulative emissions incompatible with the Paris Agreement's goal of well below 2°C. Action on short-lived GHGs and other climate forcers can not only complement this but also provide significant benefits in terms of health and food security (Shindell et al., 2012).

Land-use emissions are also important. Figure 2.15 shows the relative importance of agricultural and land use, land-use-change and forestry (LULUCF) emissions, as a percentage of total GHG emissions including LULUCF across G20 countries.[19] Argentina, Brazil and Indonesia stand out, with a large share of one or both of agricultural and LULUCF emissions. In a number of countries, the sink capacity of land use (essentially negative emissions) more than offsets agricultural emissions,[20] while for three countries, combined LULUCF and agricultural emissions comprise 15% to 20% of total GHG emissions.[21] Land-use change related to commercial agricultural expansion is one of the major sources of CO_2 emissions from deforestation (Hosonuma et al., 2012), though the share of agricultural emissions is not strongly correlated with land-use emissions.

Figure 2.14. G20 GHG emissions by sector (% of total GHG emissions excluding LULUCF)

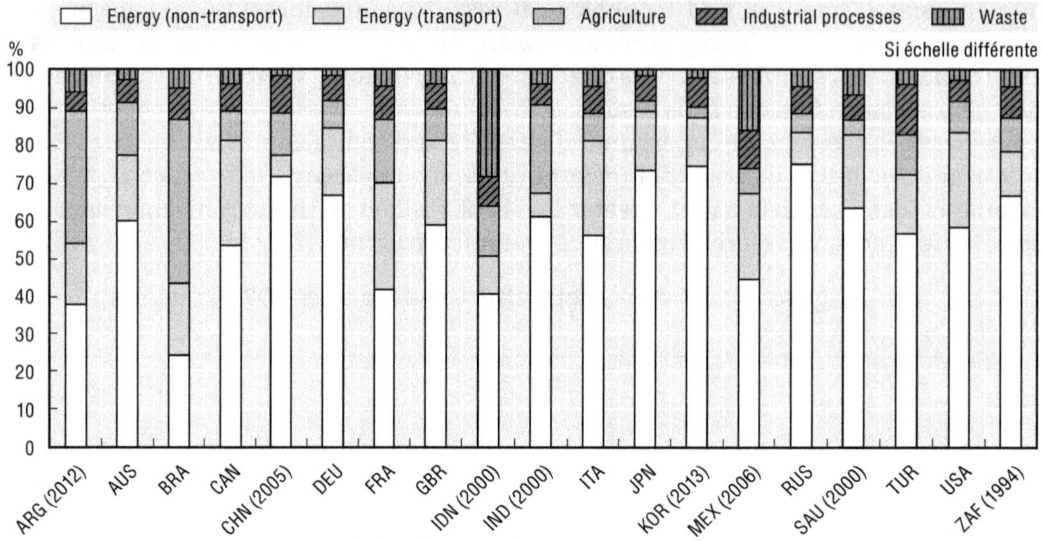

Notes: 1. 2014 or latest year available. 2. Emissions for Argentina, Mexico, and Saudi Arabia from UNFCC GHG profiles. Emissions for Brazil from MCTI, 2016.

Source: Data by sector from OECD, 2017; UNFCCC, 2014; MCTI, 2016.

StatLink http://dx.doi.org/10.1787/888933484129

Figure 2.15. G20 agriculture, land-use and forestry emissions as % of total GHG emissions

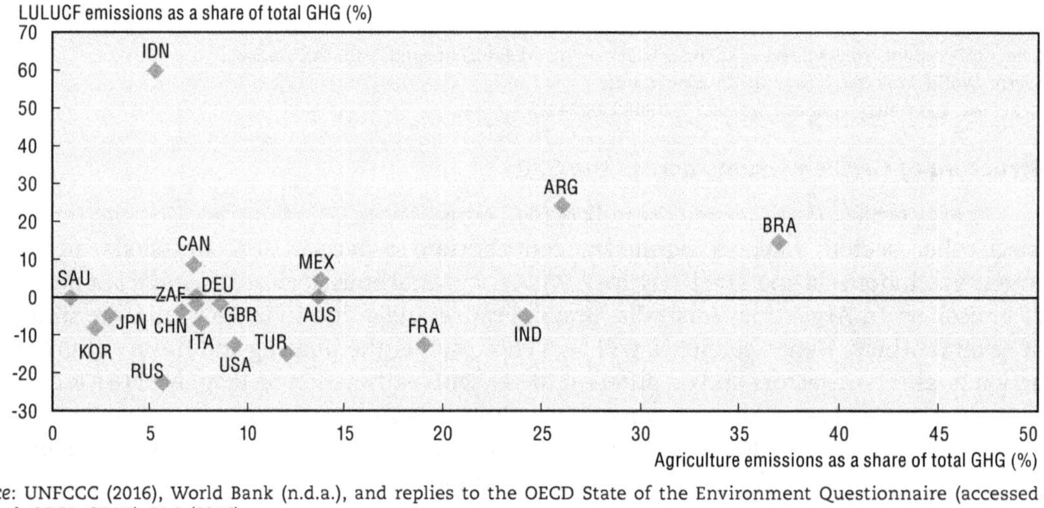

Source: UNFCCC (2016), World Bank (n.d.a.), and replies to the OECD State of the Environment Questionnaire (accessed through OECD-STAT), FAO (2016).

StatLink http://dx.doi.org/10.1787/888933484130

GDP, population growth and emissions

Future growth rates of energy-related emissions can be broken down into the growth rates of several different factors, including energy intensity of GDP and CO_2 intensity of energy (Blanco et al., 2014; Peters et al., 2017). So for a given rate of reduction in emissions, changes in GDP per person and in population together determine how quickly the other factors need to fall to keep on track to meet the Paris Agreement's goal of well below 2°C (Figure 2.16). Over the long term, GDP per capita growth rates may change as countries develop, but the current rates will influence the immediate challenges for countries in developing their low-emission, climate-resilient pathways. Countries such as Brazil that have experienced volatile economic growth rates, with sharp declines in growth in recent years, may change their relative position significantly. However, we expect the broad patterns to show some degree of stability over the period to 2030.

Countries fall broadly into three groups. In Brazil, France, Germany, Italy, Japan and Russia, recent combined growth rates in income per person and population are less than 2% per year. A second group of countries has combined growth rates between 2-4% per year, including Australia, Canada, Korea, Mexico, South Africa and the United Kingdom. A third group, including China, India, Indonesia, Saudi Arabia and Turkey, have combined growth rate in GDP per person and population of more than 4% per year.

Figure 2.16. **Growth rates of GDP per person and population in G20 countries, average 2011-15**

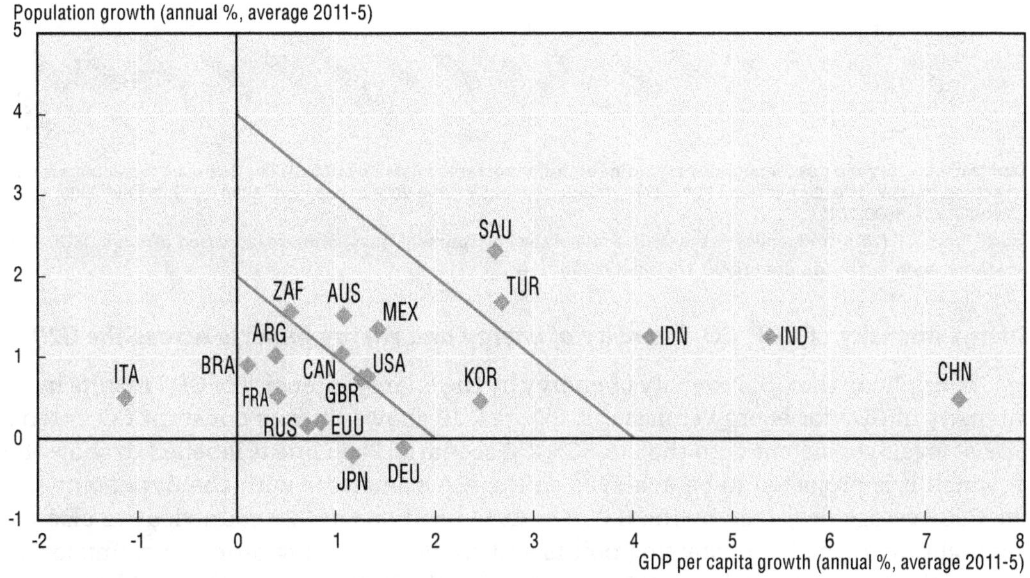

Note: Averaged over the most recent five years of data.

Source: World Bank (n.d.a).

StatLink http://dx.doi.org/10.1787/888933484141

If countries were aiming at a uniform rate of reduction in energy-related CO_2 emissions, the severity of the mitigation challenge would increase from the first to the third group. However a key element of the Paris Agreement is that countries' mitigation contributions reflect "common but differentiated responsibilities and respective capabilities in the light of different national circumstances", which is reflected in the nature and level of ambition embodied in countries' Nationally Determined Contributions (NDCs) (see section below).

However, even countries with rapid GDP or population growth can make rapid reductions in emissions per unit of GDP. GHG emissions per unit of GDP decreased in nearly all G20 countries between 1990 and 2014 (Figure 2.17). As well as structural economic changes, this progress has mainly been due to a general improvement of the energy efficiency of G20 economies rather than an improvement of the carbon intensity of the energy mix. Progress has been varied, but no country has reached the levels consistent with a 66% likelihood of staying below 2°C.[22]

Figure 2.17. **Annual % change in GHG emissions per unit of GDP for selected G20 economies**

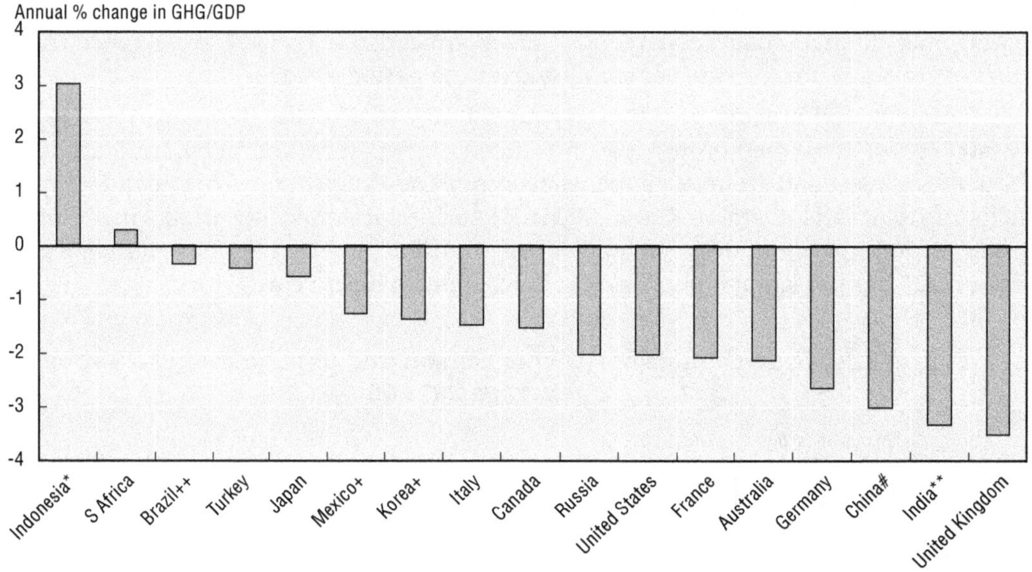

Note: Data refer to gross direct emissions excluding emissions or removals from LULUCF. The GDP used to calculate intensities is expressed in USD at 2010 prices and PPPs. The periods covered is 1990-2014 except for: *1990-2013; **1994-2000; #1994-2005; +1990-2013; ++1990-2012.

Source: UNFCCC (2016) and replies to the OECD State of the Environment Questionnaire (accessed through OECD-STAT).

StatLink http://dx.doi.org/10.1787/888933484150

Energy intensity of GDP, CO_2 intensity of energy and energy imports across the G20

Multiplying the CO_2 intensity of energy by the energy intensity of GDP results in the CO_2 intensity of GDP for energy emissions. Figure 2.18 shows lines of constant CO_2 intensity of GDP at levels consistent with the IEA 66% 2°C scenario. Each line is labelled to show the year in which it is projected to be achieved in the IEA scenario,[23] with the data point showing the G20 average projected by the IEA. The 2014 positions of G20 countries are also plotted, highlighting the different starting points and challenges facing different countries as they choose the most appropriate pathways towards the Paris Agreement's goal of well below 2°C. These lines therefore provide a clear direction of travel for country-specific levels of energy intensity and CO_2 intensity of energy. France, for example, has both a relatively low CO_2 intensity of primary energy and energy intensity of GDP, albeit not yet at the levels needed by 2050. Brazil also has a low CO_2 intensity of energy – reflecting the current large share of low-carbon power generation (like France) and the use of bioenergy – but a slightly higher energy intensity of GDP. Further improvements in such economies will require continued investment in low-carbon generation in order to avoid moving backwards, but also priority action in other CO_2-intensive sectors that are harder to decarbonise, such as transport and industry, and continued improvements in energy efficiency.

In contrast, countries like China and South Africa have both a high CO_2 intensity of energy (reflecting coal-powered generation) and a high energy intensity of GDP. Australia also has a high CO_2 intensity but slightly lower energy intensity of GDP, while Canada and Russia have a slightly lower CO_2 intensity, but are more energy-intense economies due to factors including the climate. Of course, countries may have similar levels of energy intensity or CO_2 intensity for very different reasons,[24] and different country outcomes for energy and CO_2 intensity could be consistent with the IEA 66% 2°C scenario. But the direction of travel for all is clear.

Figure 2.18. The carbon and energy intensity of G20 economies in 2014 and the path to 2050

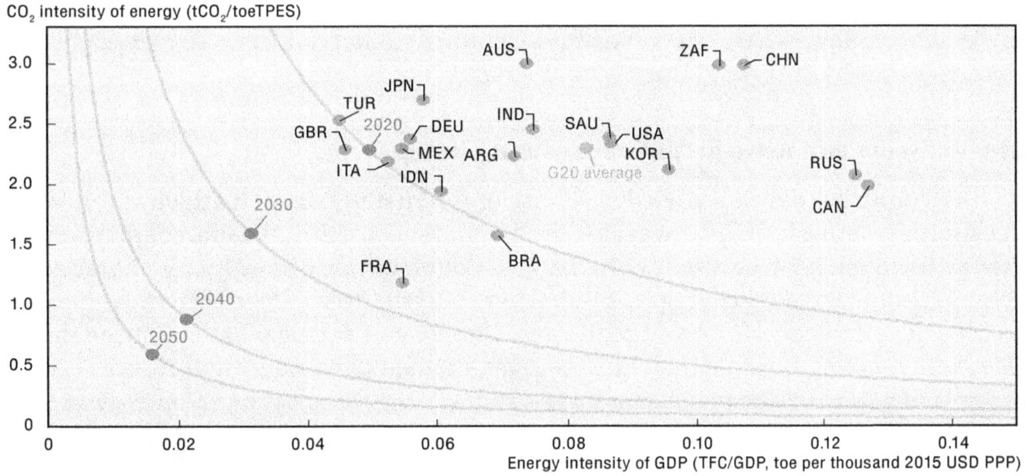

Notes: The average levels for G20 countries (excluding the European Union) refer to 2014 statistical data and the IEA 66% 2°C scenario projections for 2020, 2030, 2040 and 2050. The iso lines show other feasible combinations of CO_2 intensity and energy intensity levels. Calculations assume a constant ratio for total primary energy supply (TPES) to total final consumption (TFC). toe = tonnes of oil equivalent.

Source: Calculations based on the IEA World Indicators and IEA 66% 2°C scenario projections.

StatLink http://dx.doi.org/10.1787/888933484164

Figure 2.19. Net energy imports and CO_2 intensity of primary energy

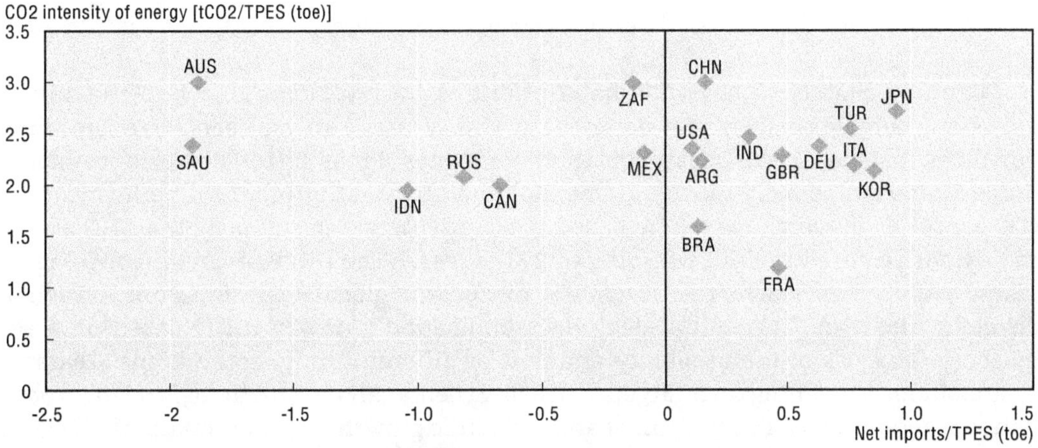

Source: World Bank World Development Indicators (database, accessed February 2017); "World Energy Balances", IEA World Energy Statistics and Balances (database, accessed February 2017).

StatLink http://dx.doi.org/10.1787/888933484176

A further important difference between countries is their position as net importers or exporters of energy (Figure 2.19). There are broadly three groups of countries. For the main net importers of energy, deploying low-carbon energy represents an opportunity in the long run to become self-sufficient in power generation, strengthening their energy security. Many of these countries also have CO_2-intensive primary energy, which means that rapid progress can be made to reduce the CO_2 intensity of electricity generation. For the second group, the main net energy exporters, the low-carbon transition represents a risk in terms of loss of export – and tax – revenues. A final group (or perhaps two sub-groups, comprising net importers and net exporters) – consists of those countries with limited net trade in energy. This may be due to the availability of significant low-carbon energy options (e.g. Brazil), but the group also includes countries with significant fossil fuel resources largely for domestic use, with limited net energy trade relative to total primary energy supply (e.g. Argentina, China, Mexico, South Africa and the United States). The challenges to decarbonisation therefore vary across countries, but are particularly significant for countries that have high CO_2 intensity of energy.

Low-emission pathways for different country groups

As countries develop their low-emission, climate-resilient pathways, an important question is whether these pathways are unique and specific to individual countries or whether groups of countries face similar challenges. Countries that have many characteristics in common could have much to gain by sharing analysis, policy development and experience as they develop their NDCs and pathways. One way of seeing what countries might have in common is to group them by income level – either Advanced (High-Income) Economies or Emerging (Middle-Income) Economies – and whether or not they are energy exporters or importers (Table 2.4).[25]

Table 2.4. **Country groupings**

Group	Advanced Exporters	Advanced Importers	Emerging Exporters	Emerging Importers
Country	Australia* Canada Saudi Arabia	France Germany Italy Japan Korea, Rep. United Kingdom United States	Indonesia Mexico Russia South Africa	Argentina Brazil China India Turkey

Source: OECD calculations. * includes New Zealand following the methodology used in the IEA 66% 2°C scenario.

Country characteristics will shape priorities in developing and implementing low-emission, climate-resilient development pathways, as can be seen by examining the outcomes of the Deep Decarbonisation Pathways Project (DDPP). This collaborative project between country modelling teams aimed to identify practical pathways that the G20 countries in which they were based could adopt, taking seriously the GHG emissions reductions required to limit warming to 2°C or less.[26] The DDPP project involved research teams from countries representing 74% of current global CO_2 emissions.[27] Each team developed its own "bottom-up" deep decarbonisation pathway (DDP) based on a sector-by-sector analysis of the feasibility and cost of different mitigation options. Teams were "autonomous in defining their targets, choosing their analytical methods, and incorporating national aspirations for development and economic growth in their scenarios" (DDPP, 2015).

Consequently, the IEA 66% 2°C scenario is more stringent than the DDPP exercise; G20 emissions are projected to fall by almost 80% by 2050 for the IEA 66% 2°C scenario, and about 50% in the DDPP exercise. Nevertheless, both the DDPP results and the IEA 66% 2°C scenario show very different energy-related CO_2 emissions pathways across the income level and energy exporter-importer country groups. Advanced Economies (Exporters and Importers) begin rapid emissions reductions from the outset and are projected to converge at very low levels by 2050. Emissions from Emerging Economies are projected to follow very different tracks.

In the IEA 66% 2°C scenario, Emerging Exporters reduce emissions from 2015 onwards, achieving a reduction of just over 60% by 2050. In the DDPP projections, however, Emerging Exporter emissions increase to 2020 before declining by a smaller 33% by 2050. Emissions from Emerging Importers grow sharply from 2010, peaking around 2017 in the IEA 66% 2°C scenario and in 2030 in the DDPP results, but then fall more rapidly than those from

Emerging Exporters. This group achieves a more than 70% reduction in emissions by 2050 in the 66% 2°C scenario, but a less than 15% reduction in DDPP, reflecting the scale of the initial increase and the differing nature of the two exercises (Figure 2.20).

Figure 2.20. **Energy-related CO_2 emissions by income-energy group**

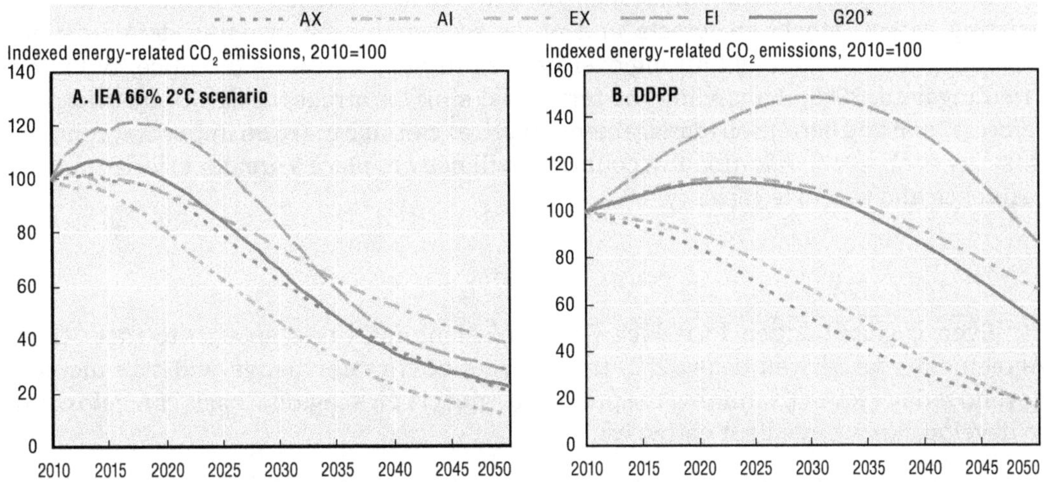

Note: AX: Advanced Exporters. AI: Advanced Importers. EX: Emerging Exporters. EI: Emerging Importers. G20 countries not included in Figure 2.20 (a) are: Argentina, Saudi Arabia, South Africa and Turkey. Australian emissions also include those for New Zealand since they are aggregated in the IEA modelling. Those not included in Figure 2.20 (b) are Argentina, Russia, Saudi Arabia and Turkey. G20* denotes the average across the countries where there is disaggregated data available for each exercise

Source: (a) IEA data underpinning IEA (2017) and OECD calculations. (b) DDPP (2015) and OECD calculations.

StatLink http://dx.doi.org/10.1787/888933484180

Another perspective can be gained by looking at emissions pathways just by income group (Figure 2.21). The joint mitigation-development challenge facing Lower Middle-Income countries is striking. The IEA scenario (LMIC reduction of 13%) would require significantly more stringent mitigation than in the bottom-up DDPP exercise (LMIC increase of 84%). Upper Middle-Income countries are projected to reduce emissions by 80% in the IEA 66% 2°C scenario but only by 36% in the DDPP results.

Figure 2.21. **Emissions pathways by income group in the IEA 66% 2°C and DDPP scenarios**

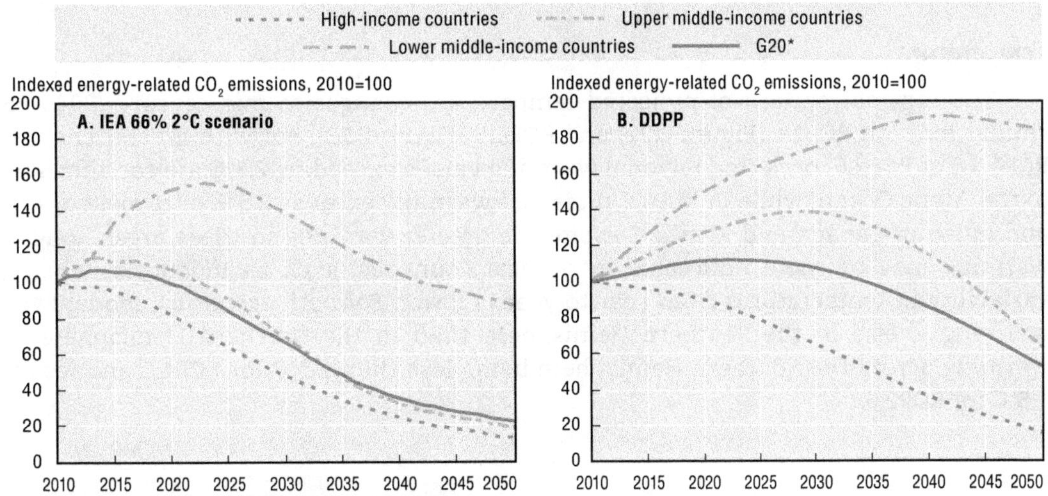

Note: For G20 country coverage, see note under Figure 2.20.

Source: a) IEA data underpinning IEA (2017) and OECD calculations. (b) DDPP (2015) and OECD calculations.

StatLink http://dx.doi.org/10.1787/888933484191

Other studies have shown potential for emissions reductions to go beyond these levels by 2050 in some emerging economies, though there remain significant challenges in doing so.[28] To keep warming well below 2°C, effective transparency, review and updating processes will clearly be essential, as well as support for climate action in developing countries.

Beyond energy-related emissions, there are clear priorities for countries to preserve existing carbon stocks in forests and other ecosystems by avoiding deforestation and forest degradation and by limiting over-use of nitrogen fertilisers (Prentice, Williams and Friedlingstein, 2015). Enhancing the terrestrial sink for atmospheric CO_2 by afforestation, reforestation and better soil management practices can also make an important contribution (Mackey et al., 2013). Additionally, countries will need to place a greater priority on building resilience and adaptive capacity.

Climate-resilient pathways reflecting regional climate change

Even if global action to reduce GHG emissions increases enough to meet the Paris Agreement goal of well below 2°C, the impacts of climate change will still increase far beyond today's level. Examining the projected impacts on a regional basis can help countries to develop climate-resilient pathways.

Projected regional climate changes

This section presents results for two different RCP scenarios simulated by a number of the climate models that informed the IPCC's Fifth Assessment Report (IPCC, 2013). The first is the RCP2.6 scenario. The second is the RCP4.5 scenario, which has mean end-of-century warming across models of 2.4°C. Both therefore have end of century warming relative to the pre-industrial time period below the level associated with the emissions pathways implied by countries' Nationally Determined Contributions (NDCs) to GHG emissions reduction post-2020, as described below. The RCP4.5 scenario is, however, broadly in line with the NDCs earlier in the century.

The following figures show maps of projected climate changes between the recent past (1986-2005) and mid-century (2046-65) for these two RCPs. The mean average change for different regions across the available climate models is shown, but individual models may give results that differ in terms of the magnitude of changes and details of the spatial patterns of change.

Temperature

The regional pattern of projected temperature changes to mid-century (2046-65) is similar for both RCP2.6 (Figure 2.22) and RCP4.5 (Figure 2.23), but with greater changes in RCP4.5. For RCP2.6, projected regional warming values exceeding 2.5°C are confined largely to the Arctic Ocean, while in RCP4.5 projected warming exceeds 2.5°C over most of Alaska and much of Canada and Russia. Despite the greater warming in these areas, long-term warming may be more noticeable in tropical countries, such as Indonesia, where the variability in temperatures from year to year is lower. For both scenarios, model-average warming is less in the Southern Hemisphere than in the Northern Hemisphere, with warming across the Southern Hemisphere being less than 2.5°C for RCP4.5 and less than 1.5°C for RCP2.6.

Figure 2.22. **Projected absolute change in annual mean surface temperature for RCP 2.6 for the period 2046-65 relative to 1986-2005**

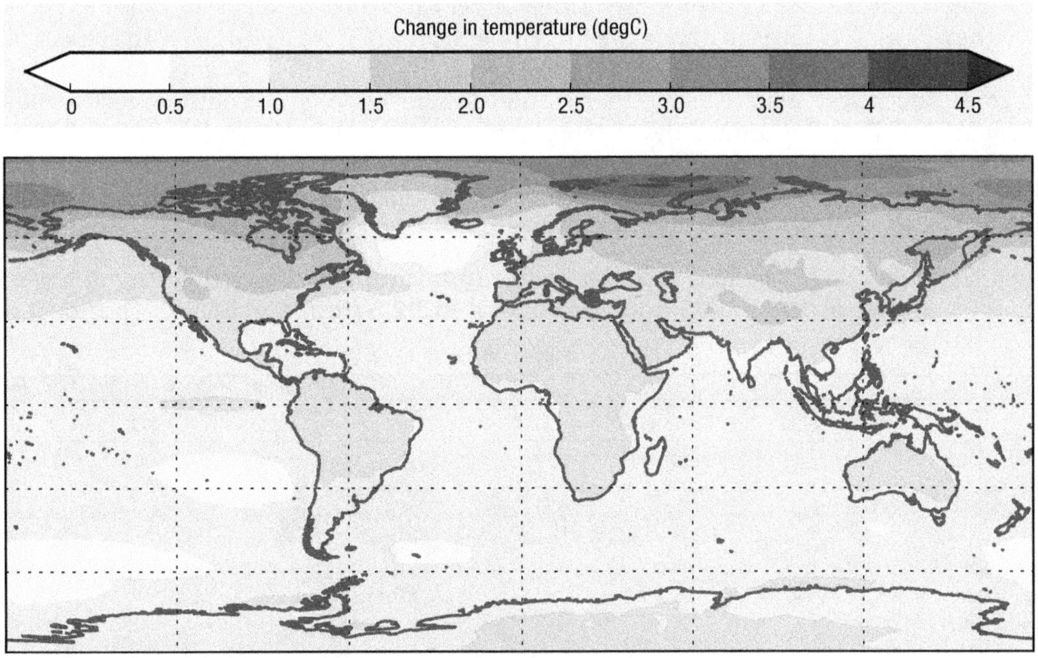

Note: Maps show average changes across available global climate model simulations.
Source: MOHC analysis.

Figure 2.23. **Projected absolute change in annual mean surface temperature for RCP 4.5 for the period 2046-65 relative to 1986-2005**

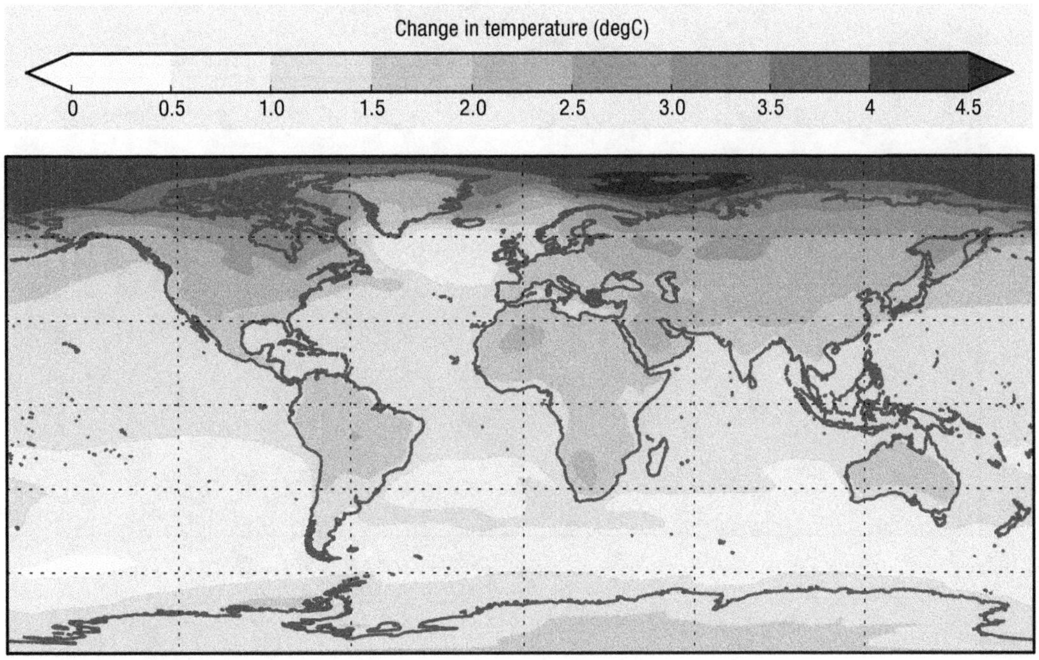

Note: Maps show average changes across available global climate model simulations.
Source: MOHC analysis.

The regional pattern of changes in extreme temperatures is quite different from that for changes in annual mean temperature. For example, those regions expected to experience the greatest increases in the temperatures of very hot days differ from those expected to see the largest increases in annual mean temperatures (Figure 2.24). For both scenarios, the maximum temperature during a year is projected to increase most over parts of continental Europe, southwest Asia, North America and inland regions of South America, such as western Brazil. As for annual mean temperatures, the increase in maximum temperature during a year is projected to be greater for RCP4.5 than for RCP2.6. For example, over parts of southeast Europe the model-average increase in maximum temperatures during a year is more than 3.0°C for RCP4.5, whereas it is less than 2.5°C under RCP2.6.

Figure 2.24. **Projected changes in the maximum temperature during a year between 1986-2005 and 2046-65 for RCP2.6 (top) and RCP4.5 (bottom)**

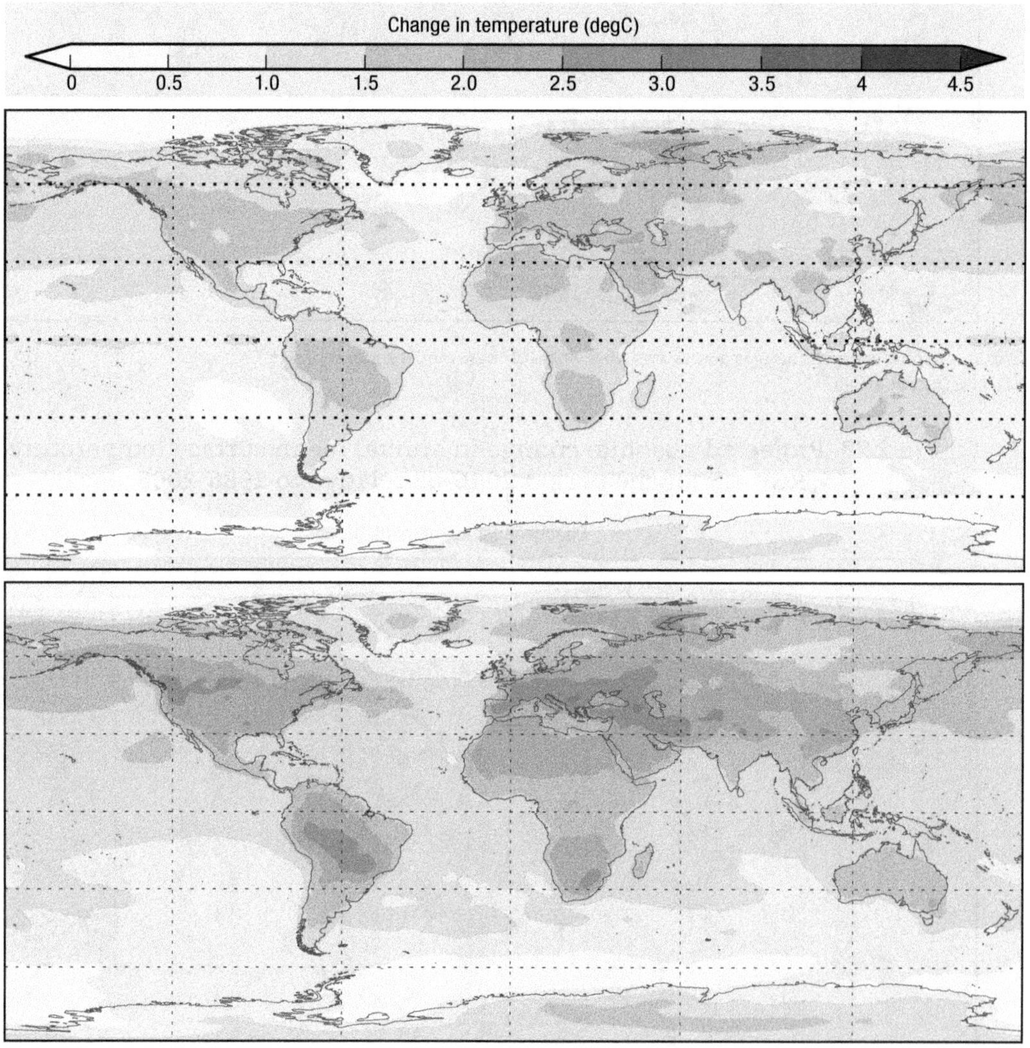

Note: Maps show average changes across available global climate model simulations.
Source: MOHC analysis.

Precipitation

In both RCP2.6 and RCP4.5, global average annual mean precipitation is likely to increase by 2-3% on average between 1986-2005 and 2046-65 (Table 2.5). Projections are highly uncertain on the country scale, however. For most of the G20 countries, some simulations show increases in precipitation while others show decreases. Nonetheless, both scenarios show the same coherent pattern of precipitation increasing in some areas and decreasing in others, particularly northern Africa, southern Europe, Central America, northern South America, southern Africa and Australia (Figure 2.25). For RCP4.5, the greatest model-average precipitation decreases for the G20 countries – of more than 6% – are projected for some of the Mediterranean countries. The same countries are projected to experience more modest precipitation decreases for RCP2.6 of around 2% or 3%. For RCP4.5, the greatest model-average precipitation increases projected for the G20 countries – of more than 7% – are for Canada and Russia.

Table 2.5. **Projected percentage changes in global average annual mean precipitation and maximum daily precipitation total during a year between 1986-2005 and 2046-65 for RCP2.6 and RCP4.5**

	Change in annual mean precipitation		Change in annual maximum daily precipitation total	
Scenario	Mean	Likely range	Mean	Likely range
RCP2.6	+2.2	+0.5 – +3.8	+5.7	+2.3 - +9.1
RCP4.5	+2.6	+1.0 – +4.1	+6.8	+1.8 - +11.8

Source: MOHC analysis.

Figure 2.25. **Projected changes in annual mean precipitation between 1986-2005 and 2046-65 for RCP2.6 (top) and RCP4.5 (bottom)**

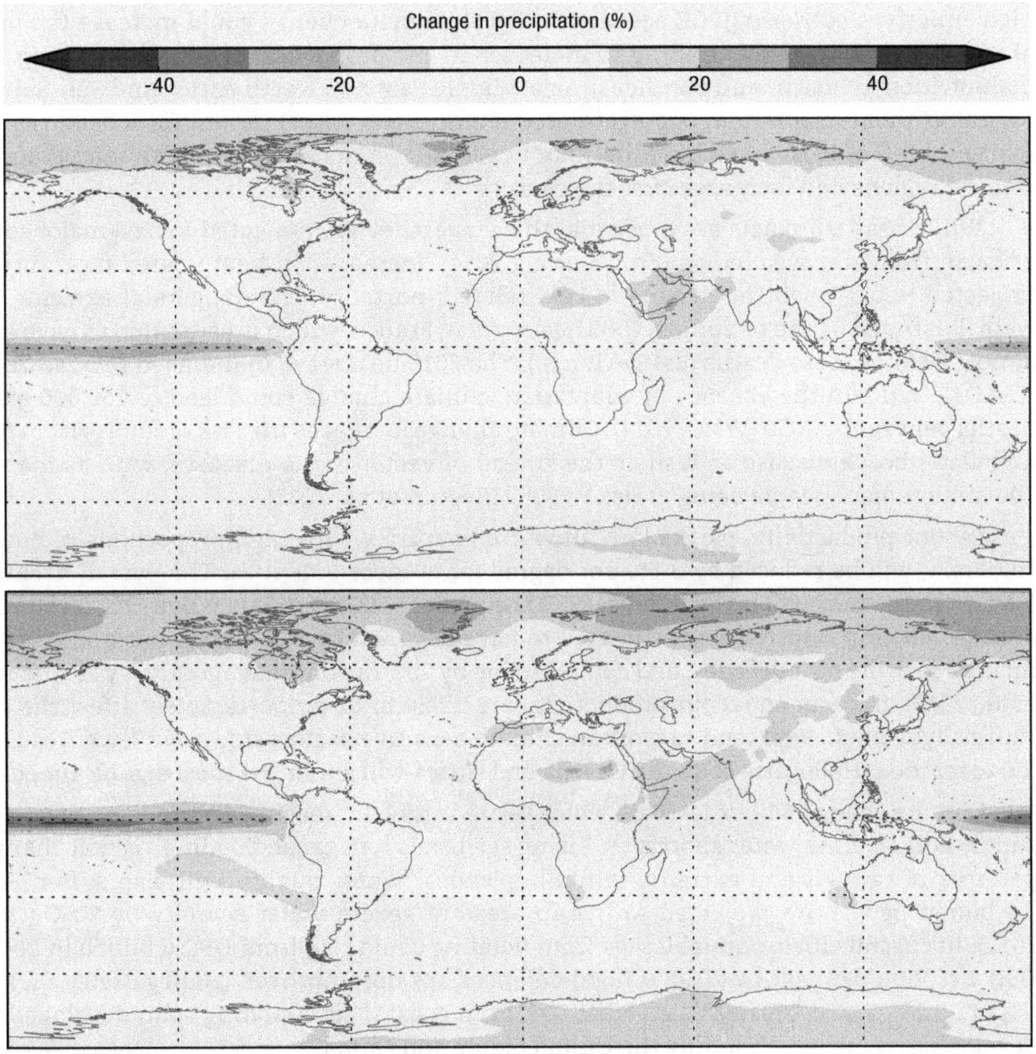

Note: Maps show average changes across available global climate model simulations.
Source: MOHC analysis.

In all G20 countries, global average extreme precipitation is expected to increase more than global average annual mean precipitation. Global average maximum daily precipitation is likely to increase by 6% on average for RCP2.6 and 7% for RCP4.5.

Climate impacts and the SDGs

The choice of development pathway will have a major influence on how climate change affects poverty levels (Hallegatte et al., 2016). In a scenario where economic growth is higher, inequality is lower and there is better provision of basic services, climate change is estimated to increase the number of people in extreme poverty in 2030 by 3 to 16 million people. By contrast, under a more pessimistic scenario, extreme poverty could increase by 35-122 million people because of climate impacts on agriculture, health, labour productivity and the incidence of natural disasters (Hallegatte et al., 2016).

Agriculture will be affected by the changes in precipitation patterns and ecosystem services that are projected to occur with climate change. IPCC (2014b) reported that negative impacts of climate change on yields of crops such as wheat and maize have been more common than positive impacts. Crop yields are projected to increase by 2050, but by less than would otherwise be the case (Ignaciuk and Mason-D'Croz, 2014). Under a very high emissions scenario (IPCC scenario RCP 8.5), climate change could increase the prices of major grains by 5-30%, leading to increases in the proportion of people suffering from malnutrition in South- and Southeast Asia, Middle East and North Africa and Sub-Saharan Africa. Without adaptation, aggregate production losses are expected for wheat, rice and maize for 2°C of local warming (Challinor et al., 2014). This applies to both temperate and tropical regions and increases over the century.

While health impacts are modest at this stage, they are projected to be a major source of harm from climate change (Smith et al., 2014). Increases in heat-related mortality are projected to outweigh the decline in cold-related mortality. The dangers of extreme heat were illustrated by the prolonged 2003 heatwave in France, which is estimated to have led to almost 20,000 excess deaths (EM-DAT, n.d.). The 2015 heat wave in India led to 2 248 deaths (EM-DAT, n.d.). In the absence of adaptation, climate change could lead to 250,000 excess deaths per year by 2050 (WHO, 2014). Climate change increases the risk of illness from food- and water-borne disease as well as the spread of vector-borne diseases, with as many as 200 million more people being at risk in 2050 (Béguin et al., 2011).

Labour productivity, particularly in warm countries with high proportions of outdoor labourers, will be reduced by 3-5% per degree for outdoor activities. The overall decline in labour productivity will be 1% in most OECD countries (OECD, 2015b). In non-OECD countries, average labour productivity is estimated to have declined by 10% during peak temperature months over the past decades, and could decline by 20% during peak months by 2050 (Dunne et al., 2013). Impacts on labour productivity are likely to disproportionately affect the poor, especially women, who tend to work in climate-sensitive sectors and have fewer resources for adaptation (Hallegatte et al., 2016). Asia and Africa will suffer the most significant effects.

Climate change will exacerbate water-related risks. Increasing demand and decreasing supply will result in water shortages. Rising sea levels will cause flooding, as will changing patterns of rainfall and extreme rainfall episodes. Water quality will also suffer. Some 3.9 billion people are projected to live in areas of severe water scarcity by 2050 (OECD, 2012). In coastal cities, annual losses from flooding could rise from USD 6 billion in 2005 to USD 1 trillion per year by 2050, if flood defences are not improved (Hallegatte et al., 2013) (Figure 2.26). The countries at greatest risk from coastal city flooding span developed and developing countries, including the United States and China.[29]

Developing climate-resilient pathways

Countries' vulnerabilities to climate change are shaped by development choices, socio-economic trends and climate effects that cross borders and will demand flexible, forward-looking approaches to decision-making.

As with mitigation action, a primary determinant of countries' ability to adapt is their GDP per capita. Richer countries will be better able to adapt to the impacts of climate change than those with lower GDP per capita; they have more resources to invest in adaptation and recovery. This can be seen in the correlation between GDP per capita and standards of protection against flooding (Hallegatte et al., 2017). There are also indirect effects: richer countries tend to have higher quality institutions, leading to more rigorous planning and better implementation of policies. More developed financial markets mean that households and businesses are better able to manage the financial consequences of extreme events.

How much rainfall countries receive – and how much this is expected to change – also affects countries' ability to adapt. Climate change is expected to reduce precipitation in regions that are already severely water-stressed. Moreover, the loss of Asian and Andean glaciers will place further stress on freshwater availability in countries in South Asia and South America. The need to reconcile supply and demand will shape the range of feasible development paths, constrain some adaptation options (such as irrigation) and increase the urgency of developing an efficient policy response.

The variability of precipitation is also a key factor for adaptation. Monthly variability in water runoff, GDP per capita and investments in water security are interconnected (Sadoff et al., 2015). River basins in high-income countries tend to have less variable runoff and higher investment in water security. In contrast, river basins in low-income countries tend to feature variable runoff and low investment in water security. As climate change makes precipitation less predictable, it will be vital to enhance investment in water security to address these fluctuations.

Figure 2.26. **The 20 cities most at risk from sea-level rise**

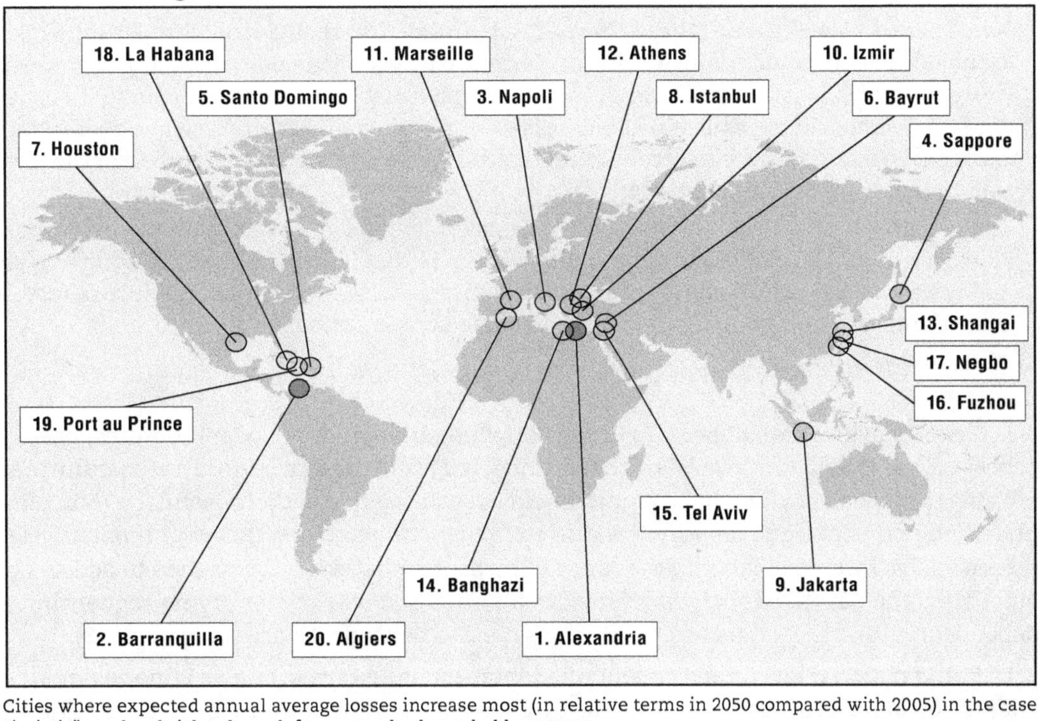

Note: Cities where expected annual average losses increase most (in relative terms in 2050 compared with 2005) in the case of "optimistic" sea-level risk, where defence standards are held constant.

Source: Hallegatte et al., 2013. Reprinted by permission from Macmillan Publishers Ltd: Nature Climate Change 3, 802–806 copyright (2013).

Political choices will also affect countries' vulnerabilities to climate change. Countries at similar levels of economic development vary widely in the levels of climate risks that they are willing to accept: New York is protected against a 1:100 year flood while Amsterdam is protected to a standard of 1:10 000. The development path that each country pursues will affect the cost and feasibility of achieving different levels of risk reduction: for example, development in low-lying coastal areas may subsequently necessitate large investments in coastal protection, or relocation to higher ground.

Countries can reduce their vulnerability to the effects of climate change by pursuing inclusive development. Poverty, marginalisation and inequality constrain people's ability to adapt to a changing climate. The poor tend to live in higher-risk areas and have less access to public services (Hallegatte et al., 2017). Moreover, the poor and marginalised have few resources with which to cushion the impact of climate shocks, with the result that such shocks can cause long-term harm, and even transform transient poverty into chronic poverty (Olsson et al., 2014). Ensuring that development is inclusive can avoid a vicious cycle between climate change and poverty.

Box 2.6. **Adaptation pathways: the Delta programme**

The Delta programme is designed to protect the Netherlands against the risk of flooding and ensure access to fresh water. An approach called "adaptation pathways" has been used to identify different sets of policy measures that could meet these objectives, given uncertainties about how the climate, the economy and society will evolve. Multiple model runs are used to project the range of potential variables over time. Based on this process, the analysis identifies tipping points where additional or different actions may be required to ensure that the objectives are met under some scenarios.

At each tipping point, there is a range of potential options – a "decision tree". Depending on the one chosen, the options available further down the track may differ. The combinations of available options offer many different pathways, which are all projected to meet the same performance criteria. These alternative pathways can then be compared using a range of qualitative and quantitative criteria. Once a pathway has been chosen, a monitoring system is established to track changes in relevant variables and change course if needed. The involvement of relevant stakeholders is essential to ensure that the right dimensions of each decision are taken into account and that there is a shared understanding of the likely consequences of different options.

This approach directly addresses the challenge of long-term planning in an environment of pervasive uncertainty. One of its main benefits is that it ensures that the actions taken today are consistent with the longer-term objectives. It also supports a flexible response, by identifying how options will open up or preclude certain actions in the future.

Source : Haasnoot et al., 2013

Since countries' circumstances differ, so will their appropriate adaptation responses. The concept of "adaptation pathways" has been pioneered to ensure that large infrastructure projects are able to respond to changing circumstances over the course of their useful life (Box 2.6). The underlying principle is to identify the range of potential outcomes that could materialise and then work backwards to identify the range of measures that would be needed to address those outcomes. The adaptation pathway provides a formalised way of identifying sequencing, path dependencies and the points where decisions need to be made (Haasnoot et al., 2013).

At the national level, the concept of adaptation pathways provides a model for viewing adaptation as a process for adjusting to changing circumstances over time. There is a succession of decision-points over time, each of which then determines the future range of opportunities that are open to decision-makers (Wise et al., 2014) (Figure 2.28). In practice, however, the process is less straightforward, because of the need to define what constitutes successful adaptation, difficulty in measuring the current state of progress and competing views about the appropriate responses to a changing climate. Nonetheless, the underlying approach of cycles of implementing actions, learning and adjusting course provides a useful description of the adaptation process.

Figure 2.28. **Iterative decision cycles**

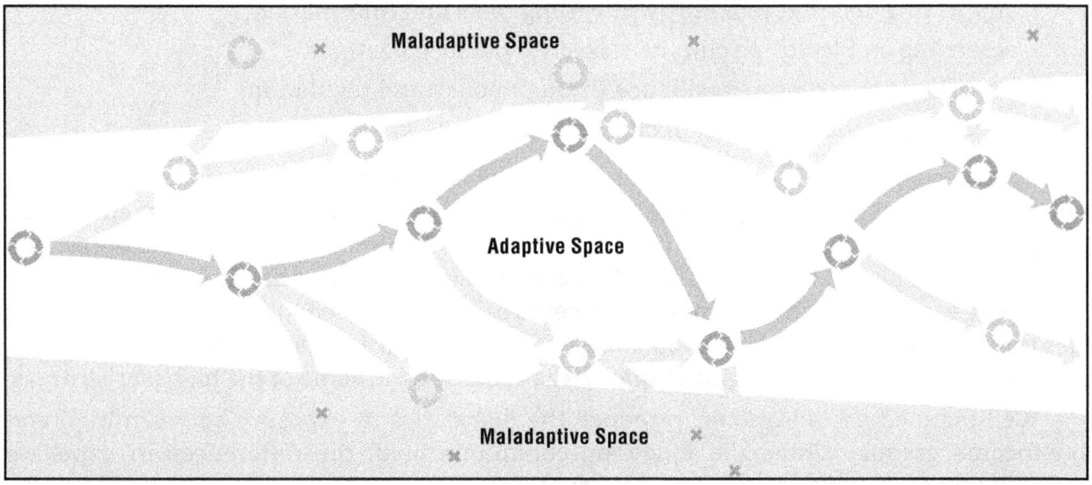

Source: Wise et al., 2014

National Adaptation Plans (NAPs) provide an important tool for communicating priorities and putting in place the key elements required to support adaptation. Adaptation will be the product of a multitude of decisions, ranging from farmers' choices of crops to urban planning, undertaken by a wide range of actors facing different sets of opportunities and constraints. Climate change will be just one of many factors that could influence how people respond to change. This means that it is neither possible nor desirable for every adaptation action to be dictated in a top-down manner. Instead, adaptation strategies such as NAPs should aim to strengthen the capacity of relevant decision makers to account for climate change. An important element of this is to influence investment decisions by demonstrating political commitment and setting the strategic direction for resilience at the national level.

The basis for effective adaptation is having access to suitable data in a usable form, combined with the tools to interpret the implications of climate change for the relevant decisions. These data should be regularly updated and reliable, which may require improvements in countries' statistical capacity.[30] Providing information is necessary, but not sufficient, to guarantee informed decision-making. The governance arrangements that determine how decisions are made may themselves need to be adapted to make them responsive to the effects of climate change. Action by governments may be required if inertia in existing governance systems means that they are no longer fit for purpose in a changing climate (Wise et al., 2014). For example, adopting a risk-based approach in the water sector requires involving a broader set of stakeholders, obtaining different information and changing the objective of the decision from meeting certain technical standards to achieving acceptable levels of risks. Regulatory reforms may be required to enable these changes to occur.

At the project level, there are clear metrics to assess progress and inform decision-making as part of an adaptation pathway. In contrast, the concept of national pathways cannot be readily quantified, because of the nature and diversity of actions that they include. For this reason, it is vital to use both quantitative and qualitative information to assess progress (OECD, 2015d). Relevant tools for doing so include national risk assessments, indicator sets and in-depth evaluations of large projects. This process is likely to be most effective when it is integrated into existing processes for monitoring and evaluation, rather than being implemented as a standalone system.

OECD analysis of infrastructure resilience shows that action is required across four policy areas (Vallejo and Mullan, 2017):

- supporting decision-making by providing tools and information;
- screening and factoring climate risks into public investments;
- enabling infrastructure resilience through policy and regulation;
- encouraging the disclosure of climate risks.

Spatial planning is another critical area for climate change adaptation, given that it can shape the location and design of new physical assets. There are two main challenges for spatial planning: ensuring that development is only permitted in lower-risk areas, and that the spatial plans are enforced. Unplanned urbanisation is a common feature of rapidly developing economies, with informal settlements being established in areas that are too risky for formal development, such as river banks and hillsides. As a consequence, the people with the fewest resources for managing climate risks are located in some of the highest risk areas.

Well-planned urbanisation can reduce the disparities in exposure between high-and low-income groups. Where the following conditions hold, the differences in exposure between income groups remain low (Revi et al., 2014):

- buildings meet construction standards;
- development is only permitted in lower-risk areas;
- infrastructure and basic services are provided to all.

Managing the effects of climate change on ecosystems will be an essential element of climate change adaptation pathways. Ecosystems are already under severe pressure as a result of deforestation, water pollution, over-fishing and other causes. The OECD Environmental Outlook to 2050 projected that biodiversity would decline in all world regions under business-as-usual policies. Climate change will place a further burden on ecosystems, as the rate of change exceeds plants and animals' abilities to adapt. There is already evidence of plants and animals having moved to new areas and changed their seasonal activities in response to climate change (Settele et al., 2014). Several policy options can be used to protect ecosystems from the impacts of climate change. The first priority is to strengthen efforts to alleviate the non-climate pressures on ecosystems. A crucial element of this is to mainstream biodiversity – and ecosystems more generally – into national and sectoral planning (OECD, forthcoming). Beyond this, several measures can be taken to lessen the effects of climate change on ecosystems (Settele et al., 2014):

- Adaptive landscape management: Ensure that landscape management strengthens resilience and capacity to adapt to change. Ensure that institutional arrangements, regulations and policies are designed with the expectation that ecosystems will change.
- Supporting biodiversity migration: Create and maintain migration "corridors" to support the process of ecosystem adaptation. In some cases, it may be necessary to move species to a new location.
- Off-site conservation: Preserve diversity through measures such as seed banks and breeding programmes. Several issues need to be resolved to ensure the successful reintroduction of preserved resources into the wild.

Ecosystem-based approaches can play an essential role in building resilience to the effects of climate change. In some cases, they can be cheaper and more flexible than hard infrastructure, and generate benefits beyond adaptation. For example, wetland protection or restoration can reduce flood risk, while also storing carbon and supporting biodiversity. Economic instruments such as Payments for Ecosystem Services should be used to enhance the provision of ecosystem services (OECD, 2010).

Linking adaptation and mitigation

Mitigation supports adaptation by delaying and reducing the scale of climate impacts. At a global level, this reduces the scale of the adaptation challenge. Mitigation also reduces the risk of encountering climate extremes that cannot be adapted to. In principle, credible commitments to a low-emission trajectory would reduce the total need for investments in climate change adaptation (OECD, 2015c). However, in practice this is not so simple (Wilbanks, 2005):

- *Dealing with uncertainty:* Adaptation decisions need to be made today based on expectations about the extent of future climate change. In terms of mitigation efforts, the question is then about expectations as well as outcomes, including the credibility of emissions reduction commitments.
- *Different time horizons:* Within the 2050 planning horizon, the differences are relatively modest between emissions trajectories but will become more severe over time. Implications for adaptation decisions will vary depending on the degree of lock-in.
- *Diverse actors:* Much adaptation is expected to be local and autonomous. Mitigation is focused on the main emitting sectors, while adaptation will take place in those that are most sensitive to the effects of climate change.
- *Distributional issues:* The benefits of adaptation are primarily local and near-term, while the primary benefits of mitigation are long-term and global.

At the level of specific adaptation measures, there are synergies and trade-offs. For example, half of the new coal power plants in China are being built in areas of high water stress (Luo et al., 2013). Replacing coal with wind or solar power would yield both mitigation and adaptation benefits. However, not all good things go together. Between mitigation and adaptation actions there are tensions as well as mutual benefits (Table 2.6). Inappropriate biofuels production, for example, could exacerbate problems with food security.

Table 2.6. **Potential synergies and trade-offs between adaptation and mitigation measures**

	Positive for mitigation	Potential trade-off with mitigation
Positive for adaptation	**Reduced deforestation:** Sequesters carbon and provides ecosystem services. **Agricultural practices (e.g. no till):** Sequesters carbon and can boost farmers' incomes. **Wetland restoration:** Carbon sequestration and reduced flood risk. **Renewable energy (wind, solar PV):** Lower water use than thermal generation.	**Desalination:** Addresses water shortage but is energy-intensive. **Increased irrigation:** Helps farmers manage variable precipitation but can be energy-intensive. **Air conditioning:** Reduces the impact of high temperatures on health, but is energy-intensive. **Construction of hard defences:** Reduces the risk of extreme events, but GHGs are embodied in the construction.
Potential trade-off with adaptation	**Inappropriate expansion of biofuels:** Could exacerbate food price shocks if biofuels displace crops. **Hydropower:** Could increase the complexity of managing water resources.	

To develop and implement effective climate policy, it is vital to ensure coherence between adaptation and mitigation policies.[31] At the level of individual projects, this means ensuring that the appraisal process takes into account the full range of relevant costs and benefits, including impacts on carbon emissions and on resources relevant for adaptation, such as water. Some projects will inevitably involve trade-offs; it is important that they are acknowledged to ensure that any negative impacts on mitigation or adaptation are justified.

Getting from here to there

Climate change is a global externality because GHG emissions in one country cause damages in other countries that are not currently adequately factored into decisions (Stern, 2007). Economic theory also tells us that a global public good such as a stable climate can only be delivered through effective collective action at the international level: each country is asked to incur costs to reduce emissions, but the benefits of these efforts are shared globally.[32] The costs and benefits of climate action are distributed unevenly across countries and over time, and are to some degree still uncertain. Mitigation costs fall early on, while the major benefits in terms of avoided impacts would be seen later in the century.[33] This provides incentives for countries to free-ride on the actions of others, either now or in terms of the damages that will face future generations.[34] Developed countries have been responsible for most of the cumulative CO_2 emissions so far, but developing countries will make up most future emissions. In the meantime, technological advances have massively reduced the costs of key renewable technologies.

This final section addresses the key question of how countries get to where they need to be. It discusses the NDCs, which are not aligned with a cost-effective path towards the Paris Agreement goal of well below 2°C. Finally, it underlines the fundamental importance of the Paris Agreement in efforts to build the trust and transparency needed to go beyond current levels of mitigation action.

The Nationally Determined Contributions

As part of the process of creating a new international climate agreement under the UNFCCC, each party submitted its proposed national climate action plan, known as its intended "nationally determined contribution" or NDC (Box 2.7). The Paris Agreement requires that parties "prepare, communicate and maintain" their NDCs.[35] In parallel, developed countries reaffirmed their commitment to support developing countries by mobilising USD 100 billion a year by 2020 from public and private sources. Emphasis was also placed on a just transition for workers, through the creation of good quality jobs in line with national development priorities.

The NDCs set out the post-2020 climate actions parties intend to take: for example, decarbonising energy supply through shifts to renewable energy, energy efficiency improvements, better land management, urban planning and low-carbon transport at the city level (see Annex 2.A1 for details of the G20 countries' NDCs). Taken together, the NDCs are a progression beyond current policies but are not enough to keep global warming below 2°C; they are more in line with emissions scenarios that keep the temperature rise to below 3°C in 2100 (UNEP, 2015).[36] Analysis of the NDCs suggests that emissions will continue rising to 2030 (UNFCCC, 2015b). Additionally, the NDCs imply significant variations in future carbon prices across countries, suggesting substantial potential gains to emissions trading.[37] To drive investment in low-emission technologies, the NDCs need to be both credible and backed by good domestic policy design, which includes flexibility to adjust (see Chapter 6) (Nemet et al., 2017).

In adopting a dynamic, hybrid approach – part bottom up, part top down monitoring and review of the adequacy of country efforts against global targets – parties to the UNFCCC have secured broad participation in international mitigation efforts, but at the (hopefully) short-term cost of environmental effectiveness and economic efficiency. The plateau in energy-related CO_2 emissions over the last three years is a positive sign, though it is still too early to claim that we are at a peak of total global emissions, let alone the subsequent rapid reductions required to keep warming "well below 2°C" (IEA, 2017).

Box 2.7. G20 countries' NDCs vary widely

The G20 countries' pledges differ in terms of the kind of emissions reduction they specify, the conditions they set, their target dates and the GHGs they cover.

An absolute emissions reduction relative to a base year. The G20 European Union countries (France, Germany, Italy and the United Kingdom) have opted for 1990 as the base year, along with the Russian Federation. This reflects the type of target and base year agreed under the Kyoto Protocol. Australia, Brazil, Canada and the United States have identified their target relative to their GHG emission levels in 2005.

A reduction in the emissions intensity of the economy relative to a base year. India, for instance, has pledged a 33-35% reduction of the emission intensity of its GDP while China aims for a 60-65% reduction. Both countries use 2005 emissions intensity of the economy as their baseline.

Emissions reduction relative to a business-as-usual scenario (without further climate policies): This is the case for the NDCs of Indonesia, Korea, Mexico, Saudi Arabia, and Turkey.

A specified emissions trajectory: South Africa has pledged a "peak, plateau and decline" of emissions, describing a path over the next 20 years. Argentina has placed an absolute cap on its 2030 emissions.

Conditionality: Several countries have set conditions for the achievement of some – or all – of their targets. These include the provision of financial, technical or capacity-building support from developed countries (e.g. for Argentina, India, Indonesia, Mexico, Saudi Arabia), the degree of the implementation of the Paris Agreement by developed countries (for South Africa). Argentina, Indonesia and Mexico have both unconditional and conditional targets, the latter requiring support from developed countries.

Target date: Most G20 countries have set 2030 as their target date. The United States and Brazil chose 2025; South Africa has target periods of 5 years going from 2020 to post-2035.

Coverage: Most G20 pledges cover the six Kyoto Protocol GHGs[38] as well as the economic sectors outlined by the IPCC.[39] Australia, Canada, the European Union, Japan, the Russian Federation, Turkey and the United States have also included nitrogen trifluoride (NF_3), added on the list of GHGs under Kyoto Phase II, in the target gases. Mexico also focuses on black carbon, while Indonesia includes only CO_2, CH_4 and N_2O.

Building on the Paris Agreement

Early efforts to forge an effective international response to climate change resulted in the 1992 UN Framework Convention on Climate Change (UNFCCC), the start of an open-ended negotiating process that led to the Paris Agreement in December 2015. The Paris Agreement aims to strengthen the international response to climate change by building on the bottom-up approach initiated at the Copenhagen COP15 meeting in 2009.[40] It also adds "an enhanced transparency framework", to help track progress of individual parties on mitigation and adaptation action as well as on support for developing countries (finance, technology and capacity-building). This framework is vital, given the evidence that trust and reciprocity are important for successful management of natural resources (Ostrom, 1990).[41] The framework will support several processes and milestones for collective stocktaking and oversight of progress made on long-term goals.[42]

An immediate priority within the UNFCCC process is to put the Paris Agreement into operation by reaching agreement on the rules and modalities for several key provisions, including those on monitoring, reporting, verification and assessing collective progress according to the timeline established at COP21.[43] Headway here is essential to build the trust needed to increase the stringency of action over time. This is the current focus of the OECD-IEA Climate Change Experts Group.

The Paris Agreement architecture has yet to demonstrate that it can catalyse the urgent and stringent mitigation action and support needed to meet the Agreement's goals. Parties must now implement their emissions limitation and reduction pledges to 2020 and their aims beyond 2020. The aggregate mitigation effect of the NDCs is inadequate, however, and countries need to scale up their efforts. Developed country support for climate action will be important, not just for mitigation but also to improve the resilience and adaptive capacity of countries facing the greatest climate challenges.

At COP21, parties were invited to communicate by 2020 the long-term low-emission development strategies they will follow up to 2050. Six countries have done so; it is crucial that more follow suit. This is an important mechanism for helping countries to align short-term actions with long-term goals and to minimise the risks of either emissions lock-in or stranded assets. One important initiative to support this and to build broader engagement and action is the 2050 Pathways Platform launched at COP22 in Marrakech, Morocco (Box 2.8).

Success will not solely depend on action at central government level. The UNFCCC process has over recent years deliberately and increasingly created mechanisms of engagement with and commitments from non-state actors, most notably under the Lima-Paris Action Agenda in the run-up to COP21, on issues as diverse as cities, private finance and forests.

Box 2.8. **The 2050 Pathways Platform**

The 2050 Pathways Platform was launched at the High-Level Event of COP22 in Marrakech. Membership is growing quickly: 22 countries, 15 cities, 17 regions and states, and 192 companies have already joined.

Short-term GHG emissions reduction targets and actions such as the NDCs need to be set and implemented consistently with the long-term global goal. Developing pathways from now until 2050 can help in envisaging the structural changes necessary to achieve net-zero GHG emissions, as opposed to incremental changes. The platform helps countries design and implement long-term deep decarbonisation strategies that will limit the average global temperature increase to well below 2°C. It does so by sharing resources (including finance and capacity building), experience and best practices. It also builds a broader constellation of cities, states, companies and investors engaged in long-term low-emission planning of their own, and in support of the national strategies. It is envisaged as a space for collective problem-solving.

Pathways to 2050 need to be socio-economic development pathways, not just GHG emission reduction pathways; adaptation is an important component. Developing 2050 pathways can help to capture the synergies between socio-economic development and climate change mitigation, for example by aligning climate action with objectives on health, innovation and food security. They are also a risk-management tool: they can avoid carbon lock-in, and therefore reduce the risk of stranded assets, by putting short-term climate actions in the context of the long-term climate transition.

Pathways to 2050 need to be co-designed – and ultimately owned – by all relevant stakeholders: not just politicians and policy-makers, but also businesses, unions, NGOs and others. They also need to be informed by the best expert knowledge and evidence. The Platform aims to leverage a range of international processes to provide: technical analysis and support; sharing lessons learned and best practices; and multi-stakeholder/cross-jurisdictional dialogues.

Source: 2050 Pathways Platform team.

Notes

1. High levels of CO_2, associated with enhanced warming, also lead to increased acidification of the ocean and impacts on corals and a wide range of marine ecosystems.
2. Yet 13 percent of the world's population lived below the international poverty line of US$1.90 per day in 2012, see World Bank (2016).
3. CO_2 contributed about 76% of global warming in 2010 (IPCC, 2013).
4. Taken here as the 1850-1900 average.
5. Scientists have more confidence in their understanding and projections of global surface temperature than of precipitation, since the latter depend on the dynamics of the atmosphere, not just on energy-balance considerations. There is also have greater confidence in projections of global or continental scale changes than at regional or local scale. Global Climate Models (GCMs) are the basis of much of the information on future climate changes presented in the IPCC's assessment reports. See Taylor, Stouffer and Meehl (2012) on the Coupled Model Intercomparison Project Phase 5 (CMIP5), which was used in IPCC AR5 (2013). Such exercises help to determine the strengths and weaknesses of the various GCMs and inform their future development.
6. The Representative Concentration Pathways (RCPs) used in the most recent IPCC AR5 report span a wide range of possible future emissions scenarios. They are used to illustrate a range of possible climate futures to 2100 (Moss et al., 2010) by specifying different concentrations of GHGs and other atmospheric constituents (such as aerosols). These scenarios are named RCP2.6, RCP4.5, RCP6.0 and RCP8.5 to reflect their impact on the net energy flows into the climate system. So RCP2.6 (4.5) would give rise to a net energy inflow to the climate system of 2.6 (4.5) Watts per square metre (Wm^2) by 2100 in the Integrated Assessment Model (IAM) used to derive them. These RCPs have been used as input to models that produce detailed simulations of the climate system.
7. In their Fifth Assessment Report, the IPCC analysed over 1 000 published emissions scenarios from integrated assessment models (IPCC, 2014a). Based on a subset of these selected for their detailed information on emissions and consistency with both historical emissions and assumptions about a feasible maximum level of negative emissions, the UK Meteorological Office Hadley Centre (MOHC) identified 39 scenarios that had a greater than 66% probability of not leading to warming above 2°C. These are shown in Figure 2.4 alongside scenarios that lead to median end of century warming of 1.75-2.0°C.
8. Estimates of the equilibrium climate sensitivity, which determines the long-run climate response to GHGs, range between 1.5°C and 4.5°C for a doubling of atmospheric CO_2 concentrations.
9. The net effect of negative emissions technologies on atmospheric concentrations is reduced by the response of the ocean and land stores of CO_2 to a reduction in atmospheric CO_2 concentration. See Mackey et al. (2013).
10. The climate effects of different GHGs relative to CO_2 are typically evaluated using the 100 year global warming potential (GWP_{100}), which also has been adopted in GHG trading schemes. However, this metric is not related to temperature outcomes, nor does it clearly highlight the need to limit cumulative CO_2 emissions (Smith et al., 2012). Indeed, there is no single metric that can equate the full climate effects of different GHGs as the appropriate metric will depend on the policy outcome sought (Shine, 2009).
11. To gain the same climatic benefit as a one-off reduction in the level of CO_2 emissions, the rate of methane emissions would need to be reduced on a permanent basis. Much of the difficulty in reducing CH_4 emissions lies in the agricultural sector and, in particular, with growing livestock numbers (Ripple et al., 2014).
12. About 70% of global N_2O emissions are due to agriculture (World Bank, 2009).
13. From the SSP Public Database Version 1.1. – see https://tntcat.iiasa.ac.at/SspDb/dsd?Action=htmlpage&page=about
14. The climate policy assumptions for SSP2 – the SSP scenario that most closely resembles historic economic and demographic trends - include some delay in establishing global action with regions transitioning to global co-operation between 2020 and 2040, making emissions in the SSP2 baseline scenario broadly consistent with the NDCs (O'Neill et al., 2015; Riahi et al., 2017).
15. The industrial process emissions are estimated from the overall carbon budget (90 $GtCO_2$ over 2015-2100) with a starting point of 2 $GtCO_2$/yr and falling to around 1 $GtCO_2$/yr by 2050, as described on p.48 of IEA (2017).
16. Modelling approaches to land-use are highly varied – see Alexander et al. (2017).
17. In Brazil, concerted public action has led to reduced deforestation over the past few years.

18. The income groups are the standard World Bank groups, notably High-Income (HIC), Upper Middle-Income (UMIC) and Lower Middle-Income (LMIC) countries. There are no low-income countries (LIC) in the G20.
19. By including LULUCF emissions in the total, emissions increase (decrease) if land-use is a net source (sink).
20. In Japan, Korea, Russia, Turkey and the United States.
21. Canada, India and Mexico.
22. Analysis of the IPCC AR5 integrated assessment scenarios, consistent with outcomes with a greater than 66% likelihood of keeping warming below 2°C, result in total GHGs emissions in 2050 between 41%- to 72% lower than in 2010 (IPCC, 2014a), which in average annual terms requires emissions reductions between of 1.3%- to 3.1% per year. If world GDP is assumed to grow at around 3% per year, this would require the sum of the total annual reductions in the emissions intensity of GDP of some 4.3% to 6.1%.
23. The IEA's average figure for the G20 is based on more disaggregated modelling, not shown in the figure.
24. For example the use of advanced technology in some countries while other countries with a similar level of energy intensity might have developed in such a way because of constraints on energy availability.
25. Using more of the indicators discussed in this chapter would provide an alternative grouping based on cluster analysis. However there would be only minor differences, in part reflecting the importance of AFOLU emissions. To match the economic analysis in Chapter 4, which does not consider AFOLU sectors, we present the results based on this more limited number of characteristics.
26. See the Executive Summary of the 2015 DDPP report at http://deepdecarbonization.org/wp-content/uploads/2015/12/DDPP_EXESUM-1.pdf .
27. The G20 countries where no results are available are: Argentine, Russia, Saudi Arabia and Turkey.
28. See for e.g. Anandarajah and Gambhir (2014), Capros et al. (2014), Gambhir et al (2013), Pye et al. (2017), and Winkler and Marquand (2009).
29. Due to their high wealth and low protection level, three American cities (Miami, New York City and New Orleans) concentrated 31% of the losses in 2005 across the 136 cities studied. Adding Guangzhou, the four top cities accounted for 43% of global losses in that same year (Hallegatte et al., 2013).
30. A number of G20 countries have invested significantly in providing access to relevant data sources, through initiatives such as the UK's Climate Impact Programme and the climate section of the United States' US Data.Gov website. The private sector is increasingly engaged in this area, through the provision of consultancy services and provision of expertise by insurance companies.
31. Interactions between mitigation and adaptation will be explored in the 2018 IPCC special report on the impacts of global warming of 1.5°C degrees (IPCC, 2016).
32. The need for international environmental agreements to be "self-enforcing" in the face of limited sanctions had the dismaying implication that participation would be inefficiently low from a global perspective precisely when such co-operation would be of greatest environmental benefit (Barrett, 1994). Concerns about "carbon leakage" by through the off-shoring of emissions-intensive industry are a further constraint on stringent mitigation action, though at current levels of carbon prices there is little evidence that carbon leakage is a major problem, except perhaps in a few fossil-intensive industries. See for example, Branger, Quirion and Chevallier (2013) and Martin et al. (2014).
33. Leading to important debates about the right discount rate to use to estimate the social cost of carbon, see Pindyck (2013) for a discussion of this and related issues.
34. See Crampton et al., 2017.
35. NDCs representing 190 parties had been submitted as of 17 January 2017.
36. Of course, whether the NDCs are consistent with a goal of well below 2°C also depends on what happens to emissions beyond the 2025-30 period for which the NDCs are applicable. A comparison of countries' pledges with emission scenarios available in the IPCC AR5 database shows that more than three quarters of the scenarios that follow a similar emission profile to that consistent with existing NDCs to 2030 give median warming values of more than 2°C in 2100 (i.e. 50% chance of warming less than 2°C), with the vast majority giving a level of median warming between 2° and 3°C.

37. Aldy and Pizer (2016) use four integrated assessment models to assess and compare the NDCs. They estimate that countries' marginal abatement costs vary by two orders of magnitude. Marginal costs rise almost proportionally with income, while total mitigation costs also reflect carbon intensity and trade in fossil fuels. See also Bataille et al. (2016) and Rogelj et al. (2016a).
38. CO_2, CH_4, N_2O, hydrofluorocarbons, perfluorocarbons, sulphur hexafluoride.
39. See Arent and Tol (2014).
40. Concerns about "top-down" approaches crystallised at the Copenhagen UNFCCC Conference of the Parties (COP15) in 2009. Outcomes at COP16 in Cancún built on the Copenhagen Accord both in terms of a new transparency regime and a formalisation of some international pledges (e.g. on climate finance). More than 90 countries, including all major emitters, put forward pledges that took a variety of forms, mostly covering the period to 2020.
41. Ostrom (1990) highlighted the significant empirical evidence of the potential for self-organising institutions successfully to manage natural resources where there is sufficient trust and reciprocity between those involved. The likelihood of co-operation was also found to increase with factors such as: (i) reliable information about short- and long-term costs and benefits; (ii) a recognition of the importance of the resource to their own achievements and a long-term view; (iii) communication between those involved; (iv) informal monitoring and sanctioning is both feasible and considered appropriate; and (v) the existence of social capital and leadership.
42. The main milestones are the Facilitative Dialogue in 2018 and the Global Stocktakes, which will take place every five years from 2023 assess collective progress towards long term goals, including mitigation and adaptation efforts and means of implementation, and will inform Parties' future actions.
43. Countries agreed in Marrakesh at the 22nd Conference of the Parties (COP22) that this "Paris rulebook" will be finalised by the end of 2018 (COP24).

References

Aldy, J. and W. Pizer (2016), "Alternative metrics for comparing domestic climate change mitigation efforts and the emerging international climate policy architecture", *Review of International Economics and Policy, Vol. 10/1, Wiley, Hoboken, pp. 3-24,* http://dx.doi.org/10.1093/reep/rev013.

Alexander, P., et al. (2017), "Assessing uncertainties in land cover projections", *Global Change Biology, Vol. 23/2, Wiley, Hoboken, pp. 767-781,* http://dx.doi.org/10.1111/gcb.13447.

Allen, M.R., et al. (2016), "New use of global warming potentials to compare cumulative and short-lived climate pollutants", *Nature Climate Change, Vol. 6, Springer Nature, London, pp. 773-776,* http://dx.doi.org/10.1038/nclimate2998.

Allen, M.R., et al. (2009), "Warming caused by cumulative carbon emissions towards the trillionth tonne", *Nature*, Vol. 458, Springer Nature, London, pp.1163-1166, http://dx.doi.org/10.1038/nature08019.

Anandarajah, G. and A. Gambhir (2014), "India's CO_2 emission pathways to 2050: What role can renewables play?", *Applied Energy, Vol. 131, Elsevier, Amsterdam, pp.79-86,* http://doi.org/10.1016/j.apenergy.2014.06.026.

Arent, D.J. and R.S.J. Tol (2014), "Chapter 10: Key economic sectors and services", In: *Climate Change 2014: Mitigation of Climate Change*, IPCC Working Group III Contribution to AR5, Cambridge University Press, Cambridge, www.ipcc.ch/pdf/assessment-report/ar5/wg3/ipcc_wg3_ar5_chapter5.pdf.

AVOID2 (2015), "AVOID2 climate change research programme", http://www.avoid.uk.net/indcs/ (accessed 12 December 2016).

Barrett, S. (1994), "Self-enforcing international environmental agreements", *Oxford Economic Papers*, Vol. 43, Oxford University Press, Oxford, pp. 878-894, www.jstor.org/stable/2663505.

Bataille, C., et al. (2016), "The need for national deep decarbonisation pathways for effective climate policy", *Climate Policy, Vol. 16, Supplement 1, Taylor & Francis, London, pp. S7-S26,* http://dx.doi.org/10.1080/14693062.2016.1173005.

Béguin, A., et al. (2011), "The Opposing Effects of Climate Change and Socio-Economic Development on the Global Distribution of Malaria", *Global Environmental Change*, Vol. 21/4, Elsevier, Amsterdam, pp. 1209-1214, http://doi.org/10.1016/j.gloenvcha.2011.06.001.

Blanco, G., et al. (2014), "Drivers, trends and mitigation", *Climate Change 2014: Mitigation of Climate Change, IPCC Working Group III Contribution to AR5. Cambridge University Press, Cambridge and New York,* www.ipcc.ch/pdf/assessment-report/ar5/wg3/ipcc_wg3_ar5_chapter5.pdf.

Branger, F., P. Quirion and J. Chevallier (2013), "Carbon leakage and competitiveness of cement and steel industries under the EU ETS: much ado about nothing", *CIRED Working Paper Series,* www2.centre-cired.fr/IMG/pdf/CIREDWP-201353.pdf.

Capros, P., et al. (2014), "European decarbonisation pathways under alternative technological and policy choices: A multi-model analysis", *Energy Strategy Reviews, Vol. 2, Elsevier, Amsterdam, pp. 231-245,* http://dx.doi.org/10.1016/j.esr.2013.12.007.

Challinor, A.J., et al, (2014), "A meta-analysis of crop yield under climate change and adaptation", *Nature Climate Change, Vol. 4/4, Springer Nature, London, pp.287-291,* http://dx.doi.org/10.1038/nclimate2153.

Ciais, P., et al. (2013), "Carbon and other biogeochemical cycles", in Stocker et al. (eds.), *Climate Change 2013: The Physical Science Basis, Contribution of Working Group I to the Fifth Assessment Report of the Intergovernmental Panel on Climate Change, Cambridge University Press, Cambridge and New York* www.ipcc.ch/pdf/assessment-report/ar5/wg1/WG1AR5_Chapter06_FINAL.pdf.

Crampton, P., et al. (eds.) (2017), *Global Carbon Pricing: The Path to Climate Cooperation, MIT Press, Cambridge, Massachusetts,* https://mitpress.mit.edu/books/global-carbon-pricing.

Creutzig, F., et al. (2015), "Bioenergy and climate change mitigation: an assessment", *Global Change Biology Bioenergy, Vol. 7/5, Wiley, Hoboken, pp.916-944,* http://dx.doi.org/10.1111/gcbb.12205.

DeAngelis, A., et al. (2010), "Observational evidence of enhanced precipitation due to irrigation over the Great Plains of the United States", *Journal of Geophysical Research*, Vol. 115, Issue D15, Wiley, Hoboken, http://dx.doi.org/10.1029/2010JD013892.

DDPP (Deep Decarbonization Pathways Project) (2015), *Pathways to Deep Decarbonization*, 2015 report – executive summary, Sustainable Development Solutions Network and Institute for Sustainable Development and International Relations, Paris, http://deepdecarbonization.org/wp-content/uploads/2015/06/DDPP_EXESUM.pdf.

Dunne, J.P., et al. (2013) "Reductions in labour capacity from heat stress under climate warming", *Nature Climate Change, Vol. 3, Springer Nature, London, pp.* 563-566, http://dx.doi.org/10.1038/nclimate1827.

EM-DAT (n.d.), *EM-DAT: The International Disaster Database*, www.emdat.be/ (accessed 9 February 2017).

FAO (2016), *FAOSTAT, Statistics Division, Food and Agriculture Organization of the United Nations, Rome Italy (accessed December 2016)*, www.fao.org/faostat/en/.

FAO (2013), "Feeding the world", in FAO (2013), FAO Statistical Yearbook 2013, Part 3, FAO, Rome, www.fao.org/docrep/018/i3107e/i3107e00.htm

Friedlingstein, P., et al. (2014), "Persistent growth of CO_2 emissions and implications for reaching climate targets", *Nature Geoscience, Vol. 7, Springer Nature, London, pp.709-715,* http://dx.doi.org/10.1038/ngeo2248.

Fuss, S., et al. (2014), "Betting on negative emissions", *Nature Climate Change, Vol. 4, Springer Nature, London, pp.* 850-853, http://dx.doi.org/10.1038/nclimate2392.

Gambhir, A., et al. (2013), "A hybrid modelling approach to develop scenarios for China's carbon dioxide emissions to 2050", *Energy Policy, Vol. 59, Elsevier, Amsterdam, pp.614-632,* http://doi.org/10.1016/j.enpol.2013.04.022.

Hallegatte, S., et al. (2017), "Unbreakable: Building the Resilience of the Poor in the Face of Natural Disasters", *Climate Change and Development Series, World Bank, Washington DC,* http://dx.doi.org/10.1596/978-1-4648-1003-9.

Hallegatte, S., et al. (2016), "Shock Waves: Managing the Impacts of Climate Change on Poverty", *Climate Change and Development Series, World Bank Group, Washington, DC,* http://hdl.handle.net/10986/22787.

Hallegatte, S., et al. (2013), "Future flood losses in major coastal cities", *Nature Climate Change*, Vol. 3, Springer Nature, London, pp. 802-806, http://dx.doi.org/10.1038/nclimate1979.

Haasnoot, M., et al. (2013), "Dynamic adaptive policy pathways: A method for crafting robust decisions for a deeply uncertain world", *Global Environmental Change, Vol. 23, pp.* 485-498, http://doi.org/10.1016/j.gloenvcha.2012.12.006.

Hosonuma, N., et al. (2012), "An assessment of deforestation and forest degradation in developing countries", *Environmental Research Letters, Vol. 7/4, IOP Publishing, Bristol,* http://dx.doi.org/10.1088/1748-9326/7/4/044009.

Huitema, D., et al. (2016), "The governance of adaptation: choices, reasons, and effects: Introduction to the Special Feature", *Ecology and Society, Vol. 21/3, Resilience Alliance, Wolfville, Nova Scotia,* http://dx.doi.org/10.5751/ES-08797-210337.

IEA (2017), "IEA finds CO_2 emissions flat for third straight year even as global economy grew in 2016", Press release, 17 March 2017, https://www.iea.org/newsroom/news/2017/march/iea-finds-co2-emissions-flat-for-third-straight-year-even-as-global-economy-grew.html.

IEA (2017), Chapters 1 and 2 in *Perspectives for the Energy Transition: Investment Needs for a Low-Carbon Energy System, International Energy Agency,* OECD Publishing, Paris, www.energiewende2017.com/wp-content/uploads/2017/03/Perspectives-for-the-Energy-Transition_WEB.pdf.

IIASA (n.d.), *SSP Database*, International Institute for Applied Systems Analysis, Laxenburg, Austria, https://tntcat.iiasa.ac.at/SspDb/dsd?Action=htmlpage&page=about./.

Ignaciuk, A. and D. Mason-D'Croz (2014), "Modelling adaptation to climate change in agriculture", *OECD Food, Agriculture and Fisheries Papers*, No. 70, OECD Publishing, Paris, http://dx.doi.org/10.1787/5jxrclljnbxq-en.

IPCC (2016), *Decision IPCC/XLIV-4, Sixth Assessment Report (AR6) Products, Outline of the Special Report on 1.5°C,* 44th session of the IPCC Panel, Bangkok (Thailand), 17-20 October 2016, Intergovernmental Panel on Climate Change, Geneva, www.ipcc.ch/meetings/session44/l2_adopted_outline_sr15.pdf.

IPCC (2014a), *Climate Change 2014: Mitigation of Climate Change, Contribution of Working Group III to the Fifth Assessment Report of the Intergovernmental Panel on Climate Change, IPCC, Cambridge University Press, Cambridge, and New York,* http://www.ipcc.ch/report/ar5/wg3/.

IPCC (2014b), "Summary for policymakers", in Field et al. (eds.) *Climate Change 2014: Impacts, Adaptation, and Vulnerability, Part A: Global and Sectoral Aspects, Contribution of Working Group II to the Fifth Assessment Report of the Intergovernmental Panel on Climate Change, Cambridge University Press, Cambridge and New York, pp.* 1-32.

IPCC (2014c), *Climate Change 2014: Synthesis Report, Contribution of Working Groups I, II and III to the Fifth Assessment Report of the Intergovernmental Panel on Climate Change,* IPCC, Geneva, www.ipcc.ch/report/ar5/syr/.

IPCC (2013), *Climate Change 2013: The Physical Science Basis, Contribution of Working Group I to the Fifth Assessment Report of the Intergovernmental Panel on Climate Change"*, Cambridge University Press, Cambridge and New York, www.ipcc.ch/report/ar5/wg1/.

Lenton, T.M. et al. (2008), "Tipping elements in the Earth's climate system", *Proceedings of the National Academy of Sciences, Vol. 105/6, National Academy of Sciences of the United States of America, Washington DC, pp.*1786-1793, http://dx.doi.org/10.1073/pnas.0705414105.

Lowe, J.A., et al. (2009), "How difficult is it to recover from dangerous levels of global warming?", *Environmental Research Letters, Vol. 4/1, IOP Publishing, Bristol,* http://dx.doi.org/10.1088/1748-9326/4/1/014012.

Luo, T., et al. (2013), *Majority of China's proposed coal-fired power plants located in water-stressed regions,* World Resources Institute, Washington, www.wri.org/blog/2013/08/majority-china%E2%80%99s-proposed-coal-fired-power-plants-located-water-stressed-regions.

Mackey, B., et al. (2013), "Untangling the confusion around land carbon science and climate change mitigation policy", *Nature Climate Change, Vol. 3/6, Springer Nature, London, pp.*552-557, http://dx.doi.org/10.1038/nclimate1804.

Martin, R., et al. (2014), "On the empirical content of carbon leakage criteria in the EU Emissions Trading Scheme", *Ecological Economics, Vol. 105, Elsevier, Amsterdam, pp.78-88,* http://doi.org/10.1016/j.ecolecon.2014.05.010.

MCTI (2016), *Annual Estimates of GHG Emissions in Brazil,* Ministry of Science, Technology and Innovation, Brasilia, Brazil, http://seeg.eco.br/2014/12/06/seeg-2014-engl/.

Moss, R.H. et al. (2010), "The next generation of scenarios for climate change research and assessment", *Nature, Vol. 463, Springer Nature, London, pp. 747-756,* http://dx.doi.org/10.1038/nature08823.

Myhre, G. and D. Shindell (2013), "Anthropogenic and Natural Radiative Forcing – Supplementary Material", In Stocker et al. (eds.), *Climate Change 2013: The Physical Science Basis, Contribution of Working Group I to the Fifth Assessment Report of the Intergovernmental Panel on Climate Change, Cambridge University Press, Cambridge and New York,* www.ipcc.ch/pdf/assessment-report/ar5/wg1/supplementary/WG1AR5_Ch08SM_FINAL.pdf.

Nemet et al. (2017), "Addressing policy credibility problems for low-carbon investment", *Global Environmental Change, Vol. 42, Elsevier, Amsterdam, pp.474-57,* http://doi.org/10.1016/j.gloenvcha.2016.12.004.

NOAA (2017), *Trends in atmospheric carbon dioxide,* U.S. National Oceanic and Atmospheric Administration, Silver Spring, Maryland, www.esrl.noaa.gov/gmd/ccgg/trends/.

OECD (forthcoming), *Biodiversity and Development: Mainstreaming and Managing for Results,* OECD Publishing, Paris.

OECD (2017), «Air and climate: Greenhouse gas emissions by source», *OECD Environment Statistics* (database).DOI: http://dx.doi.org/10.1787/data-00594-en (accessed 28 March 2017).

OECD (2016a), *Policy Guidance on Resource Efficiency,* OECD Publishing, Paris, http://dx.doi.org/10.1787/9789264257344-en.

OECD (2016b), *Summary record of the 8th OECD Food Chain Analysis Network Meeting – 23-24 June 2016, OECD, Paris* www.oecd.org/site/agrfcn/meetings/FCAN%20June%202016%20Summary%20Report_for%20website.pdf.

OECD (2015a), *Climate Change Mitigation: Policies and Progress,* OECD Publishing, Paris, http://dx.doi.org/10.1787/9789264238787-en.

OECD (2015b), *The Economic Consequences of Climate Change,* OECD Publishing, Paris, http://dx.doi.org/10.1787/9789264235410-en.

OECD (2015c), *Climate Change Risks and Adaptation: Linking Policy and Economics,* OECD Publishing, Paris, http://dx.doi.org/10.1787/9789264234611-en.

OECD (2015d), *National Climate Change Adaptation: Emerging Practices in Monitoring and Evaluation,* OECD Publishing, Paris, http://dx.doi.org/10.1787/9789264229679-en.

OECD (2015e), "Key joint messages emerging from the OECD-France Conference on Agriculture and agricultural soils facing climate change and food security challenges: public policies and practices", 16 September 2015, OECD, Paris, www.ag4climate.org/keymessages/.

OECD (2014), *Green Growth Indicators for Agriculture: A Preliminary Assessment,* OECD Publishing, Paris, http://dx.doi.org/10.1787/9789264223202-en.

OECD (2012), *OECD Environmental Outlook to 2050: The Consequences of Inaction,* OECD Publishing, Paris, http://dx.doi.org/10.1787/9789264122246-en.

OECD (2010), *Paying for Biodiversity: Enhancing the Cost-Effectiveness of Payments for Ecosystem Services, OECD Publishing, Paris,* http://dx.doi.org/10.1787/9789264090279-en.

Olivier, J.G.J., et al. (2016), *Trends in global CO_2 emissions: 2016 Report, European Commission, Joint Research Centre (JRC), Directorate C – Energy, Transport and Climate; PBL Netherlands Environmental Assessment Agency, The Hague,* http://edgar.jrc.ec.europa.eu/news_docs/jrc-2016-trends-in-global-co2-emissions-2016-report-103425.pdf.

Olsson, L., et al. (2014), "Livelihoods and poverty", in *Climate Change 2014: Impacts, Adaptation, and Vulnerability, Part A: Global and Sectoral Aspects, Contribution of Working Group II to the Fifth Assessment Report of the Intergovernmental Panel on Climate Change*, Cambridge University Press, Cambridge and New York, pp. 793-832, http://danida.vnu.edu.vn/cpis/files/IPCC/wg2/pdf/WGIIAR5-Chap13_FINAL.pdf.

O'Neill, B.C., et al. (2015), "The roads ahead: Narratives for shared socioeconomic pathways describing world futures in the 21st century", *Global Environmental Change*, Vol.42, Elsevier, Amsterdam, pp. 169-180, http://dx.doi.org/10.1016/j.gloenvcha.2015.01.004.

Ostrom, E. (2009), "A General Framework for Analyzing Sustainability of Social-Ecological Systems", *Science, Vol. 325/5939, American Association for the Advancement of Science, New York, pp.* 419-422, http://dx.doi.org/10.1126/science.1172133.

Ostrom, E. (1990), *Governing the Commons: The Evolution of Institutions for Collective Action, Cambridge University Press, Cambridge and New York.*

Peters, G.P., et al. (2017), "Key indicators to track current progress and future ambition of the Paris Agreement", Nature Climate Change, Vol. 7/2, Springer Nature, pp.118-122, http://dx.doi.org/ 10.1038/nclimate3202.

Pindyck, R.S. (2013), "Climate change policy: What do the models tell us?", *Journal of Economic Literature, Vol. 51/3, American Economic Association, Pittsburgh, pp.*860-872, http://dx.doi.org/10.1257/jel.51.3.860.

Popp, A., et al. (2017), "Land-use futures in the shared socioeconomic pathways", *Global Environmental Change*, Vol. 42, Elsevier, Amsterdam, pp. 331-345, http://dx.doi.org/10.1016/j.gloenvcha.2016.10.002.

Prentice, I.C., Williams, S. and P. Friedlingstein (2015), "Biosphere feedbacks and climate change", *Grantham Institute Briefing Paper No.12, Imperial College London, London,* https://www.imperial.ac.uk/media/imperial-college/grantham-institute/public/publications/briefing-papers/Biosphere-feedbacks-and-climate-change-Briefing-Paper-No-12v2.pdf.

Pye, S., et al. (2017), "Achieving net-zero emissions through the reframing of UK national targets in the post-Paris Agreement era", *Nature Energy, Vol. 2, Springer Nature, London,* http://dx.doi.org/10.1038/nenergy.2017.24.

Revi, A., et al. (2014), "Urban Areas", in Field, C.B. et al. (eds.), *Climate Change 2014: Impacts, Adaptation, and Vulnerability: Part A: Global and Sectoral Aspects, Contribution of Working Group II to the Fifth Assessment Report of the Intergovernmental Panel on Climate Change*, Cambridge University Press, Cambridge and New York, pp. 535-612, http://ar5-syr.ipcc.ch/ipcc/ipcc/resources/pdf/WGII/WGIIAR5-Chap8_FINAL.pdf.

Riahi, K., et al. (2017), "The shared socioeconomic pathways and their energy, land use, and greenhouse gas emissions implications: an overview", *Global Environmental Change, Vol. 42, Elsevier, Amsterdam,* pp.153-168, http://dx.doi.org/10.1016/j.gloenvcha.2016.05.009.

Ripple, W.J., et al. (2014), "Ruminants, climate change and climate policy", *Nature Climate Change*, Vol. 4/1, *Springer Nature, pp.*2-5, http://dx.doi.org/10.1038/nclimate2081.

Rittel, H.W. and M.M. Webber (1973) "Dilemmas in a general theory of planning", *Policy Sciences*, Vol. 4/2, *Springer Nature, London, pp.*155-169, http://dx.doi.org/10.1007/BF01405730.

Rogelj, J., et al. (2016a), "Paris Agreement climate proposals need a boost to keep warming below 2°C", *Nature*, Vol. 534/7609, *Springer Nature, London, pp.* 631-639, http://dx.doi.org/10.1038/nature18307.

Rogelj, J., et al. (2016b), "Differences between carbon budget estimates unravelled", *Nature Climate Change*, Vol. 6/3, Springer Nature, London, pp. 245-252, http://dx.doi.org/10.1038/nclimate2868.

Rose, S. K., et al. (2014), "Bioenergy in energy transformation and climate management", *Climatic Change*, Vol. 123/3, *Springer Nature, pp.* 477-493, http://dx.doi.org/10.1007/s10584-013-0965-3.

Sadoff, C.W., et al. (2015) *Securing Water, Sustaining Growth: Report of the GWP/OECD Task Force on Water Security and Sustainable Growth*, University of Oxford, Oxford, www.water.ox.ac.uk/wp-content/uploads/2015/04/SCHOOL-OF-GEOGRAPHY-SECURING-WATER-SUSTAINING-GROWTH-DOWNLOADABLE.pdf.

Schiermeier, Q. (2007), "The new face of the article", *Nature, Vol. 446/7132, Springer Nature, London, pp.* 133-135, http://dx.doi.org/10.1038/446133a.

Settele, J., et al. (2014), "Terrestrial and inland water systems", in Field, C.B. et al. (eds.), *Climate Change 2014: Impacts, Adaptation, and Vulnerability, Part A: Global and Sectoral Aspects, Contribution of Working Group II to the Fifth Assessment Report of the Intergovernmental Panel on Climate Change, Cambridge University Press, Cambridge and New York, pp.* 271-359, www.ipcc.ch/pdf/assessment-report/ar5/wg2/WGIIAR5-Chap4_FINAL.pdf.

Shepherd, T.G. (2014), "Atmospheric circulation as a source of uncertainty in climate change projections", *Nature Geoscience, Vol. 7/10, Springer Nature, pp.703-708,* http://dx.doi.org/10.1038/ngeo2253.

Shindell, D., et al. (2012), "Simultaneously mitigating near-term climate change and improving human health and food security", *Science, Vol. 335/6065, American Association for the Advancement of Science, New York, pp. 183-189,* http://dx.doi.org/10.1126/science.1210026.

Shine, K.P. (2009), "The global warming potential—The need for an interdisciplinary retrial", *Climatic Change, Vol. 96/4, Springer Nature, pp.467-472,* http://dx.doi.org/10.1007/s10584-009-9647-6.

Smith, P. (2016), "Soil carbon sequestration and biochar as negative emission technologies", *Global Change Biology, Vol. 22/3, Wiley, Hoboken, pp.1315-1324,* http://dx.doi.org/10.1111/gcb.13178.

Smith, P., et al. (2015), "Biophysical and economic limits to negative CO_2 emissions", *Nature Climate Change, Vol. 6, Springer Nature, London, pp.42-50,* http://dx.doi.org/ 10.1038/nclimate2870.

Smith, P., et al. (2014), "Agriculture, Forestry and Other Land Use (AFOLU)", in Edenhofer et al. (eds.), *Climate Change 2014: Mitigation of Climate Change, Contribution of Working Group III to the Fifth Assessment Report of the Intergovernmental Panel on Climate Change, Cambridge University Press, Cambridge and New York,* www.ipcc.ch/pdf/assessment-report/ar5/wg3/ipcc_wg3_ar5_chapter11.pdf.

Smith, P., et al. (2013), "How much land-based greenhouse gas mitigation can be achieved without compromising food security and environmental goals?", *Global Change Biology, Vol. 19/8, Wiley, Hoboken, pp.2285-2302,* http://dx.doi.org/10.1111/gcb.12160.

Smith, K.R., et al. (2014), "Human health: impacts, adaptation, and co-benefits", in Field et al. (eds.), *Climate Change 2014: Impacts, Adaptation, and Vulnerability, Part A: Global and Sectoral Aspects, Contribution of Working Group II to the Fifth Assessment Report of the Intergovernmental Panel on Climate Change, Cambridge University Press, Cambridge and New York,* www.ipcc.ch/pdf/assessment-report/ar5/wg2/WGIIAR5-Chap11_FINAL.pdf.

Smith, S.M., et al. (2012), "Equivalence of greenhouse-gas emissions for peak temperature limits", *Nature Climate Change, Vol. 2/7, Springer Nature, London, pp.535-538,* http://dx.doi.org/ 10.1038/nclimate1496.

Solomon, S., et al. (2009), "Irreversible climate change due to carbon dioxide emissions", *Proceedings of the National Academy of Sciences, Vol. 106, National Academy of Sciences of the United States of America, pp.* 1704-1709, http://dx.doi.org/10.1073/pnas.0812721106.

Steffen, W., et al. (2004), *Global Change and the Earth System: A Planet Under Pressure, Springer-Verlag, Berlin and Heidelberg.*

Stern, N.H. (2007), The economics of climate change: the Stern review, Cambridge University Press, Cambridge.

Taylor, K., R. Stouffer and G. Meehl (2012), *An Overview of CMIP5 and the Experiment Design*, AMS, http://dx.doi.org/10.1175/BAMS-D-11-00094.1.

UNEP (2015), *The Emissions Gap Report 2015: A UNEP Synthesis Report, United Nations Environment Programme,* http://web.unep.org/emissionsgapreport2015.

UNFCCC (n.d.), *NDC Registry (interim) (accessed 9 February 2017),* www4.unfccc.int/ndcregistry/Pages/All.aspx.

UNFCCC (2016), *National Inventory Submissions 2016: CRF tables.* http://unfccc.int/national_reports/annex_i_ghg_inventories/national_inventories_submissions/items/9492.php.

UNFCCC (2015a), *Paris Agreement,* http://unfccc.int/files/essential_background/convention/application/pdf/english_paris_agreement.pdf.

UNFCCC (2015b), *Synthesis report on the aggregate effect of the intended nationally determined contributions,* FCCC/CP/2015/7, http://unfccc.int/focus/indc_portal/items/9240.php.

UNFCCC (2014), *GHG emission profiles for non-Annex I Parties* (accessed 29 March 2017). http://unfccc.int/ghg_data/ghg_data_unfccc/ghg_profiles/items/4626.php.

Vallejo, L. and M. Mullan (2017), "Climate-resilient infrastructure: Getting the policies right", *OECD Environment Working Papers*, No. 121, OECD Publishing, Paris, http://dx.doi.org/10.1787/02f74d61-en.

Wigley, T.M., R. Richels and J.A. Edmonds (1996), "Economic and environmental choices in the stabilization of atmospheric CO_2 concentrations", *Nature, Vol. 379/6562, Springer Nature, London, pp.240,* http://dx.doi.org/10.1038/379240a0.

Wilbanks, T. (2005), "Issues in developing a capacity for integrated analysis of mitigation and adaptation", *Environmental Science and Policy, Vol. 8/6, pp. 541-547,* http://dx.doi.org/10.1016/j.envsci.2005.06.014.

Winkler, H. and A. Marquand (2009), "Changing development paths: From an energy-intensive to low-carbon economy in South Africa", *Climate and Development, Vol. 1/1, Taylor & Francis, London, pp.47-65,* http://hdl.handle.net/11427/16928.

Wise, R.M., et al. (2014), "Reconceptualising adaptation to climate change as part of pathways of change and response", *Global Environmental Change*, Elsevier, Amsterdam, Vol. 28, pp. 325-336, http://doi.org/10.1016/j.gloenvcha.2013.12.002.

World Bank (n.d.a), *World Development Indicators Database,* (accessed February 2017), World Bank, Washington DC, http://data.worldbank.org/data-catalog/world-development-indicators.

World Bank (n.d.b.), *World Governance Indicators Database,* (accessed February 2017), World Bank, Washington DC, http://info.worldbank.org/governance/wgi/#home.

World Bank (2016), "Highlights: Featuring the Sustainable Development Goals", in *2016 World Development Indicators*, World Bank, Washington DC, http://databank.worldbank.org/data/download/site-content/wdi-2016-highlights-featuring-sdgs-booklet.pdf.

World Bank (2009), *World Development Indicators 2008,* World Bank, Washington DC, http://documents.worldbank.org/curated/en/587251468176971009/World-development-indicators-2008.

WHO (World Health Organisation) (2014), "Quantitative risk assessment of the effects of climate change on selected causes of death, 2030s and 2050s", WHO, Geneva, www.who.int/globalchange/publications/quantitative-risk-assessment/en/.

WMO (World Meteorological Organization) (2017), *Statement on the State of the Global Climate in 2016, WMO-No. 1189, WMO, Geneva,* http://library.wmo.int/opac/doc_num.php?explnum_id=3414.

Zhang, X., et al. (2015), "Managing nitrogen for sustainable development", *Nature*, Vol. 528/7580, Springer Nature, London, pp.51-59, http://dx.doi.org/10.1038/nature15743

Annex 2.A1. Summary of G20 countries' (I)NDCs

Summary of G20 countries' (I)NDCs

G20 ECONOMY	TYPE	BASE YEAR	TARGET DATE	CONDITIONALITY	GHG EMISSIONS MITIGATION TARGET	SECTORS COVERED	GASES COVERED	MITIGATION MEASURES	ADAPTATION MEASURES
ARGENTINA	Emission ceiling	n/a	By 2030	Unconditional	To not exceed 483 $MtCO_2eq$	Economy-wide, including energy, industrial processes, agriculture (including cattle rearing), LULUCF, waste	CO_2, CH_4, N_2O, HFC, PFC, SF6	Action on sustainable forestry, energy efficiency, biofuels, nuclear power, renewable energy and transport modal shift	1. Early warning mechanisms and systems for response and recovery from climate disasters 2. Sustainable management of native forests 3. Water resource management 4. Crop management 5. Management of the health impacts of climate change 6. Implementation of measures to face extreme events 7. Ecosystem-based biodiversity conservation and adaptation
	Emission ceiling			Conditional on financial, technical and capacity-building support from abroad	To not exceed 369 $MtCO_2eq$				
AUSTRALIA	Absolute reduction from base year emissions	2005	By 2030	Unconditional	Reduction of 26-28%	Economy-wide, including energy, industrial processes and product use, agriculture, LULUCF, waste	CO_2, CH_4, N_2O, HFC, PFC, SF6, NF_3	1. Emissions Reduction Fund – provides incentives for emissions reduction activities across the economy. 2. Safeguard Mechanism – sets emissions limits for facilities emitting >100 000 tonnes per year. 3. Renewable Energy Target of 23% of electricity supply to come from renewable sources by 2020. 4. National Energy Productivity Plan to achieve a 40% improvement in energy productivity by 2030. 5. Grants for research, development, demonstration and deployment of clean energy technologies.	1. Work to build climate resilience and support adaptation to climate change. 2. Develop a National Climate Resilience and Adaptation Strategy.
BRAZIL	Absolute reduction from base year emissions	2005	By 2025	Unconditional	Reduction of 37%	Economy-wide, including emissions from forest managed areas (conservation units and indigenous lands)	Not specified	Not specified	Outlined in the National Adaptation Plan, which focuses on risk areas, housing, basic infrastructure (especially in the areas of health, sanitation and transportation).
	Absolute reduction from base year emissions	2005	By 2030	n/a	Reduction of 43% (indicative value, for reference only)				
CANADA	Absolute reduction from base year emissions	2005	By 2030	Unconditional	Reduction of 30%	Economy-wide (although excludes emissions from natural disturbances)	CO_2, CH_4, N_2O, HFC, PFC, SF6, NF_3	Regulation measures in the transport and energy sectors and with regards to renewable fuels	Not specified
CHINA	Peaking of emissions	n/a	By 2030	Not specified	n/a	Economy-wide	Not specified	ii) Increase the share of non-fossil fuels in primary energy consumption to approx. 20%. iii) Increase forest stock volume by 4.5 cubic meters compared with 2005 levels.	i) Enhance mechanisms and capacities for climate vulnerable sectors. ii) Strengthen early warning and emergency response systems.
	Emission intensity of GDP	2005	By 2030	Not specified	Reduction of 60-65%				

Summary of G20 countries' (I)NDCs *(cont.)*

G20 ECONOMY	TYPE	BASE YEAR	TARGET DATE	CONDITIONALITY	GHG EMISSIONS MITIGATION TARGET	SECTORS COVERED	GASES COVERED	MITIGATION MEASURES	ADAPTATION MEASURES
EUROPEAN UNION	Absolute reduction from base year emissions	1990	By 2030	Not specified	Reduction of 40%	Economy-wide, including energy, industrial processes and product use, agriculture, waste, LULUCF	CO_2, CH_4, N_2O, HFC, PFC, SF6, NF_3	Not specified	Not specified
INDIA	Emission intensity of GDP	2005	By 2030	Dependent on financial, technical and capacity-building support from developed countries	Reduction of 33-35%	Economy-wide	Not specified	*i)* Achieve 40% cumulative electric power installed capacity from non-fossil fuel based energy sources by 2030. *ii)* Create an additional carbon sink of 2.53 billion tCO_2eq through afforestation by 2030. *iii)* Introduce cleaner, more efficient technologies in thermal power generation. *iv)* Promotion renewables and increase the share of alternative fuels in the country's fuel mix. *v)* Reduce emissions from the transport and waste sectors. *vi)* Promote energy efficiency. *vii)* Fully implement India's afforestation programmes.	*i)* Enhance investment in development programs in vulnerable sectors. *ii)* Develop climate-resilient infrastructure. *iii)* Enhance climate-resilience more generally.
INDONESIA	Emission reduction relative to BAU baseline	BAU scenario of 2.869 $GtCO_2$-eq in 2030	By 2030	Unconditional	Reduction of 29%	Energy (including transport), industrial processes and product use, agriculture, LULUCF, waste	CO_2, CH_4, N_2O	LULUCF: Reducing deforestation and forest degradation, restoring ecosystem functions, sustainable forest management. Energy: 23% of energy coming from new and renewable energy by 2025 Waste: Enhance management capacity of urban wastewater, reduce landfill waste, using waste for energy production	Outlined in the National Action Plan on Climate Change Adaptation. Includes local capacity strengthening, improved knowledge management, identifying synergies between the adaptation and disaster risk reduction agendas, application of adaptive technologies.
	Emission reduction relative to BAU baseline			Dependent on financial, technical and capacity-building support from developed countries	Reduction of 41%				
JAPAN	Absolute reduction from base year emissions	FY 2013	By FY 2030	Not specified	Reduction of 26%	Economy-wide: Energy (incl. CO_2 from transport), industrial processes and product use, agriculture, LULUCF, waste)	CO_2, CH_4, N_2O, HFCs, PFCs, SF6 and NF_3	INDC includes a 2030 target for *i)* final energy consumption: 326 M kl, *ii)* Japan's energy mix: 22-24% renewable energy (dominated by solar and hydropower), 2220% nuclear, 26% coal, 27% LNG, 3% oil. A detailed list of the policy measures considered for each sector is included as an annex to the NDC.	Not specified

Summary of G20 countries' (I)NDCs *(cont.)*

G20 ECONOMY	TYPE	BASE YEAR	TARGET DATE	CONDITIONALITY	GHG EMISSIONS MITIGATION TARGET	SECTORS COVERED	GASES COVERED	MITIGATION MEASURES	ADAPTATION MEASURES
KOREA	Emission reduction relative to BAU baseline	BAU of 850.6 $MtCO_2$-eq in 2030	By 2030	Not specified	Reduction of 37%	Economy-wide, excluding LULUCF (energy, industrial processes and product use, agriculture and waste)	CO_2, CH_4, N_2O, HFCs, PFCs, SF6	Use of carbon credits to achieve 2030 mitigation target. Other mitigation measures used include i) an emissions trading scheme for the industrial sector (launched in 2015), ii) renewable energy regulations for the power sector, iii) a Green Building Standards Code and a system for the Performance Evaluation of Eco-friendly Homes for the buildings sector, iv) low-carbon standards for fuel efficiency and tax incentives to purchase electric vehicles in the transport sector.	Outlined in the National Climate Change Adaptation Plan (2010). *i)* strengthening infrastructure for climate change monitoring, forecasting and analysis; *ii)* developing a management system for disaster prevention and stable water supply; *iii)* developing a climate-resilient ecosystem; *iv)* making a systemic transition to a climate-resilient social and economic structure; and *v)* enhancing the system for the management of negative impacts of climate change on health.
MEXICO	Emission reduction relative to BAU baseline	BAU scenario projecting economic growth in the absence of climate policies	By 2030	Unconditional	Reduction of 25%	Nation-wide (Energy, Industrial processes and product use, agriculture, waste, LULUCF)	CO_2, CH_4, N_2O, HFCs, PFCs, SF6, Black carbon	Mexico's mitigation commitment includes an unconditional reduction of GHG emissions of 22% by 2030. The target increases to 25% when Black Carbon is included.	*i)* Strengthen the adaptive capacity of at least 50% the most vulnerable municipalities. *ii)* Establish early warning systems and risk management at every level of government. *iii)* Reach a rate of 0% deforestation by the year 2030.
	Emission reduction relative to BAU baseline		By 2030	Conditional on financial, technical and capacity-building support from abroad, and international agreement on carbon price, carbon border adjustments.	Reduction of 40%				
RUSSIAN FEDERATION	Absolute reduction from base year emissions	1990	By 2030	Conditional on the maximum absorbing capacity of forests	Reduction of 25-30%	Economy-wide (Energy, industrial processes and product use, agriculture, LULUCF, waste)	CO_2, CH_4, N_2O, HFCs, PFCs, SF6 and NF_3	Not specified	Not specified
SAUDI ARABIA	Emission reduction relative to BAU baseline	BAU scenario projecting economic growth in the absence of climate policies	By 2030	Conditional on the provision of technical assistance and capacity-building	Avoid 130 $MtCO_2eq$	Not specified	Not specified	*i)* Improve energy efficiency via the expansion of the Saudi Energy Efficiency Program, which currently only focuses on industry, buildings and transport sectors. *ii)* Develop renewable energy and natural gas exploitation to diversify the energy mix. *iii)* Develop CCUS via plans to construct a CCU plant able to capture 1 500 tCO_2 per day. *iv)* Recover methane and minimise flaring.	Focus on: *i)* water and wastewater management, *ii)* urban planning, *ii)* marine protection, *iv)* reducing desertification, *v)* developing integrated coastal zone management planning, *vi)* further developing early warning systems, and *vii)* develop integrated water management planning.

Summary of G20 countries' (I)NDCs *(cont.)*

G20 ECONOMY	TYPE	BASE YEAR	TARGET DATE	CONDITIONALITY	GHG EMISSIONS MITIGATION TARGET	SECTORS COVERED	GASES COVERED	MITIGATION MEASURES	ADAPTATION MEASURES
SOUTH AFRICA	Emissions peak, plateau and decline (starting 2020 year-end)	Peak, plateau, decline	Peak: 2020-2025, Plateau 2025-2035, Decline as of 2035	Conditional on the degree of implementation of the Convention by developed countries	Peak: 398-614 $MtCO_2$-eq	Economy-wide (Energy, industrial processes and product use, agriculture, LULUCF, waste)	CO_2, CH_4, N_2O, HFCs, PFCs, SF6 (with a particular focus on CO_2, CH_4, N_2O)	Carbon tax, desired emission reduction outcomes for sectors, company-level carbon budgets, regulatory standards and controls.	*i)* Develop a National Adaptation Plan *ii)* Take climate considerations into account in development policy frameworks. *iii)* Build institutional capacity for climate change response planning and implementation *iv)* Develop early warning systems for climate vulnerable sectors *v)* Develop vulnerability assessment and adaptation needs framework *vi)* Communicate on adaptation efforts
TURKEY	Emission reduction relative to BAU baseline	BAU scenario projecting economic growth in the absence of climate policies	By 2030	Not specified, although NDC mentions that the country will receive financial, technological, and capacity-building support from abroad.	Reduction of up to 21%	Economy-wide (Energy, industrial processes and product use, agriculture, LULUCF, waste)	CO_2, CH_4, N_2O, HFCs, PFCs, SF6 and NF_3	Use of carbon credits from international market mechanisms. An exhaustive list of the measures planned by sector is listed in the INDC.	Not specified
UNITED STATES	Absolute reduction from base year emissions	2005	By 2025	Not specified	Reduction of 26-28%	Economy-wide (energy, industrial processes and product use, agriculture, LULUCF, waste)	CO_2, CH_4, N_2O, HFCs, PFCs, SF6 and NF_3	Does not intend to use international market mechanisms to reach 2025 target. Regulatory measures mitigate emissions include: *i)* fuel economy standards for light-duty and heavy-duty vehicles; *ii)* energy conservation standards for building appliances/equipment as well as building codes for building envelopes; *iii)* regulation to cut emissions from existing power plants; *iv)* methane-specific standards for landfills and the oil and gas sector *v)* the Significant New Alternatives Policy program (targets HFCs).	Not specified

Source: UNFCCC, n.d.

Investing in Climate, Investing in Growth

Chapter 3

Infrastructure for climate and growth

Infrastructure investment is vital to underpin economic growth and development, but current levels of investment are inadequate. Meeting the Paris Agreement's mitigation and adaptation objectives will also require a radical shift in the world's infrastructure base. This chapter considers the current gap in infrastructure investment, the infrastructure and technology transformations needed to shift onto low-emission, climate-resilient pathways, and the incremental capital costs involved. It then looks at the energy sector as an indicative assessment of progress in aligning infrastructure investment plans for the transition, before exploring how governments might better align short-term investment strategies with long-term decarbonisation and resilience goals.

Choices made today about the types, features and location of infrastructure will heavily influence the extent of the impacts of climate change and the vulnerability or resilience of societies to it. Creating low-emission, climate-resilient pathways compatible with the Paris Agreement, as described in Chapter 2, requires a radical shift in our infrastructure bases, mainly for energy, mobility services and buildings. Sustainable infrastructure – infrastructure that is socially, economically and environmentally sound – is a key foundation for economic activity and for reaching the Sustainable Development Goals (SDGs). Since the financial crisis, however, infrastructure of all kinds has suffered from chronic underinvestment.

The first section of this chapter documents the current gap in infrastructure investment required to sustain growth and development. The inconsistencies between current investment trends and climate goals, and the infrastructure investment and technology transformations needed to shift G20 governments onto low-emission, climate-resilient pathways are then addressed. The chapter then focuses on the energy sector as an indicative assessment of progress in aligning infrastructure investment plans for the transition, highlighting the risks of locking in emissions and stranding assets that come with continued investment in fossil-fuel infrastructure. Finally, the chapter concludes with guidance to G20 countries on how they could better align short-term investment strategies with long-term, low-emission decarbonisation goals, and the need to enhance resilience to climate impacts.

Scaling up infrastructure investment to sustain growth and development

Infrastructure in sectors such as energy, transport, water and telecommunications is the backbone of our economies, essential for sustained, inclusive growth and for meeting the SDGs. But current levels of investment in infrastructure are generally too low to sustain growth, and often of insufficient quality. Ensuring affordable and reliable access to basic services remains a major challenge in lower and middle-income countries, while advanced economies are struggling with chronic underinvestment in their ageing infrastructure. Infrastructure investment in the G20 countries needs to be significantly scaled up to fill this gap.

Current levels of infrastructure investment are insufficient to sustain growth and development

Effective energy and transport infrastructure underpins almost all economic activity. Many studies have underscored the positive relationship between high-quality public infrastructure and economy-wide productivity in the long run (e.g. Berg et al., 2012; Ghazanchyan and Stotsky, 2013; Calderon and Serven, 2014). Infrastructure investment is also a way of stimulating demand in the short term: after the financial crisis, many G20 countries devoted a major share of their fiscal stimulus to infrastructure investment (see Chapter 4). On average, emerging and developing economies devoted 40% of their stimulus packages to infrastructure spending, while advanced economies devoted 21% (ILO and IILS, 2011).

Infrastructure investment can also have an impact on promoting inclusive development and fighting income inequality. Inclusive growth, human well-being and poverty reduction depend critically on the type, extent and quality of the infrastructure that supports key services: food, energy, water supply, safe and resilient cities, and sustainable industrialisation (Bhattacharya et al., 2016a). For example, SDG7 ("*Ensure access to affordable, reliable, sustainable and modern energy for all*") requires considerable investment in energy infrastructure in urban and rural areas. Investments in sustainable infrastructure can boost growth and employment and contribute to "*promote sustained, inclusive and sustainable economic growth, full and productive employment and decent work for all*" (SDG8). Transport infrastructure – such as roads, railways, ports and airports – connects home to work, and rural areas to domestic and regional markets, contributing to economic development and the goal of *"ending poverty in all its forms everywhere"* (SDG 1). Infrastructure choices also affect our natural environment and the sustainable use of natural assets such as air, water, terrestrial ecosystems and forests (SDGs 13, 14 and 15).

Despite the links between infrastructure investment and growth and development, underinvestment in infrastructure has been chronic over the past decades. The stock of public capital relative to GDP decreased by 15% globally in the past 30 years (Bhattacharya et al., 2016b; IMF, 2014). Over the past two decades, global infrastructure investment has averaged 3.5% of world GDP (Woetzel et al., 2016).

In advanced G20 economies, public investment fell from 5% of GDP in the late 1960s to 3% in the mid-2000s. Despite increased infrastructure investment following the recent financial crisis, spending remains at a historic low, resulting in an ageing and poorly maintained infrastructure stock in many G20 countries. In the United States, for instance, the National Association of Manufacturers rates transport-related land-based infrastructure as mediocre to poor, with US bridges on average 42 years old, and 1 in 9 structurally deficient. In addition, 65% of roads in 2013 were in "less than good condition", a significant factor in 30% of road fatalities (National Economic Council and the President's Council of Economic Advisers, 2014).

In emerging and low-income economies, public investment fell from 8% of GDP in the late 1970s to 4-5% in the mid-2000s, rising again to 6-7% in 2012. This increase has been led by China, which in 2014 accounted for USD 1.3 trillion of the USD 2.2 trillion invested in infrastructure in developing and emerging economies. This is not only more than all other developing countries, but also more than all developed countries combined (Bhattacharya et al., 2016b).

The quality of infrastructure is critical for development. Many middle-income economies – such as Brazil, India, Russia and South Africa – are left with infrastructure bases of low quality, which constrains medium- and near-term growth. In South Africa, for instance, only 46% of households had piped water of good quality in 2012 and only 71% of households had access to sewerage networks. One-fifth of South African firms identified unreliable electricity supply as a major constraint to doing business (Development Bank of Southern Africa, 2012). Even in China, despite sustained investment in the past decades, the quality of urban infrastructure is not always adapted to the challenges faced by rapidly growing cities (Pan, 2016). Some suggest that China has in fact overinvested in infrastructure and highlight a need to reallocate investments towards more productive infrastructure (Ansar et al., 2016).

Unprecedented levels of infrastructure investment are needed to *i)* maintain and upgrade ageing infrastructure in high-income countries; and *ii)* achieve universal access to basic services in middle-income economies. G20 countries face different priorities in improving infrastructure quality and access (Figure 3.1). Rapid rates of urbanisation and population growth require an expansion of transport and electricity infrastructure, especially in developing countries. By 2050, the global population is expected to increase to 9 billion people, 66% of which will be urban, compared with 54% in 2014. Demand for urban mobility is expected to nearly double between now and 2050, with most of this growth concentrated in developing countries (OECD/ITF, 2017). One in 8 people still live in extreme poverty, nearly 800 million suffer from hunger, 1.1 billion live without electricity, and water scarcity affects more than 2 billion (UN, 2016). Countries that are caught in a low-growth trap could use this opportunity to boost their growth in the short-term, capitalising on the current environment of low interest rates, or optimise the taxation-spending balance to increase infrastructure spending (see Chapter 4).

The importance of infrastructure quality for sustainable growth and well-being can be seen by looking at both access to basic services and at a measure of the quality of the underlying infrastructure (Figure 3.1). For example, while many high-income and middle-income countries boast near-universal access to electricity, in many cases the quality of electricity supply is mediocre, with important consequences for both economic activity and well-being.

Figure 3.1. **Quality of infrastructure and access to basic services in G20 countries, by income and growth groups**

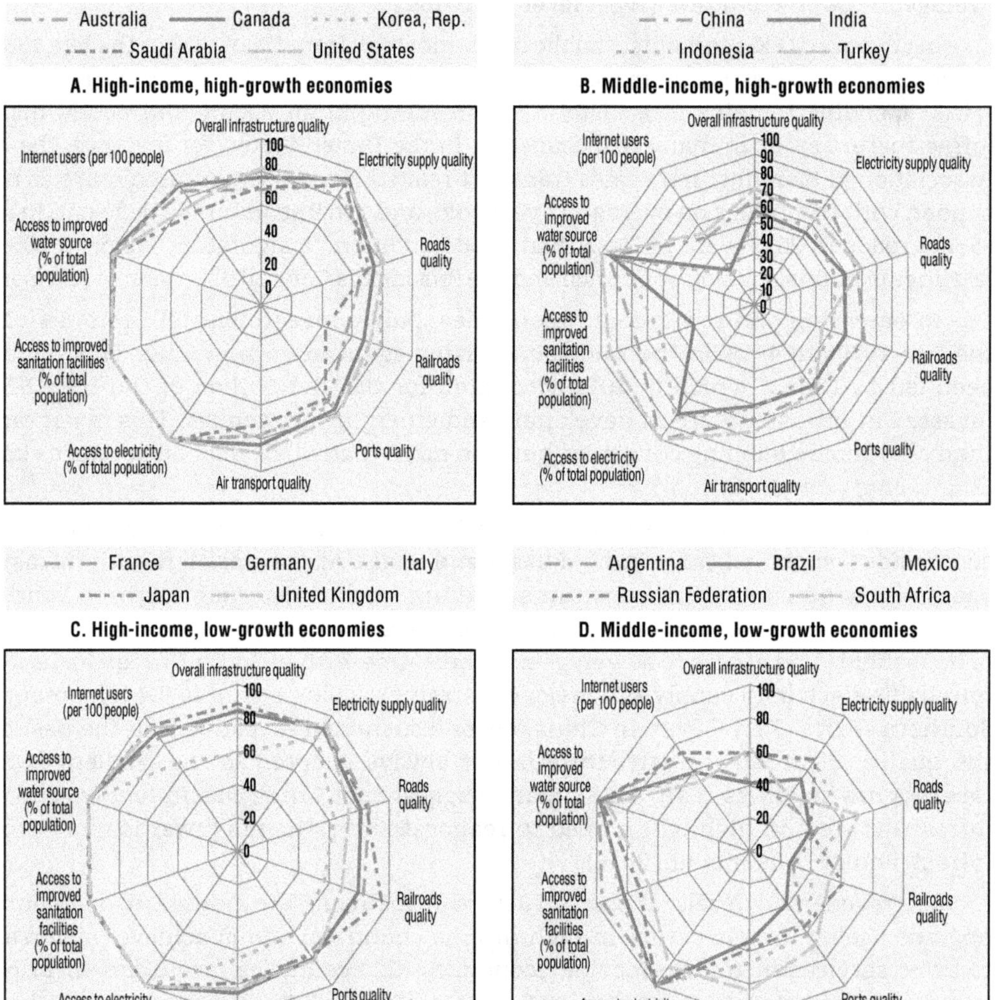

Note: The growth groups are based on the 2010-15 average of GDP growth, population growth and gross capital formation as a share of GDP.

Source: Authors, based on WEF (2015) and World Bank (n.d.a.) (accessed on 28 February 2017).

StatLink http://dx.doi.org/10.1787/888933484204

The infrastructure investment gap

The OECD estimates that around USD 95 trillion of investments will be needed between 2016 and 2030 in energy, transport, water and telecommunications infrastructure to sustain growth, or around USD 6.3 trillion per year, even if governments take no further action on climate change (Table 3.1). This number is to be compared with current infrastructure spending of around USD 3.4 to USD 4.4 trillion (IEA, 2017; IEA, 2016b; Woetzel et al., 2016; Bhattacharya et al., 2016b). Middle-income countries are expected to represent around 60% to 70% of future infrastructure needs (Pardee Centre, n.d; NCE, 2016; Bhattacharya et al., 2016b) (Figure 3.2). The majority of infrastructure investments are required in transport and power, two critical sectors that are also at the heart of decarbonisation strategies (Figure 3.3). However, all infrastructure estimates need to be read with caution (Box 3.1).

Table 3.1. **Global estimates of infrastructure investment needs 2016-30, by sector (before taking into account climate considerations)**

USD 2015 trillion		Annual average	Cumulative
Energy supply	Power and Transmission & Distribution (T&D)	0.7	11.2
	Fossil fuel supply chain	1.0	14.3
Energy demand		0.4	6.6
Transport infrastructure	Road	2.1	31.8
	Rail	0.4	6.4
	Airports and ports	0.2	2.7
Water and sanitation		0.9	13.6
Telecoms		0.6	8.3
TOTAL		**6.3**	**94.9**

Sources: IEA (2017) for energy supply and demand; IEA (2016d) for road and rail infrastructure; OECD (2012) for airports and ports; McKinsey (Woetzel et al., 2016) for telecoms. The water and sanitation estimate is an average of estimates from: Booz Allen Hamilton (2007), McKinsey (Woetzel et al., 2016) and OECD (2006). See technical note on estimating infrastructure investment needs for further details on methodology (http://oe.cd/g20climatereport).

Figure 3.2. **Evolution of infrastructure investment needs by income groups in the G20**

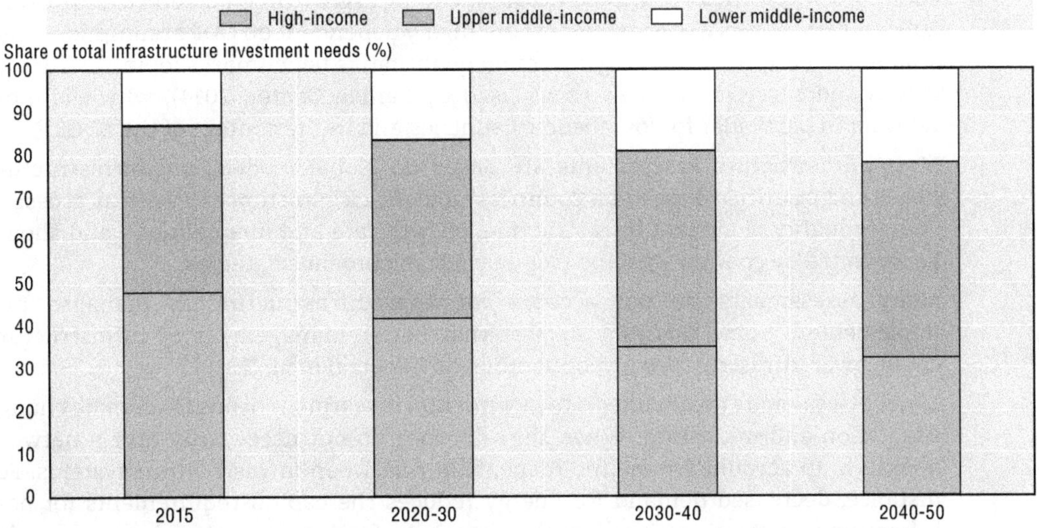

Source: Pardee Center (n.d. accessed February 2017).

StatLink http://dx.doi.org/10.1787/888933484216

Figure 3.3. **Global investment needs by sector, 2016-30**

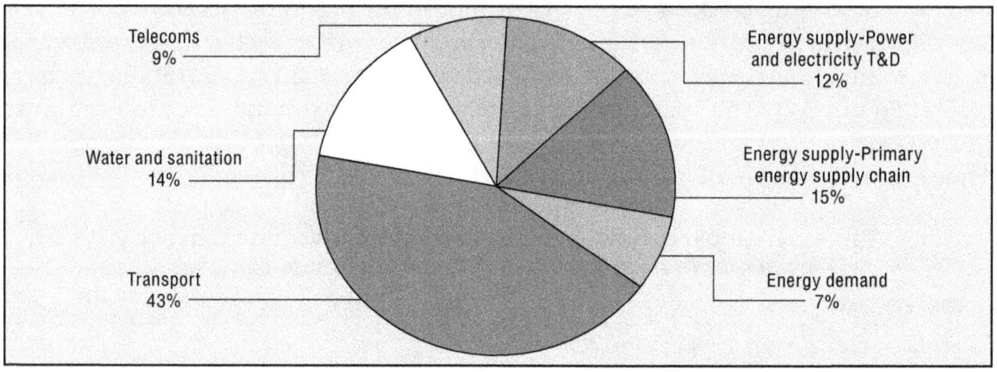

Source: As per Table 3.1.

StatLink http://dx.doi.org/10.1787/888933484229

Box 3.1. **The challenges of estimating infrastructure investment needs**[1]

There have been several attempts to provide estimates on infrastructure investment needs (WEF, 2013; NCE, 2014; Bhattacharya et al., 2016; Woetzel et al., 2016; Kennedy and Corfee, 2012). Each projection is highly uncertain as it combines several distinct sources, each with different underlying assumptions:

- Projections attempt to take as a starting point existing infrastructure investment, but there is a lack of comprehensive data on investments across countries, including G20 countries (AsDB, 2017, Bhattacharya et al., 2016b). There is a need for national and international agencies to gather more comprehensive, better quality data on infrastructure investment.
- Most infrastructure needs assessments are based on projected GDP growth and country-level elasticity of infrastructure spending to growth (Woetzel et al., 2016; NCE, 2016), which results in estimates that are highly dependent on GDP assumptions. Few studies are based on achieving minimum quantitative benchmarks for infrastructure stocks and services (such as those used by Pardee Center, 2014), which is more relevant in particular for low-income countries and in the context of the SDGs.
- Most infrastructure assessments are based on global models, but infrastructure needs and priorities depend on countries' specific circumstances – such as access to energy, quality of current infrastructure, growth rate and inequalities – and should be informed by country-specific long-term development strategies.
- Many assessments do not account for how infrastructure is managed and implemented. Some analysts suggest that better management of infrastructure could lower infrastructure investment needs (Woetzel et al., 2016).
- Many assessments do not integrate incremental investment needs for climate change adaptation and mitigation. When they do, they do not necessarily take a network approach, to account for the interdependency between infrastructure systems. For instance, decreased demand for energy reduces the capital requirements for new infrastructure in oil, gas and coal, potentially freeing up rail and port capacity (Kennedy and Corfee-Morlot, 2012).

The figures presented here offer an up-to-date estimate based on the sources listed in Table 3.1. The new estimate in this report is around USD 4.9 trillion per year for energy, transport, water and telecommunications infrastructure, reflecting a recent reevaluation of investment needed in transport (IEA, 2016d). This estimate is of a similar order of magnitude

Box 3.1. The challenges of estimating infrastructure investment needs[1] *(cont.)*

to figures presented in other analyses. The New Climate Economy (NCE) (2014) estimated that the world needed to invest USD 57 trillion (USD 3.8 billion per year) in infrastructure between 2014 and 2030, or around USD 96 trillion (USD 89 trillion in 2010 dollars) including primary energy generation and energy efficiency. More recent estimates by Bhattacharya et al. (2016b) anticipate larger needs: USD 75-86 trillion (or USD 5.4 trillion a year), excluding primary energy and energy efficiency – USD 1.6 trillion more per year than the NCE.[2] McKinsey (Woetzel et al., 2016) estimates cumulative needs of USD 49 trillion (or USD 3.3 trillion per year) for the period 2016-30 (Table 3.2). The Pardee Center (2014) estimates that annual spending in infrastructure will be on average USD 4.3 trillion per year between 2014 and 2050.

Table 3.2. Selected estimates of infrastructure investment needs, 2016-30 – annual averages in 2015 USD trillion per sector

	Energy supply					
	Power and T&D	Primary energy use supply chain	Energy demand/ efficiency	Transport	Water and sanitation	Telecoms
OECD (2017)	0.7	1.0	0.4	2.7	0.9	0.6
Bhattacharya et al. (2016b)	1.5	0.8	1.6	2.0	0.9	1.0
McKinsey (Woetzel et al., 2016)	1.0	not included	not included	1.2	0.5	0.6
NCE (2014)	0.7	0.9	1.7	1.0	1.5	0.5

Note: See technical note on estimating infrastructure investment needs for further details (http://oe.cd/g20climatereport).

Sources: NCE, 2014; 2016; Bhattacharya et al., 2016b; Pardee Center, 2014; Woetzel et al., 2016.

Shifting infrastructure investment for low-emission, climate-resilient pathways

Low-emission, climate-resilient pathways will require an unprecedented transformation of our infrastructure system. Most existing energy and transport infrastructure was designed and built for a world of cheap and abundant fossil fuels, contributing to economic growth in many regions but also to GHG emissions. As a result, around 60% of GHG emissions are hard-wired into existing infrastructure (NCE, 2016; IPCC, 2014). In an effort to keep average global warming well below 2˚C, the Paris Agreement stipulates that a "balance" between anthropogenic sources and sinks of GHGs must be reached by 2050-2100, so that there are zero net emissions to the atmosphere in the second half of the century (see Chapter 2). In many cases, it will be important to shift as much investment as possible towards zero-emission (rather than low-emission) options, given that some difficult-to-decarbonise sectors will still have residual emissions.

In addition to being responsible for more than 80% of energy-related CO_2 emissions (IEA, 2016a), G20 countries represent around two-thirds of global investment needs in infrastructure. This share is expected to raise to 75% of infrastructure needs between 2016 and 2030 (Pardee Center, n.d.). G20 country choices are critical to the world's ability to mitigate climate change and will also dictate the resilience of G20 infrastucture to climate change impacts. The infrastructure required for the low-emission transition is also integral to meeting many of the SDGs beyond SDG13 on climate change (Figure 3.4).

Figure 3.4. **The links between low-carbon, climate-resilient infrastructure and the SDGs**

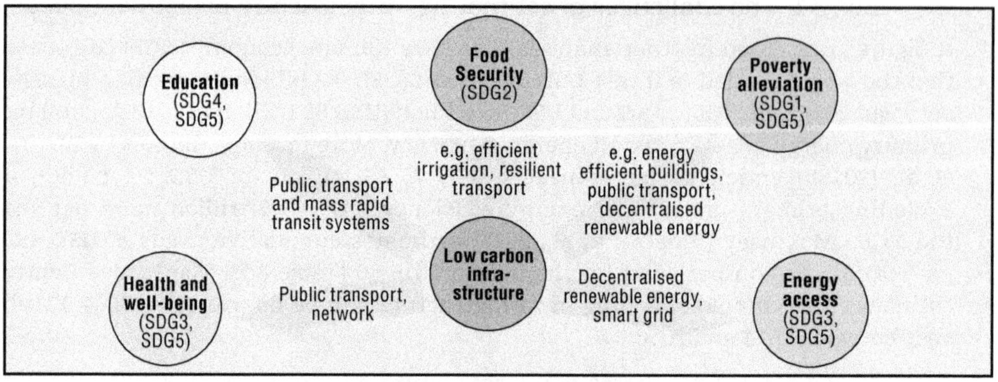

Infrastructure and technology shifts for low-emission pathways

Achieving low-emission, climate-resilient pathways requires strategies spanning infrastructure, technology development and innovation in the energy, land-use and agriculture sectors. This section examines the implications for infrastructure and technology of the shift to zero net emissions across these different categories (Table 3.3).

Table 3.3. **Examples of infrastructure and technologies needed for a low-emission transition**

	Strategies	Infrastructure needs		Technologies
Transport	Improve carbon intensity of vehicles Shift to more efficient transport modes Avoid carbon intensive mobility when possible	Passenger	Charging infrastructure for electric cars and fueling infrastructure for hydrogen cars Intelligent Transport Systems Smart grids Rail Mass rapid transit systems (light rail, metro, bus rapid transit lanes) Infrastructure for walking, cycling	Electric cars Advanced biofuels and biojet (algae) for air and maritime transport Hydrogen aircrafts Batteries
		Freight	Hinterland rail infrastructure	Electrification of trucks Advanced biofuels, hydrogen for shipping Investment in agriculture research (yields)
Energy	Decarbonise the power sector Electrification of end-uses Energy efficiency	Energy and power generation	Renewable energy (wind, solar, thermal energy, tidal, waves) Smart grids Infrastructure for CO_2 transport and storage	Energy storage (thermal cycle, power to gas, batteries) Tidal, thermal energy CCS (large-scale demonstration)
		Buildings	Retrofitting of the building stock Energy-efficient new build Heat supply	Zero energy or positive energy buildings Alternative material for steel and cement
Heavy industries	Energy efficiency in industrial processes Material efficiency Capture of emissions		Energy efficiency in industrial processes Infrastructure for CO2 transport and storage	CCS (large-scale demonstration of industrial CCS applications) Hydrogen in steel making
Land use	Improve carbon sequestration by land Minimise emissions from food production, including livestocks	Negative emissions	Infrastructure for CO_2 transport and storage	CCS Direct air capture and storage BECCS (deployment at commercial scale) Biochar Ocean liming
		Agriculture	Restoration of degraded grassland	Research on yields improvements Innovative agricultural practices to improve productivity

Source: Authors.

Infrastructure for low-emission energy and transport systems

Energy production and use accounts for around two-thirds of all anthropogenic GHG emissions, mostly in the form of CO_2 from the combustion of fossil fuels (IEA, 2017). Creating low-emission pathways requires radical changes in infrastructure, not only to reduce the carbon intensity of energy supply, but also to create less energy-intensive behaviours and to reduce energy use in transport, buildings and industry. The main elements of infrastructure-related changes needed to reshape energy supply and use are described here, with the main technological breakthroughs needed covered in Box 3.2.

Key to the energy transition is the decarbonisation of electricity, including phasing out inefficient coal-fired power plants and unabated coal, the widespread deployment of renewable energy sources, further development of nuclear power according to country choices, and potentially the development of negative emissions technologies (NETs) such as bio-energy with carbon capture and storage (BECCS) (IEA, 2017). Significant investments in smart grids will be needed to help manage demand and support increased penetration of intermittent renewable energy. On the demand side, reducing energy use in transport and buildings will be key.

Transport produces roughly 23% of global CO_2 emissions and is the fastest-growing source globally. Without further policy action, CO_2 emissions from transport could double by 2050 (OECD/ITF, 2017). Reducing emissions from transport is not only crucial for a low-carbon transition: it also reduces air pollution and congestion. The strategies necessary will depend on each country's circumstances, for example to what extent cities have already been developed around car ownership, and where opportunities exist to use urban planning to reduce the need for personal vehicles (OECD, 2015a). In general, ambitions will only be fulfilled with integrated policy action to:

- *avoid* unnecessary travel and reduce the demand for total motorised transport activity;
- promote the *shift* to low-emission and even zero transport modes; and
- *improve* the carbon intensity and energy efficiency of fuels and vehicle technologies. Significant advances have been made recently, notably in the electrification of transport via battery and fuel cells vehicles that are now on the market.

Building sector energy use was responsible for 9% of CO_2 emissions in 2013 in G20 countries. Increasing energy efficiency in buildings has not been sufficient to offset large increases in energy demand driven by the growth in population, energy-intensive appliances, and heating and cooling of buildings (IEA, 2016c). This is despite the availability of technologies that could lead to widespread decarbonisation of buildings through immediate widespread uptake. In developing and emerging economies, the building sector tends to be dominated by new construction and demolition of older buildings as cities expand. Integrating energy efficiency principles early in construction is therefore more important than retrofitting existing buildings. In mature economies, 75-90% of today's buildings will most likely still be in service by 2050. Many of these buildings are not built to the standards of today's energy efficiency codes and do not benefit from the latest energy-saving technologies; as a result, 30% of current buildings will need to be retrofitted by 2030 (IEA, 2017). Energy demand and efficiency of the appliances contained in buildings also has a major impact (Climate Policy Initiative, 2013). Managing policy decisions in tandem with investment decisions on heating, cooling, and power transmission and distribution infrastructure could enable additional cost reductions.

Box 3.2. Which technological innovations are needed for a low-carbon economy?

Many of the technologies needed to decarbonise the economy are known and available at a commercial scale, even though ongoing R&D will likely see further cost reductions: electric vehicles, renewable electricity generation and advanced building insulation techniques are all examples. However, to achieve pathways consistent with the Paris goals, many new technological breakthroughs will be required. Twenty-one technological innovation priorities were identified for this project that are crucial to achieving a low-carbon economy but have not yet been deployed at commercial scale and therefore still require significant R&D. Some key examples are described here.

Carbon capture and storage (CCS)

Current scenario projections rely heavily on CCS to meet emission targets. In the IEA scenario consistent with a 66% chance of reaching the Paris Agreement's 2°C goal, CCS contributes around 15% of emissions reductions by 2050 (IEA, 2017). In industry, it accounts for one-fourth of cumulative CO_2 emissions savings by 2050 relative to the New Policies Scenario. Furthermore, negative emissions technologies (NETs) such as bioenergy with CCS (BECCS) would benefit from the advancement of conventional CCS. While the components of carbon capture, transport, injection and storage have been demonstrated individually at commercial scale (Florin and Fennell, 2010), large-scale demonstration is an urgent priority to overcome the challenges of whole systems integration across the CCS chain (LCICG, 2014). The main research priorities are: (1) developing advanced adsorption and membrane processes; (2) advanced processes such as Ca-looping; and (3) improved modelling of CO_2 storage, including optimal injection scenarios and expected leakage (IEA, 2012; UKCCSRC, 2015).

The cost of CCS for power generation is estimated at USD 43-80/tCO_2 (IEA, 2012). CCS applied to industrial processes is less well developed and is generally more challenging, but has the potential to be cheaper than CCS for power generation. Each process and site is unique and will likely require bespoke equipment and plant design. Current cost estimates are USD 15-138/tCO_2 for cement and USD 51-64/tCO_2 for steel (Fennell et al., 2012). Research priorities for industrial CCS include: (1) improving heat and flow integration; (2) testing the impact of impurities on the capture process; and (3) developing novel sorbents optimised for industrial operating conditions.

Industrial sector (energy use and process emissions)

The industrial sector accounts for one-third of global emissions. Of this, steel, cement and chemicals together make up over 70% (IEA, 2010). Energy efficiency improvements will not be able to reduce industrial emissions as needed. The other options for achieving low (or zero) emissions from industrial processes are: switching from fossil fuels to biomass or hydrogen; electrification; and CCS. With the exception of biomass usage in certain applications, all these options are still in the concept phase. There is an urgent need to develop breakthrough processes (e.g. steel production based on hydrogen or electrolysis) that could result in a step-change in emissions reductions. Development of alternative building materials to steel and cement could reduce emissions from both industry and the built environment. Alternative cement chemistries (i.e. not based on limestone) could provide a low-carbon solution for cement, but extensive testing would be required to gain wide-scale acceptance in the construction industry.

Aviation sector

CO_2 emissions from aviation amounted to 700 MtCO_2 in 2013, or around 2% of global CO_2 emissions (Elgowainy et al., 2012). With demand expected to rise by around 5% per annum, emissions could be as high as 3 100 MtCO_2 by 2050 (ATAG, 2014). In the medium term, radical new aircraft designs (e.g. the "blended wing" concept) could improve fuel efficiency by 25% compared with the most efficient planes today (DfT, 2007). In the short term, options for low (or zero)

Box 3.2. **Which technological innovations are needed for a low-carbon economy?** *(cont.)*

carbon airplanes are extremely limited. Biofuels present the most viable alternative but are limited to those that meet industry standards and are interchangeable with conventional fuels. New engine designs that can cope with the low aromatics composition of biofuels could open the aviation sector up to cheaper biofuels supply options. Hydrogen-powered planes should not be ruled out. In 2016, the first four-seater hydrogen fuel-cell powered plane took flight (Pultarova, 2016). While this is promising, significant technical challenges need to be overcome for commercial-scale hydrogen powered planes to become a reality. In particular, the low energy density of hydrogen requires a large storage volume, which will require major design modification. A starting point for hydrogen in aviation may be for use during taxiing. EasyJet is exploring this idea (Carrington, 2016).

These alternative fuels for aviation, as well as other sectors, will rely on cost-effective and scaled-up supply chains. Researching and designing new plant strains optimised for biofuel production would increase crop yield and reduce the cost of biofuel supply. Other promising avenues for investigation include cellulosic biomass, algae and halophytes (Epstein, 2014). Hydrogen supply from electrolysis, which requires a large amount of electricity, could be superseded by new technologies such as photocatalytic water splitting (Hisatomi et al., 2014; Moniz et al., 2015) or microbial processes (Magnuson et al., 2009), reducing the amount of electricity required per unit of hydrogen produced.

Negative emission technologies (NETs)

There are five main NETs: direct air capture, the lime-soda process, augmented ocean disposal, biochar and bioenergy with CCS (BECCS), the best known. Cost estimates for NETs are USD 59-155/tCO_2e (Workman et al., 2011). With the exception of BECCS, all NETs are in a very early stage of technical development. BECCS relies on a sustainable source of biomass; given competing pressures for bioenergy across different sectors, it is unlikely that BECCS alone will be adequate. The main research priorities are: (1) developing novel sorbents to reduce the energy input for direct air-capture technologies and the soda/lime process; (2) optimising the design of pyrolysis plants for biochar production (3) integrated testing of CCS with 100% biomass-firing; (4) improving liquefaction processes for artificial trees; and (5) systematic studies of biochar effectiveness, focusing on repeatability and side-effects (Gurwick et al., 2013; Workman et al., 2011).

Electricity storage

Electricity storage is required to accommodate high levels of intermittent renewable generation. Beyond 2050, scenarios limiting global warming to 2°C have a share of generation from intermittent renewables greater than 50%. A rule of thumb is that for every GW of intermittent renewables, 1 GWh of storage is required (Budischak et al., 2013). The research priorities for electrical batteries include new cell chemistries emerging from the lithium-ion family, such as lithium-air (Grande et al., 2015) and lithium-sulphur (Fotouhi et al., 2016), or other metals such as sodium and magnesium (Erickson et al., 2015). These could improve power and charge density (Zhang, 2013), decreasing the cost per unit of energy stored. Improved manufacturing techniques and efficient management of battery packs could provide evolutionary cost and performance improvements. Capital costs of lithium-ion batteries of around USD 193-254 per kWh of storage capacity are possible (Darling et al., 2014) and new cell chemistries could offer further reductions to reach the USD 150/kWh thought to be the threshold for commercialisation of battery technologies for battery electric vehicles (Nykvist and Nilsson, 2015). Less mature electricity storage technologies, such as redox flow batteries, molten salt batteries, flywheels, and power-to-gas could also play an important role in balancing supply and demand over different timescales (from seconds to months), and different scales (distributed and centralised) (Brandon et al., 2016).

Source: Napp, T. (forthcoming).

The importance of innovation in land use sectors

Chapter 2 highlighted the importance of agriculture, foresty and land use (AFOLU) for low-emission pathways, accounting for around 25% of global anthropogenic GHG emissions, mainly deforestation (9-10% of emissions) and agriculture (10-12%, mainly methane and nitrous dioxide) (IPCC, 2014). In some countries, proportions are much higher: land use and agriculture were responsible for 48% of emissions in Indonesia, 46% in Brazil, 31% in Argentina, and 27% in Australia (FAO, n.d.). By 2050, land will have to supply 60% more food than it does today to feed a growing population (Alexandratos and Bruinsma, 2012). It will have to do so in a way that does not further harm the climate. AFOLU sectors are expected to play a significant role in low-emission development pathways through carbon sequestration and sustainable approaches to managing land and livestock, and climate adaptation.

While crucial for low-emission pathways, AFOLU sectors differ from other sectors of the economy in the sense that infrastructure is not central to low-emission strategies (Box 3.3), at least in the short term. In the long term, infrastructure investments will be needed to increase resilience of agriculture (for example through access to on site renewable energy sources), to optimise the transport of produced goods, and to further develop ship and rail freight (Box 3.3).

Innovation is central to low-emission, climate-resilient land-use strategies. Although agricultural emissions of methane (CH_4) and nitrous oxide (N_2O) are notoriously difficult to reduce, technological innovation offers possible paths. This includes improving crop and livestock productivity (e.g. by developing crop varieties that are resilient to local hazards and that inhibit the production of nitrous oxides); more efficient fertiliser use; improved soil management; and practices aimed at reducing CH_4 emissions from ruminants, rice paddies and manure management. Better agricultural practices that increase the productivity of arable land in a sustainable manner would also help to halt and reverse deforestation and widespread land degradation, which is estimated to cost USD 100 billion per year (Delgado et al. 2015).

Figure 3.5. **Government spending on agricultural knowledge and innovation systems in 2012-14 in selected G20 countries, as a share of agricultural value added**

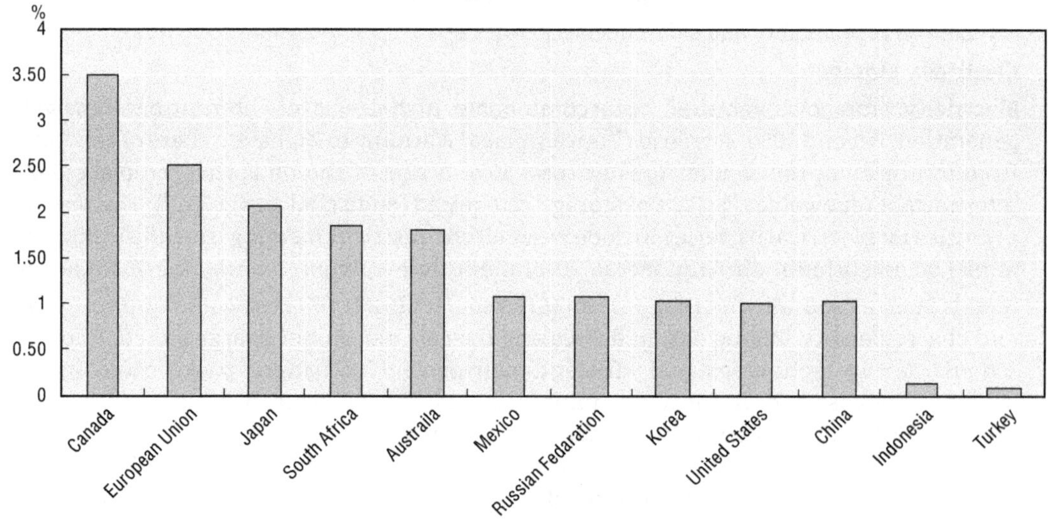

Note: a. Government spending on agricultural knowledge and innovation systems includes funding of agricultural research, agricultural education, training and extension services for farmers. b. Exchange rates used in the OECD Producer and Consumer Support Estimates database have been applied here: http://www.oecd.org/agriculture/agricultural-policies/producerandconsumersupportestimatesdatabase.htm. c. Data for other G20 countries are not available.

Source: OECD (2016b).

StatLink http://dx.doi.org/10.1787/888933484231

Agricultural innovation is not only about technological improvements but also about education, training and organisational improvements. Further investment in research and development and education is hence central to spur agricultural innovation that can improve sustainable productivity growth (Ignaciuk, 2015). Indeed, the level of technological development and innovation in agriculture has a direct impact on its capacity to produce adequate and sustainable supplies of food and feed (OECD, 2014). Given the importance of sustainable productivity growth for achieving ambitious mitigation targets, G20 countries can be encouraged to increase their spending in agricultural knowledge and innovation systems (Figure 3.5).

Box 3.3. Investing in innovation and infrastructure for resilient agriculture

Ensuring access to a secure water supply will be one of the main challenges of the land use sector – particularly agriculture – in the years to come. Climate change is expected to reduce crop yields in some areas. Coupled with increased demand for food from a growing population with increasingly rich diets, this will impose serious strains on agricultural systems, threatening food security in the most vulnerable countries.

Strategies to adapt agricultural systems are varied. Much can already be achieved by increasing the sector's reliance on on-site renewable energy sources, as well as optimising the transport of produced goods by shrinking the distance food is transported, and developing ship and rail freight. Technology also has a considerable role to play, via such measures as:

- developing new crop varieties that are drought-resistant and better adapted to higher temperatures; and
- improving water efficiency via the widespread dissemination of pressurised irrigation systems (e.g. sprinklers and drip irrigation), which decrease water demand while increasing the efficiency of water use.

Significant investment in R&D will be required to increase the resilience of agricultural systems to climate change. In OECD member countries, annual adaptation costs in agricultural R&D and in improved irrigation technology are estimated at USD 16-20 billion by 2050. In the short term, most of this investment is likely to come from public sources, although by 2050 the private sector is likely to invest more in this area than the public sector (Ignaciuk and Mason-D'Croz, 2014). Governments could facilitate private investment by lowering investment barriers that impede R&D, ensuring that private knowledge is disseminated, and encouraging public-private partnerships for R&D, where appropriate (Ignaciuk, 2015).

Incremental investment needs: mitigation

Assessment of the incremental capital requirements for putting the world on track to meet the mitigation objectives of the Paris Agreement depends on a number of factors, including the interpretation of the target (e.g. well below 2°C or efforts towards 1.5°C, likelihood of reaching the target); assumptions concerning decarbonisation strategies chosen (e.g. with or without nuclear, accounting or not for behavioural changes such as modal shifts in transport); and assumptions made on several factors such as the evolution of GDP, population, and technology costs.[3]

Consistent with the global pathways analysis in Chapter 2, this section takes as its core the IEA scenario consistent with a 66% likelihood of keeping the global average surface temperature increase to below 2°C throughout the century (IEA 66% 2°C scenario, IEA 2017).

The OECD estimates that around USD 103 trillion of cumulative investment between 2016 and 2030 would be required for the IEA 66% 2°C scenario, or 10% more than in a scenario where no further action is taken to mitigate climate change. The major shift of energy supply investments towards low-emission alternatives and significant scaling-up of demand-side investments for energy efficiency assumed by the scenario would require 29% more investment in the energy sector alone (IEA, 2017). Annual investment needs in transport, water and sanitation, telecommunications and energy supply and demand would be around USD 6.9 trillion over the next 15 years, versus USD 6.3 trillion a year with no further action (Figure 3.6, left-hand panel).

The incremental capital cost of shifting investments for the IEA 66% 2°C scenario is therefore significant, but not prohibitive; furthermore, incremental costs would be offset by fuel savings of up to USD 1.7 trillion per year through 2030 (Figure 3.6, right-hand panel). Factoring in modal shifts in transport could also lower overall investment needs for low-emission pathways, due to reduction in vehicle ownership and less investment needed in parking space (IEA, 2016d). Finally, provided low-emission infrastructure investment is pursued in an integrated way with climate-consistent, growth-enhancing policies, it could form an integral part of a new growth model for low-carbon growth, offsetting incremental costs entirely (Chapter 4).

Figure 3.6. **Global annual infrastructure investment needs for a 66% scenario 2°C, and fuel savings, 2016-30, USD 2015 trillion**

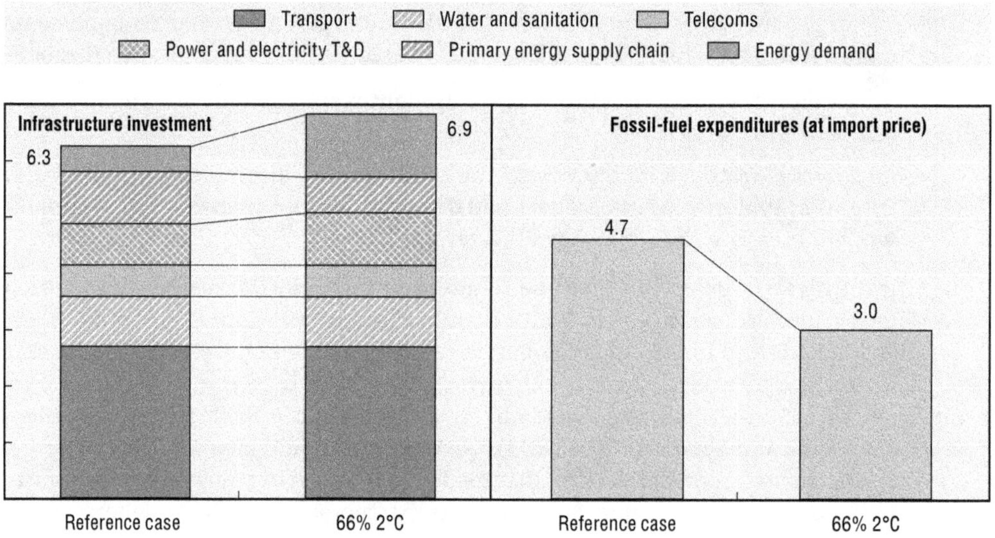

Notes: Reference case assumes no further action by governments to mitigate climate change.

Sources: IEA (2017) and IEA (2016a) for energy supply and demand; IEA (2016d) for road and rail infrastructure; OECD (2012) for airports and ports; McKinsey (Woetzel et al., 2016) for telecommunications. The water and sanitation estimate is an average of estimates from: Booz Allen Hamilton (2007), McKinsey (Woetzel et al., 2016) and OECD (2006). See technical note on estimate of infrastructure investment needs for further details on methodology (http://oe.cd/g20climatereport).

StatLink http://dx.doi.org/10.1787/888933484243

The global infrastructure investment needs estimate presented here is higher than in previous exercises, partly because many past estimates were based on a less ambitious scenario with a lower chance of limiting warming to below 2°C. NCE (2016) and Kennedy and Corfee-Morlot (2012), for example, estimated that incremental capital costs could increase by as little as 5% compared to a business as usual scenario in a low-emissions future. The impact on investment needs of increasing the level of ambition is not just incremental and linear: it implies a radical reorientation of investments and measures to decarbonise sectors that are harder and more expensive to decarbonise (transport, aviation, industry). For instance, cumulative global investments increase by 13% in the IEA 66% 2°C scenario compared with a scenario with a 50% chance of meeting 2°C, mainly due to increased investment in low-emission electricity supply and end uses (IEA, 2017).

There are many uncertainties associated with those estimates. Further research is required to understand the impact of the digitalisation of energy on telecommunication infrastructure, for example. Deployment of BECCS may generate significant investments in CO_2 pipelines (Chapter 2). There are also many remaining uncertainties on the impact of a low carbon future on future demand in infrastructure beyond energy. Between 2010 and 2015, fossil fuels represented between 11% and 18% of the value of international trade in goods (UN, n.d.). Fossil fuels accounted for an average of 42% of total maritime traded volumes between 2011 and 2015 (UNCTAD, 2016). In the long term, a world less reliant on fossil fuels is likely to require fewer port capacities, oil and gas tankers, and hinterland railways to transport coal (Kennedy and Corfee-Morlot, 2012). Specific country contexts will also influence investment needs. Encouraging more efficient transport modes from the outset in developing and emerging economies where infrastructure continues to be built could generate significant savings, reducing the need for road and parking spaces, which in many non-OECD countries are more costly than the additional investments required in public transport infrastructure (IEA, 2016d).

Box 3.4. **Investment needs for low-emission urban mobility**

Cities have a major role to play in strategies to decarbonise transport (see Chapter 2). It is essential to integrate transport and land-use planning to reduce overall demand and facilitate the shift from individual cars to mass transit systems. The International Transport Forum undertook a modelling exercise to assess transport investment needs in G20 countries between 2015 and 2050 under three different scenarios for urban development (see OECD/ITF (2017) for more details) (Figure 3.7).

In the baseline scenario (BASE), no additional measures to reduce travel demand and CO_2 emissions are implemented. The combined effects of urban extension, population and income growth will result in a surge in motorised mobility. Road traffic – the sum of car-km and motorcycle-km – will increase globally by 91%. Most of the increase comes from G20 countries, with 7 600 billion additional vehicle-km out of a total of 11 100 billion. In the G20, this increases CO_2 emissions by 10%.

In the Integrated land-use and transport planning scenario (LUT), stringent policies targeting land-use planning, development of public transport and restriction of car use significantly mitigate CO_2 emissions. In G20 countries, transport emissions decrease by 34%.

In the strong investment scenario (INVEST), budgetary constraints on transit infrastructure are removed, increasing investment in mass transit infrastructure – urban rail, underground and tramways – especially in middle-income countries. This leads to a decrease of 50% in CO2 emissions.

Overall, aggregate infrastructure investment needs are smaller in the transit-oriented scenarios (USD 9 trillion in LUT and USD 13 trillion in INVEST) than in the baseline (USD 14 trillion). However, the results differ by income groups. High-income economies need to frontload urban transport investment towards light rail systems in the next 10 years. Middle-income countries can significantly decrease overall investment needs by 2050 by shifting investments in the next 10 years to rail.

Box 3.4. **Investment needs for low-emission urban mobility** *(cont.)*

Figure 3.7. **Investment in urban infrastructure in G20 countries, 2016-50, road and rail – ITF projection**

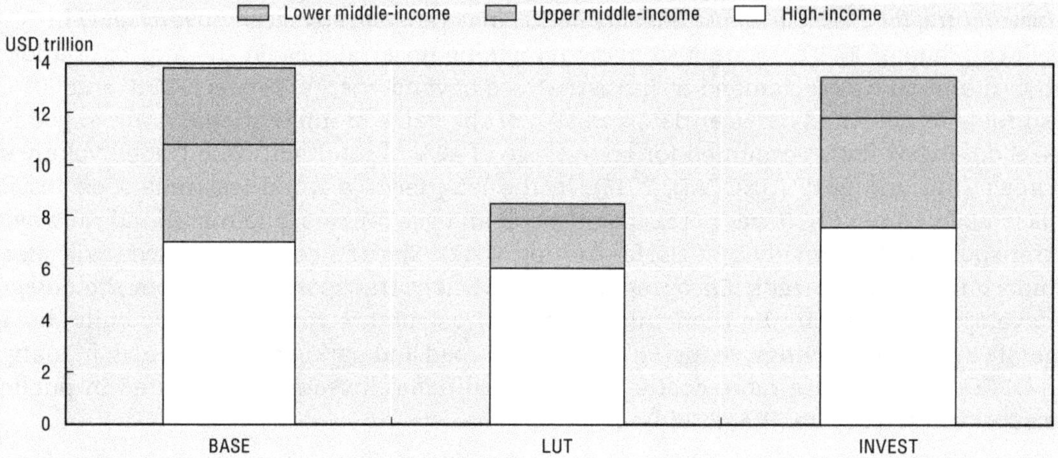

Source: Based on ITF data (accessed on 28 February 2017).

StatLink http://dx.doi.org/10.1787/888933484251

Incremental investment needs: adaptation

Estimates of the additional funding required for infrastructure adaptation depend on specific definitions of what constitutes "infrastructure adaptation", including which sectors are included (Box 3.5). In practice, costs are very context-specific, adding to the challenge.

Box 3.5. **Defining adaptation investments**

Adaptation investments can be considered across three areas:

Adaptation investments that create an enabling environment, such as investing in climate information, awareness raising and capacity building, and adapting governance systems to better account for the projected changes and deep uncertainty regarding climate change. If private stakeholders are sufficiently aware of climate risks, some adaptation investments make economic sense without public support.

Adaptation investments that "climate proof" infrastructure, reducing the exposure or vulnerability of an infrastructure asset or network, whether from the outset or as part of a retrofitting process. Such investment can take the form of engineering work with clearly identifiable additional costs, such as building a bridge higher than would otherwise be the case or building to higher design standards. It can also mean considering reduced exposure when siting or designing, often without incurring additional costs, for example siting back-up power generators to avoid them being flooded or modifying operational routines. It can also consist of pursuing a different approach to provide the same service, for example expanding green spaces to absorb rainfall in urban areas, instead of investing in larger drainage pipes.

Adaptation investments that fill gaps in infrastructure provision, particularly in developing countries, where infrastructure can be insufficient even for addressing current climate challenges.

Source: Agrawala and Fankhauser (2008).

Several estimates of the global costs of adaptation feature a category on infrastructure adaptation. These tend to estimate the costs of "climate proofing" infrastructure by applying an adaptation cost mark-up to future investment plans to take account of future climate change. Such investments are estimated to be small compared with other factors that may influence the future costs of infrastructure. The cost of adapting infrastructure has been estimated at no more than 1-2% of the total cost of providing that infrastructure (Hughes, Chinowsky and Strzepek, 2010).

Other estimates take into account adaptation investments that fill gaps in infrastructure provision. Below are three recent estimates:

- The UNFCCC (2007) estimated that by 2030, the world would be spending USD 8–130 billion more each year on *new* infrastructure than would otherwise be needed in response to impacts associated with climate change, with two-thirds of the investment in OECD countries. This estimate excludes operating and maintenance costs, as well as the costs of adapting *existing* infrastructure, and any additional investment needed in water supply infrastructure (USD 11 billion, 85% of which will be needed in non-Annex 1 Parties) or housing.
- The UNFCCC estimates were criticised for failing to account for the infrastructure deficit in low and middle-income countries (LMICs), the investments in governance and technical capacity needed to maintain infrastructure in those countries, as well as the "residual" losses that cannot be prevented even with adaptation. With these elements taken into account, adaptation infrastructure investments in LMICs are eight times higher than the high-bound UNFCCC estimate (Parry et al. 2009.).
- Infrastructure accounts for a significant share of the USD 70-100 billion in annual global adaptation costs, according to a 2010 World Bank study on the costs between 2010 and 2050 of adapting to an approximately 2°C warmer world. Infrastructure adaptation is estimated to require USD 13-27.5 billion per year, depending on wetter or drier climate scenarios (Figure 3.8). Urban infrastructure (drainage, public buildings) accounts for over half of these costs, followed by railways (18%) and roads (16%), with costs highest in East and South Asia. This amount does not account for coastal zone adaptation, water supply or flood protection.

Figure 3.8. **World Bank estimates of global adaptation investment needs 2010-50**

USD 13-27.5 billion per year

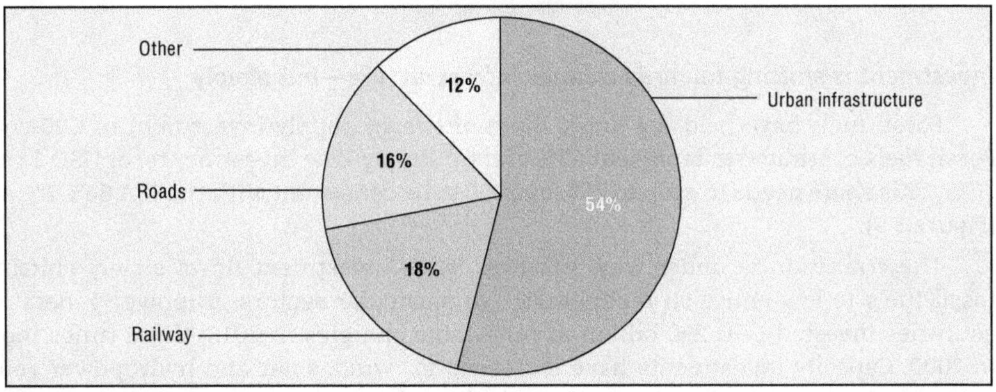

Note: The estimate provided above does not account for adaptation in coastal zone adaptation, water supply or flood protection.

Source: World Bank (2010).

StatLink http://dx.doi.org/10.1787/888933484263

A transition is under way, but not at the required pace

The estimated global carbon budget consistent with a 66% likelihood of limiting global warming to below 2°C (described in Chapter 2) equates to 15 to 30 years of fossil fuel-related CO_2 emissions at current rates. Given the slow rate of capital stock turnover (Table 3.4), the infrastructure investment choices countries make over the next 15 years will be pivotal in determining the extent of global climate change. If governments continue to invest in fossil-fuel infrastructure, they risk locking in even higher levels of GHG emissions for decades to come, and they will enhance the risk of stranded assets. Long operational lives also make infrastructure vulnerable to the impacts of climate change in the coming decades. Overall, unless global emissions peak by around 2030 and fall to zero by 2100, serious climatic disruption could draw up to 720 million people back into extreme poverty (Granoff et al., 2015).

Information on infrastructure projects is not always complete or available to the level of detail required to allow meaningful analyses on progress in shifting investment in line with the Paris Agreement's goals. Energy is the only sector where information is more complete, as surveys and commercial databases track information on power plant capacity announced, at pre-construction stage, under construction, cancelled or in operation. This section therefore focuses on the energy sector as an indicative assessment of progress in aligning infrastructure investment plans for the transition, using the IEA 66% 2°C scenario as a benchmark.

Table 3.4. **Typical lifespans of selected infrastructure and equipment**

	Lifespan
Water infrastructure (dams, reservoirs, sanitation facilities)	30-200 yr
Transportation (port, bridges)	30-200 yr
Buildings, housing (insulation, windows, buildings)	30-150 yr
Power plants (coal-fired, gas-fired, nuclear)	20-60 yr
Cars	15-20 yr
Building appliances	10-20 yr
Industrial boiler	10-30 yr
Cities, urbanisms, land use planning	> 100 yr

Source: Corfee-Morlot et al. (2012).

Investment is shifting towards cleaner infrastructure – but slowly

Fossil fuels have held the lion's share of energy supply investment in G20 countries. Fossil fuels continued to represent 63% of total supply-side investments, or USD 1 trillion in 2015. This share needs to drop to 26% by 2050 to be consistent with the IEA 66% 2°C scenario (Figure 3.9).

The transition is under way, however, with investment flows slowly shifting from fossil fuels to low-emission technologies in particular sectors. In power generation, G20 countries invested USD 290 billion in renewable energies in 2015, three times more than in 2000. Capacity investments have increased for wind, solar and hydropower generation in particular (IEA, 2016b). Since 2011, these technologies have captured approximately 40% of total annual investments in power generation (IEA, 2016b). This increase in total renewables capacity investment is even more impressive given that the cost of production of the technologies has decreased in the past few years: since the end of 2009, solar PV module prices have fallen by around 80% and wind turbine prices by 30-40% (IEA, 2017).

Figure 3.9. **G20 investment in energy supply 2000-15, and investment needs in the 66% 2°C scenario**

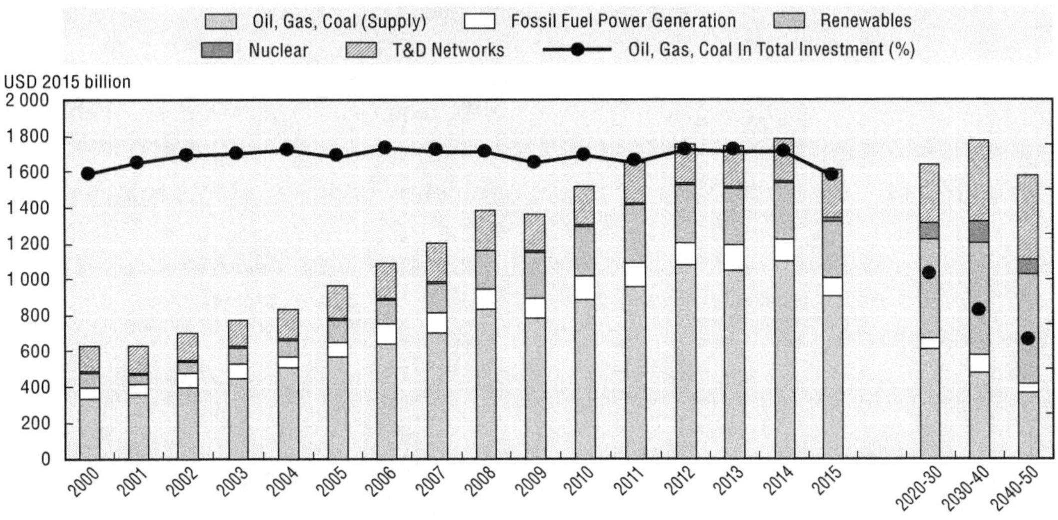

Sources: IEA, 2017; IEA, 2016b.

StatLink http://dx.doi.org/10.1787/888933484273

In the transport sector, most of the investment in G20 countries has targeted road transport since 2000, but the share of rail infrastructure investment – important to help promote the shift from emissions-intensive road transport – has been growing steadily, from 20% in 2000 to 26% in 2014, with a peak at 31% in 2010 (Figure 3.10). From a low of USD 250 billion in 2003, investment has more than doubled in size to reach USD 650 billion in 2014 (OECD/ITF, 2017). Investment in rail needs to increase significantly in the coming years to help fully decarbonise the economy.

Figure 3.10. **Road and rail infrastructure investment in G20 countries, 2000-14**

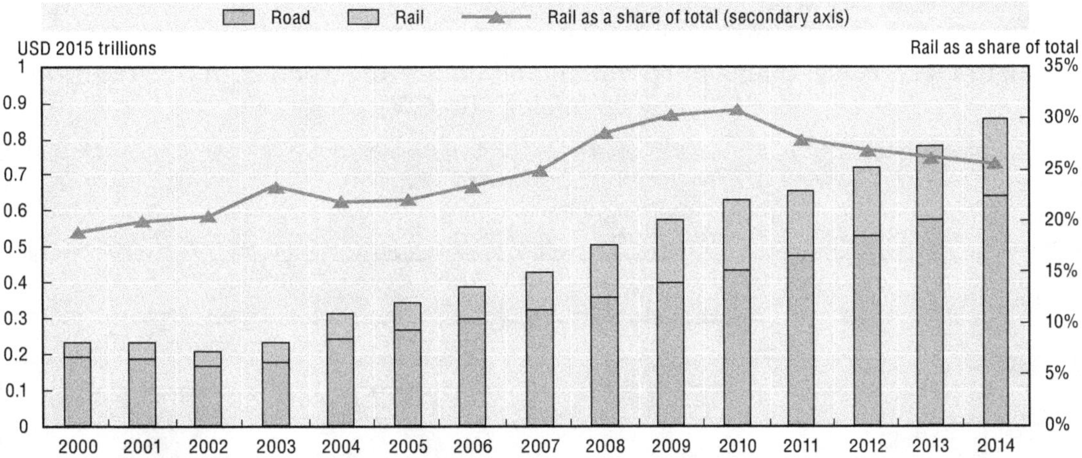

Source: OECD/ITF, 2017.

StatLink http://dx.doi.org/10.1787/888933484286

Investment plans are not yet aligned with the Paris Agreement's objectives

How, then, do current investment patterns and national energy sector infrastructure plans match up with the trajectory needed to achieve Paris objectives? In the power sector, the current capacity mix in G20 countries is still far from that required by the IEA 2050 scenario (Figure 3.11, left-hand panel). However, the plants under construction and planned for the next five years paint a different picture. The right-hand panel of Figure 3.11 compares this pipeline with required additions up to 2025 in the IEA 66% 2˚C scenario. The share of zero-carbon capacity additions is close to that required under the scenario (72% renewables and nuclear, versus 76% required). Solar and wind represent 84% of renewable generation capacity under construction, versus 36% for the plants in operation (Figure 3.12). However, the share of coal is much greater than the required level (22% of planned additions, versus 8% required). So, across the G20, the real challenge facing the power sector is accelerating the phase-out of coal-fired power generation.

Figure 3.11. **Current capacity and current pipeline of power plants relative to those required in a 66% 2°C scenario**

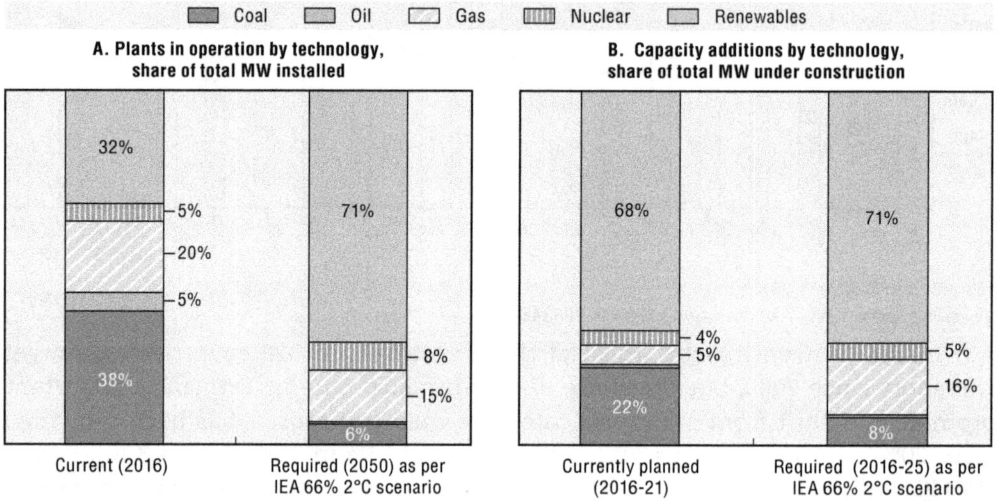

Note: Results are presented as share of total gigawatts and refer to power generation in operation in G20 countries in 2016, the energy mix in 2050 in the IEA 2°C 66% scenario, capacity additions in G20 countries for the period 2015-21, and global capacity additions in the IEA 2°C 66% scenario in the period 2016-25.

Source: Authors' analysis from *i)* Platts WEPP (2017) for oil and gas under construction; *ii)* the Global Coal Plant Tracker (2017) for coal under construction; *iii)* IAEA (2016) for nuclear under construction; *iv)* IEA (2016c) for renewable energy under construction; and *v)* IEA (2017) for capacity additions in the IEA 2°C 66% scenario.

StatLink http://dx.doi.org/10.1787/888933484293

Figure 3.12. **Power plants in operation and under construction in G20 countries, by technology (in GW)**

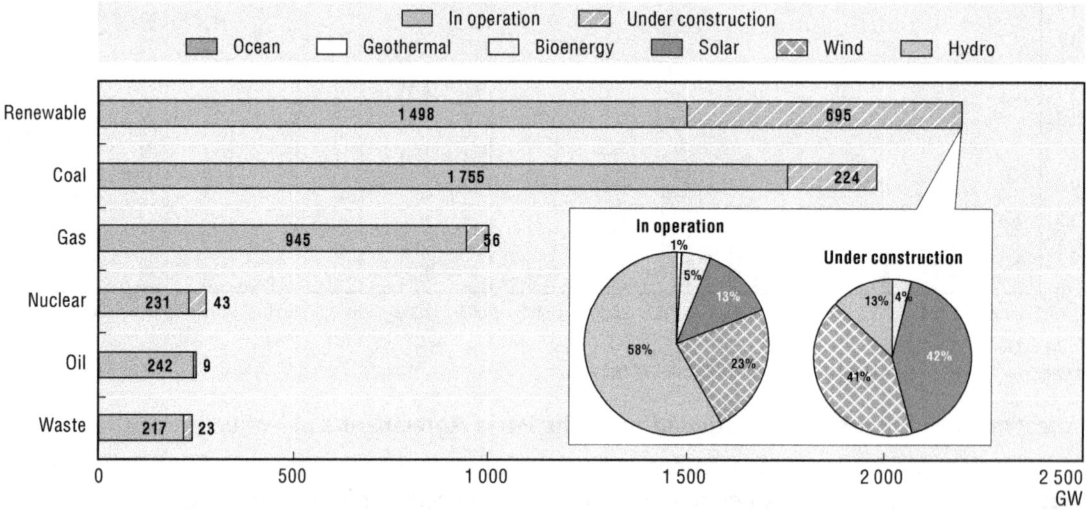

Source: Authors' analysis from *i)* Platts WEPP (2017) for oil and gas under construction (accessed March 2017); *ii)* the Global Coal Plant Tracker (2017) (accessed on 28 February 2017) for coal under construction; *iii)* IAEA (2016) for nuclear under construction (November 2016); and *iv)* IEA (2016c) for renewable energy under construction.

StatLink http://dx.doi.org/10.1787/888933484305

The level of coal plants in the pipeline is high despite recent falls in global capacity under development, mainly due to shifting policies and economic conditions in China and India – which account for 86% of coal power built globally between 2006 and 2016 – together with a reduction in overall power demand (Box 3.6). Pre-construction activity decreased by 48% from January 2016 to January 2017. Construction starts dropped 62%, and ongoing construction decreased by 19%. Coal plant retirements are taking place at an unprecedented pace, with 64 GW of retirements in the past two years, mainly in the European Union and the United States (Shearer et al., 2017). Nevertheless, the proportion of overall G20 capacity investment that is coal based could increase in the future, as 416 GW of coal plants are in pre-construction, and 543 GW are "on hold" (Figure 3.13).[4] Considerable further efforts are therefore needed. These efforts will not only be domestic. G20 economies also influence the type of infrastructure that is built outside of their borders, and especially in developing countries through development finance and export credits (Box 3.7)

Box 3.6. Recent reductions of the coal project pipelines in China and India

In China, over 300 GW of projects in various stages of development were put on hold in 2016 until after the 13th Five Year Plan (2016-20), including 55 GW of projects that were already under construction. According to a survey by Greenpeace, the amount of new coal power capacity authorised for construction in 2016 in China was 22 GW, a decline of 85% from the 142 GW authorised in 2015.

In India, the draft National Energy Plan, released in December 2016, states that no further coal power capacity beyond that currently under construction will be needed until at least 2027; but there is already 177 GW in the pipeline before that date. Moreover, India is in the midst of a solar power revolution, with bids as low as Rs 2.97 (USD 0.044) per kilowatt-hour, and government proposals to install 215 GW of renewables (biomass, small hydro, wind, distributed solar PV, and utility scale solar PV) by 2027. Although some policy and financial challenges need to be addressed to reach the ambitious goals set by the government, the combination of the current low capacity utilisation rate of several coal power plants and the declining cost of renewables has caused many financial backers of coal projects to withdraw support. Construction activity is now on hold for 31 coal plant units at 13 sites totalling 12 725 MW of capacity, mainly due to frozen financing.

Source: Extract from Shearer et al. (2017).

Figure 3.13. Coal power plants under construction, 2015-21, top five G20 countries

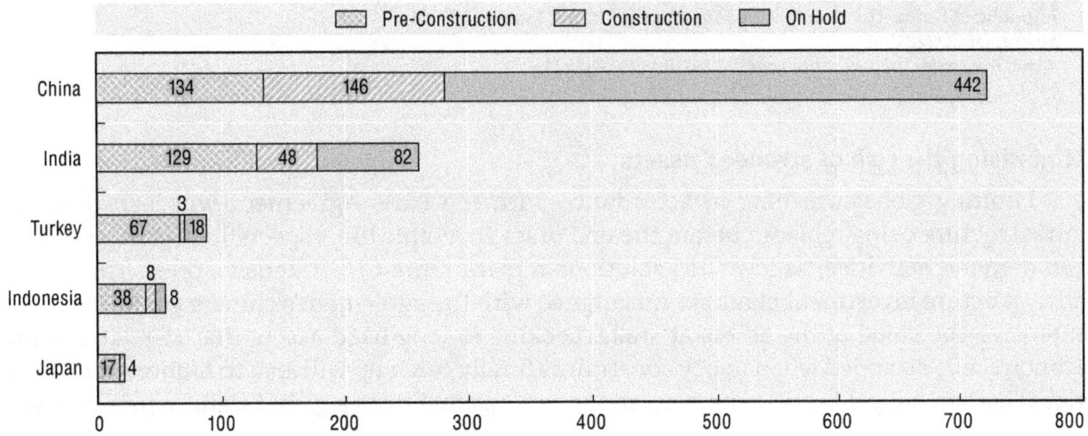

Source: Authors' analysis based on the Global Coal Plant Tracker (2017) (accessed on 5 April 2017).

StatLink http://dx.doi.org/10.1787/888933484310

Box 3.7. **Aligning ODA and export credits for infrastructure investment with the Paris Agreement's objectives**

The G20 includes the biggest aid providers globally – roughly 77% of ODA and ODA-like flows come from G20 countries, according to the OECD-DAC statistical system – and while aid supports only a small share of infrastructure investment overall, it plays a critical role in low-income countries where it is difficult to mobilise domestic and external finance. Export credits – commercially motivated support linked to a country's trade strategy – also play an important role in financing infrastructure. For example, 20% of external finance for infrastructure projects in Sub-Saharan Africa is provided by China EXIM Bank alone (Gutman, Sy and Chattopadhyay, 2015).

An analysis of export credits in support of power generation from G20 countries that report to the OECD shows that the overwhelming majority of these credits supported fossil fuel technologies over the last decade (Figure 3.14). Export credits provided by G20 countries for coal power generation specifically amounted to USD 13.1 billion. Most signatories to the OECD's Arrangement on Export Credits have agreed to begin limiting export credits related to coal.

Figure 3.14. **Official export credits for power generation projects**

(Share per sector, G20 members reporting to the OECD Working Party on Export Credits and Credit Guarantees)

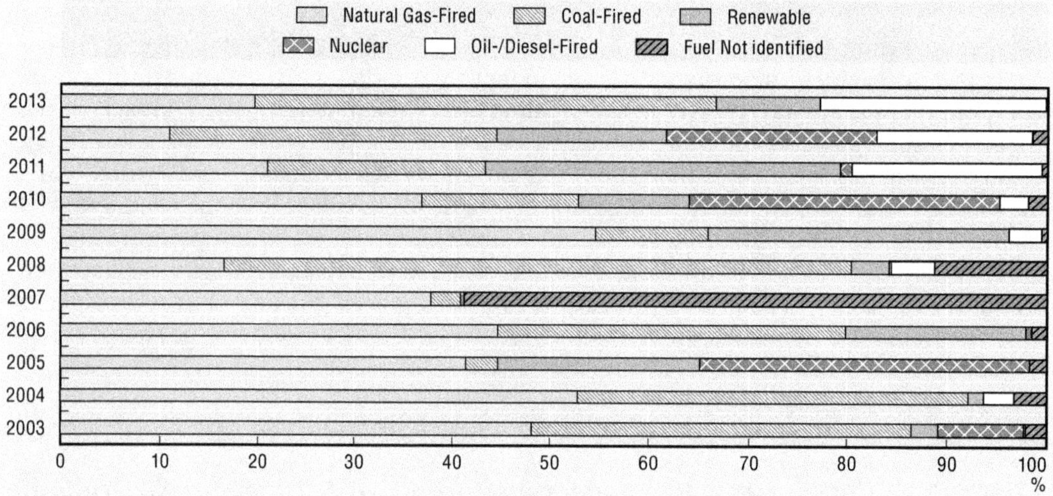

Note: G20 countries that report to the export credit committee are Canada, France, Germany, Italy, Japan, Republic of Korea, the United Kingdom and the United States.

Source: OECD (2015a).

StatLink http://dx.doi.org/10.1787/888933484324

Minimising the risk of stranded assets

Limiting global warming in accordance with the Paris Agreement will lead to some infrastructure being replaced before the end of its economic life, especially in energy supply and demand activities, as low-GHG solutions replace more GHG-intensive ones. The longer infrastructure investment plans are misaligned with the agreement's climate goals, the more extensive the value of the assets at stake. Locking in long-lived assets that risk later being economically stranded when policy constraints finally catch up will lead to higher costs if the global carbon budget is still to be met, and is sub-optimal from a global welfare perspective.[5]

Stranded assets are a common feature of market economies that spur reallocation of capital as some firms are outcompeted by others (Caldecott et al., 2017). A range of approaches has been used to define and quantify the climate-related risk for assets (Box 3.8). Similar assets may also face different financial risks depending on their contribution to their country's emission profile, making the identification of the exact magnitude of assets at risk more challenging. A natural gas power plant, for example, can play a positive role if it replaces low-efficiency coal or balances variable sources of power generation, or a negative role if it slows the penetration of renewables.

Box 3.8. Climate-related risks for assets: clarifying the terms of the discussion

Many different definitions have been used in the debate on the impact of climate policy and climate change on assets.

- **Stranded assets:** Assets whose investment cannot be fully recouped as the result of climate policy (e.g. a coal power plant closing before it has recouped investment as its electricity is no longer competitive, whether because of a carbon price, other forms of support to low-carbon generation, or on pure financial grounds). More precisely, if the revenues of an asset are lower than its capital expenditure minus operating costs, the difference is the estimate of the stranded asset.
- **Assets at risk under climate change:** Infrastructure at risk of being destroyed or made unusable as the result of local climate changes (flooding, sea-level rise, typhoons, droughts). Dietz et al. (2016) provide a first estimate of value at risk, estimated at 1.8% of global financial assets in their central estimate (USD 2.5 trillion), rising to 16.9% in a 99% percentile scenario (USD 24 trillion).
- **Foregone revenues:** Revenues lost as lower volumes of fossil fuels are sold, and sold at a lower price than would otherwise be the case without climate mitigation policies (also known as the "carbon bubble"). IEA argues that the foregone revenues can be larger than stranded assets as the former include profits, even if these are discounted.
- **Capital value loss:** The capital value that a company loses as its activity is impaired by climate policy (and possibly climate change damages), as used by IRENA for its upstream fossil fuel estimates of stranded assets (IRENA, 2017b). There is much overlap between foregone revenues and the capital value of an energy company, although much depends on how the company is managed, and how quickly it can diversify its portfolio (e.g. a company that produces oil exclusively versus an oil and gas company with a renewable energy branch and ownership in electricity distribution).
- **Unburnable carbon:** Fossil fuel resources that are not used due to climate mitigation policies, but that would be burned if there were no constraint on emissions, usually expressed in energy amounts (Carbon Tracker Initiative, 2013).

IEA (2017) and IRENA (2017a) represent the latest estimates of energy-related assets at risk; both use the notion of stranded assets, although their methelodgy, sectoral coverage and assumptions about the future energy mix differ. Assuming an orderly transition to meet the Paris Agreement objectives, the IEA 66% 2°C scenario estimates stranded assets at USD 852 billion between 2014 and 2050, distributed as follows:

- USD 320 billion for power (96% of which are coal-fired power plants), with about half of the stranded assets occurring before 2030.

- USD 532 billion for production facilities, including coal mines, oil and gas wells and processing plants, that fail to recover their capital investment as a result of climate policy (USD 120 billion for gas, USD 400 billion for oil and USD 12 billion for coal).

A less orderly transition – for example, a delay followed by abrupt action – is likely to have more deleterious effects. The IEA considers a "disjointed transition case", in which climate policy would change abruptly in 2025, shifting from weaker action to a more ambitious trajectory, allowing the world to stay within the carbon budget of the 66% 2°C scenario. This would mean a change in investors' and market expectations, with investments previously committed to fossil fuel-based production that would eventually be stranded following the change in policy. Stranded assets would then amount to USD 2.1 trillion, with the brunt of the additional assets in oil (USD 1 trillion) and gas (USD 300 billion). The "delayed action" scenario in Chapter 4 builds on these numbers.

IRENA provides a different set of estimates of asset risks based on a renewable energy-driven low-carbon transition scenario, REmap (IRENA, 2017a). In terms of sectoral coverage, IRENA differs from the IEA in including heavy industry and buildings, in addition to oil and gas.[6] Among other differences, while the same emission budget as the IEA is used, IRENA projects renewables to provide 65% of total primary energy by 2050, against 47% for the IEA scenario.[7] Results for the delayed action case are indicated in parentheses, confirming the much higher financial impact of an abrupt adjustment in mitigation policy:

- The capital value loss for the oil, gas and coal sector is estimated at USD 3.8 trillion (USD 7 trillion in a Delayed Policy Action case).
- Stranded assets in power generation are estimated at USD 200-300 billion for a low assumption of plants economic lifetimes and USD 1.2 trillion with longer lifetimes (USD 1.9 trillion in a Delayed Policy Action case).
- Stranded assets in industry are estimated at USD 220 billion (USD 740 billion in the Delayed Policy Action case).

A combination of IEA and IRENA estimates indicate that stranded assets could amount to USD 1.06 trillion for the energy supply and industry sectors – using IRENA's low range for industrial assets economic lifetime – a number that would nearly triple under a delayed action scenario. These amounts are significant for sectors at stake. However, they appear manageable when compared with the global infrastructure investment needs over the same period to 2050 – i.e. USD 244 trillion, particularly if exits are well planned and impacts on the work force are mitigated (Chapter 6).

Possible ripple effects through the financial system also need to be taken into account. Stranded assets can be viewed as the primary effect of what may be broader effects on the financial situation of companies and sectors in the low-carbon transition. As the value of physical investment in energy production assets that will not be recovered becomes visible to investors, they should reassess publicly listed companies' value, taking into account future earnings. How companies would anticipate, and adapt to, a more stringent climate policy environment is highly uncertain at this stage, and estimates of capital value losses therefore carry more uncertainty than stranded assets. In general, because capital value loss casts a wider net than stranded assets, capital value loss ought to be higher, unless the company has diversified its activities or changed business model, which cannot be evaluated ex ante. Financial stability concerns add to the case for swift action (Carney, 2015).

Stranded assets are not only about energy. A changing climate also weighs on crop yield productivity, which calls for sustainable agriculture investment to taper volatility of future earnings (Morel et al., 2016). The risk of stranding is particularly high in countries like Brazil and Malaysia where deforestation gives way to agriculture (Rautner et al., 2016).

Aligning short-term infrastructure investment plans with long-term, low-emission, climate-resilient development strategies

Barriers to accelerating investment in low-emission and resilient infrastructure include a lack of long-term infrastructure planning that integrates climate mitigation and resilience from the outset, and a lack of a pipeline of bankable and sustainable projects that internalise positive and negative externalities over the lifetime of infrastructure. In order to overcome these barriers, G20 countries should first develop clear infrastructure investment plans that consider mitigation and adaptation as part of their work on developing pathways to 2050.

This section looks at how countries have framed long-term plans, before considering how governments might improve the transparency of infrastructure project pipelines, both to improve the alignment of short-term infrastructure investment with long-term, low-emission, climate-resilient development strategies and to enhance investment flows to that end. The other barriers to accelerate low-emission and resilient infrastructure investment are discussed in Chapter 5.

Develop long-term low-emission strategies to reconcile short-term actions and long-term decarbonisation goals

The Paris Agreement invites parties to communicate by 2020 long-term, low-emission development strategies to 2050 as one of its mechanisms to support strengthening of the international response to climate change. In addition to helping to scale up the ambition of the NDCs, which remain inadequate to reach the Paris Agreement's goals (Chapter 2), such strategies are vital to assist countries in reconciling short-term actions with long-term climate goals. Aligning short-term infrastructure investment plans with long-term, low-emission development strategies will help minimise the risk of both emissions lock-in and stranded assets. Long-term infrastructure investment planning is equally important to ensure flexible, forward-looking investments in resilience, to minimise future impacts from climate change and related economic damage and social hardship.

Post-2030 decarbonisation pathways require different infrastructure, technologies and industrial bases. Countries need to prepare in the next 15 years the technologies and infrastructure necessary to overcome the fossil fuel bias of our economies. In addition, what is considered to be "low-carbon" may differ across countries and over time. Not all "low-carbon" infrastructure is necessarily consistent with the trajectory to a carbon neutral society by the second half of the century; what could be considered as low-carbon in the next five years in some places may not be considered low-carbon elsewhere or on a different timescale.

To date, six countries have submitted mid-century long-term plans to the UNFCCC: Bénin, Canada, France, Germany, Mexico and the United States (Box 3.9). Many other countries are in the process of developing such plans; it is vital that they follow suit. China, India, Russia and the G7 countries have all indicated their intent to develop such strategies before 2020. The 2050 Pathways Platform initiative launched at the UN Climate Change Conference in Marrakech (COP22) represents an important complementary initiative (see Box 2.9).

G20 leaders recognised at the 2014 G20 Summit in Brisbane a lack of a clear pipeline of bankable infrastructure projects as one barrier to infrastructure investment. The lack of information on the pipeline of infrastructure projects makes it difficult to match investment needs and investors, including for low-emission, climate-resilient infrastructure. Providing detailed, comprehensive information on infrastructure projects is key to sending the right signals to private stakeholders to invest in the transition. The lack of information also makes it difficult to carry out a cross-country assessment of consistency of infrastructure plans with long-term mitigation and adaptation goals.

This challenge is particularly important for transitional or "bridge" technologies. Switching from oil or coal to natural gas, for example, will reduce GHG emissions and help countries achieve their 2030 targets and NDCs. But in the mid-term it may generate infrastructure that is costly to replace as further decarbonisation is necessary. There would then be a choice either to let the asset become stranded or to lock in its emissions and accept a continued dependence on fossil fuels that could prevent countries from achieving 2050 targets.

Retrofitting infrastructure post-construction, or stranding assets before the end of their economic life, can be very costly – more costly than designing infrastructure from the outset to take into account climate considerations (Corfee-Morlot et al., 2012; NCE, 2016). To minimise the scale of such problems, each country needs to define now which low-emission options and technologies are consistent with its low-emission pathway to 2050 and beyond, as well as the timing with which new and existing assets need to be deployed and/or phased out. Given the uncertainties associated with the deployment of technologies that are necessary for low-emission pathways (e.g. BECCS), there is a need for a continual reassessment of ambition, as set out in the Paris Agreement.

How do strategic infrastructure plans match up with long-term mitigation and adaptation goals?

At the 2014 G20 Summit in Brisbane, G20 leaders recognised that "tackling global investment and infrastructure shortfalls is crucial to lifting growth, job creation and productivity" and endorsed the Global Infrastructure Initiative (GII), a multi-year work programme to improve the quality of public and private infrastructure investment. In 2015, the G20 Investment and Infrastructure Working Group (IIWG) conducted a voluntary survey to compile information on countries' investment strategies, including the main challenges being addressed, policy priorities, and the policy context of these strategies. This section draws on that work, which remains in progress, in reviewing the extent to which current investment plans and pipelines of infrastructure projects are consistent with climate goals in G20 countries (Table 3.5).

Box 3.9. Examples of mid-century long-term plans under the Paris Agreement

France has committed to reducing carbon emissions by 40% by 2030, compared with 1990 levels, and by 75% by 2050. This means that annual emissions reductions must accelerate from 8 megatonnes of carbon dioxide equivalent ($MtCO_2eq$) per year to 9-10 $MtCO_2eq$. Sectoral targets are spelled out for three "carbon budget" periods – 2015-18, 2019-23 and 2024-28 – followed by a long-term target to be achieved by 2050. The national low-carbon strategy is founded on two pillars: including carbon footprint reductions as a key consideration in all economic decisions; and redirecting investments to support the energy transition, through interventions such as environmental quality labels, guaranteeing public funds, and gradually increasing carbon taxes without increasing the overall tax burden.

The United States has committed to reducing its GHG emissions by 26-28% below its 2005 levels by 2025, making every effort to reach a 28% reduction (including LULUCF). It considers this target to be in line with a straight-line emission reduction pathway from 2020 to deep, economy-wide emissions reduction of 80% or more by 2050. To reach these targets, the government has set out three pillars for action:

- shifting to a low-carbon energy system, while putting a particular emphasis on *i)* increasing the energy efficiency of buildings, vehicles and plug-in appliances, *ii)* decarbonising electricity, and *iii)* shifting to clean electricity and low-carbon fuels in transport, buildings and industry;
- carbon sequestration and removal, taking advantage of the country's natural land resources and their capacity to continue to act as a net carbon sink;

Box 3.9. **Examples of mid-century long-term plans under the Paris Agreement** *(cont.)*

- reducing emissions from non-CO_2 gases, notably via the introduction of *i)* stringent standards and incentives to limit CH_4 emissions from oil and gas production and from landfills; and *ii)* new technologies and best practices for livestock agriculture.

Germany's Climate Action Plan 2050 (adopted in November 2016) sets out to obtain extensive GHG neutrality by 2050, which implies reducing total GHG emissions by 80-95% from 1990 levels. The strategy includes a mid-term target of 55% emissions reduction by 2030, and provides several strategic measures, including:

- sector-specific emissions reduction targets for 2030 that will undergo an impact assessment and possibly be revised in 2018;
- a road map towards an almost climate-neutral building stock;
- a commission for growth, structural change and regional development, which will bring together stakeholders from different levels of government, business, industry and various regions, in order to develop strategies for implementation of the Climate Action Plan by the end of 2018.

Canada's Mid-Century Long-Term Low-Greenhouse Gas Development Strategy sets out to cut GHG emissions by 80% by 2050 from 2005 levels. The strategy is not policy prescriptive, but seeks to inform the Pan-Canadian Framework on Clean Growth and Climate Change, and more generally the conversation on how Canada can achieve a low-carbon economy. It describes modelling analyses that illustrate various scenarios towards deep emissions reductions and outlines potential GHG abatement opportunities. Furthermore, it identifies the areas in which emissions reduction will be more challenging, thus requiring an increased policy focus. The Pan-Canadian Framework has four pillars: *i)* pricing carbon pollution; *ii)* complementary measures to further reduce emissions across the economy; *iii)* measures to adapt to the impacts of climate change and build resilience; and *iv)* actions to accelerate innovation, support clean technology, and create jobs.

Sources: FMESDE (n.d.); GFMoENBN (2016); Government of Canada (2016); Government and Provinces of Canada (2016); White House (2016); UNFCCC (2015).

Table 3.5. **Overview of infrastructure plans in the G20 countries**

			ENERGY						TRANSPORT												WATER						AFOLU					
			Fossil fuels			Renewables			Rail/road			Electric v.			Airports			Ports			WSS			Flood			Agriculure			Forests		
COUNTRY	Adaptation strategy incl. in infrastructure plan	Mitigation strategy incl. in infrastructure plan	P	B	T	P	B	T	P	B	T	P	B	T	P	B	T	P	B	T	P	B	T	P	B	T	P	B	T	P	B	T
Argentina																																
Australia*	✓	✓																														
Brazil	✓	✓																														
Canada*	✓	✓																														
China	✓	✓																														
France		✓																														
Germany*		✓																														
India*	✓	✓																														
Indonesia*	✓	✓																														
Italy		✓																														
Japan	✓	✓																														
Korea		✓																														
Mexico																																
Russia		✓																														
Saudi Arabia																																
South Africa	✓	✓																														
Turkey																																
UK	✓	✓																														
US**																																

Notes: a. The table does not provide an overview of the countries' emissions reduction targets, as stated in the NDCs. These are reflected in Annex 2.A1. This table provides a snapshot of the coverage of national infrastructure plans in G20 countries, and the extent to which adaptation and mitigation are mainstreamed into those plans. b. Abbreviations: WSS: Water supply and sanitation ; AFOLU: Agriculture, forestry and other land use; c. Shaded cells indicate the availability of an infrastructure plan (P), associated budget (B), and target (T); d. Infrastructure plans (P) must include details on the location, type (new construction vs renewal), and the action(s) to be performed in each project (e.g. capacity strengthening, efficiency measures or safety enhancing measures) to be captured in the table; e. Outcome-based targets (T) included are those expressed as, inter alia, kilometres (road and rail), megawatts (energy), number of units (EVs), a percentage (the share of renewable energy in total energy supply or the share of roads that will be upgraded), or hectares (afforestation). Sectoral emissions reduction targets are not taken into account in the table, as these alone provide no information about the kind of infrastructure measures that will be implemented. * These are federal states, where infrastructure projects to a large extent are planned at a state level. The table reflects national plans only; ** The US has stated that it will issue an infrastructure plan during 2017; however, this was not available at the time the information for this table was compiled.

Mainstreaming climate mitigation and adaptation in infrastructure plans

Low-emission growth and economic development are often presented, erroneously, as competing priorities. While there will always be trade-offs and competing objectives between different goals for infrastructure investment, many climate-friendly infrastructure options also provide relief from problems like congestion, air pollution and access to energy in rural locations that have to date lacked easy answers (Box 3.10). This can be a boost to mainstreaming of climate considerations into infrastructure plans. As with any large-scale investments, the essential task is to ensure that all the costs and benefits are considered coherently at the outset, taking into account the time frames during which the infrastructure will be operated.

Table 3.5 shows that only 9 countries – less than half of the G20 – have integrated both mitigation and adaptation considerations into infrastructure planning. An additional four countries only mention mitigation. Five mention neither climate mitigation nor adaptation. In addition, only seven G20 countries have made available a detailed plan of infrastructure projects covering at least three of the four economic sectors of primary concern from a climate perspective (e.g. transport, energy, water and AFOLU, addressed below). The majority cover only one of these areas, or have not communicated infrastructure plans in these areas at all. There is therefore considerable scope for G20 countries to heighten their efforts to both align infrastructure plans across key economic sectors with climate mitigation and adaptation goals, and communicate those plans.

For transport, five G20 countries have provided detailed plans for road, rail, ports and airport infrastructure. Five more have an overall target specific to road and rail. Many countries that do not have a detailed plan tend to either have specific targets (e.g. Turkey) and/or allocated a budget for infrastructure (e.g. India). While these are promising signs, there is a need to better shape and define the future nature of transport in these countries for the transition. China, Russia and the United States are yet to communicate targets, budgets and plans for transport infrastructure. Infrastructure to facilitate the deployment of electric vehicles – such as public charging stations – is also important to the transition in the transport sector. However, to date, G20 infrastructure plans make no mention of concrete charging station infrastructure.

For energy, 17 G20 countries have defined renewable energy targets. Most, however, have not communicated a pipeline of projects for the years to come. Further, Table 3.5 also indicates that fossil-fuel related energy is still prevalent in many governmental plans. Ten G20 countries have targets for fossil fuel energy.

Water and AFOLU receive little attention in national infrastructure plans. For water supply and sanitation, only five countries have defined infrastructure plans. One additional country has set aside an envelope of funding for this issue. As for AFOLU, three countries have defined a pipeline of projects in agriculture. A further three have either established a budget or a target but are yet to provide information on the specific projects involved. In terms of forestry, information is even more scarce: targets exist in only three G20 countries, and one country has identified a budget to invest in this sector. Given the importance of these two sectors in transiting to low-emission, climate-resilient economies, there is scope for G20 countries to develop more robust plans, budgets and targets in their strategies in these areas.

Box 3.10. Examples of co-benefits between low-carbon infrastructure and other SDGs

Air pollution

Improved air quality is one of several co-benefits of climate action that have positive implications for human health. The OECD estimates that in 2010, 3 million people died prematurely because of outdoor air pollution. Unless policies become more stringent, projections suggest 6-9 million people will die prematurely each year by 2060. These deaths are largely projected to take place in densely populated regions with high concentrations of PM2.5 (particulate matter 2.5 micrometers or less in diameter) and, to a lesser extent, ozone (especially China and India) and in regions with aging populations, such as China and Eastern Europe.

In addition, increasing concentrations of PM2.5 and ozone are projected to lead to substantially more cases of illness. This will imply more hospital admissions, greater health expenditure, a higher number of lost working days and limitations on normal daily activities. Air pollution-related healthcare costs are projected to increase from USD 21 billion in 2015 (using constant 2010 USD and PPP exchange rates) to USD 176 billion in 2060, reflecting both a large number of additional cases of illness due to air pollution, and a projected increase in healthcare costs per illness. While a reduction in the burning of fossil fuels is likely to decrease the risk of heart and lung diseases, such as lung cancer, as well as neurologic disorders, other measures also provide clear benefits for human health. For example, replacing cars by more active forms of transport such as walking and cycling can reduce obesity, lung disease, heart disease, breast cancer and depression (Armstrong, 2012).

If climate change mitigation and air pollution policies are integrated, air quality could improve to a point where 40% of the global population currently exposed to dangerous PM levels would breathe air that meets World Health Organisation clean air quality guidelines. At the same time, expenditure on air pollution control will be reduced by EUR 250 billion in 2050. According to the estimates provided by the study, one-third of the total financial co-benefits by 2050 will occur in China, while annual cost savings of EUR 35 billion are estimated for the European Union, provided the current air pollution legislation and climate policies are adopted in parallel (Rafaj et al., 2012).

Reducing congestion

A number of governments have implemented Bus Rapid Transit (BRT) systems to reduce local air pollution and improve health. National railway systems have also reduced congestion, while improving access to remote, small or low-income communities, and supporting economic development and trade (Ang and Marchal, 2013). By improving connectivity and reducing congestion, these policies can boost the contribution of urban centres to productivity growth (OECD, 2015b).

Sources: OECD (2015b; 2016a); Armstrong (2012); Rao et al. (2016); Rafaj et al. (2012).

Improving the transparency of infrastructure project pipelines

Infrastructure development plans and project pipeline information that are inaccessible, incomplete or poorly aligned with long-term climate mitigation and adaptation goals are likely to hinder the flow of infrastructure investment in support of climate goals. Several mechanisms are available to help governments improve the transparency of infrastructure project pipelines.

The Global Infrastructure Hub (GI Hub) launched by the G20 in 2014 could prove a useful tool to increase transparency and strengthen the global pipeline of private and public infrastructure investment opportunities. It showcases investment-ready projects to multilateral banks and private investors. As of February 2017, the project pipeline consisted of 44 projects from eight countries, with a total value of more than USD 29 million (although several early-stage projects have not yet disclosed their values) (GI Hub, n.d.). Out of the eight countries that have contributed to the GI Hub Project Pipeline, only four are G20 countries. The participation of more G20 countries in the Hub would provide a more complete and transparent picture to investors of the direction of infrastructure plans as a whole.

Other global initiatives also help to improve the transparency of infrastructure project pipelines. These can be divided into influencers, mobilisers and tool providers (Mercer and IDB, 2016). *Influencers* – such as the OECD Centre on Green Finance and Investment, the New Climate Economy and the Global Infrastructure Investor Association – provide research and leadership to align infrastructure investment plans with sustainability targets. *Mobilisers*, such as the GI Hub, assist *i)* governments in developing bankable projects and *ii)* investors in funnelling funds into those projects. *Tool providers* – such as the IRENA Navigator and the World Bank's REFINe – aim at facilitating the integration of environmental and social components of infrastructure projects into investment decisions (Mercer and IDB, 2016).

Other platforms provide information on public-private partnerships (PPPs) for infrastructure projects, with the aim of matching investors to projects. For example, the World Bank's Private Participation in Infrastructure (PPI) Project Database contains data on 6 400 infrastructure projects in 139 low- and middle-income countries (World Bank, n.d.b). The World Bank also provides a range of other resources on PPPs for infrastructure, including regional and sectorial updates on overall infrastructure investments through PPPs, as well as sample agreements, checklists, risk matrices, standard bidding documents and other material facilitating the establishment of PPPs, notably in developing countries (World Bank, n.d.c; n.d.d). Strengthening those existing tools to improve the data quality on existing infrastructure investments and future plans and needs is a key priority for G20 countries, and critical to gain the confidence of private sector investors in low-carbon, climate-resilient infrastructure (Chapter 5).

Notes

1. All estimates were converted to 2015 USD for comparability.
2. Bhattacharya et al. (2016b) explain that such an increase is the result of a different methodological approach, and argue that previous estimate failed to reflect the increase in infrastructure spending over the past decade, mainly in middle-income countries. Batthacharya et al.'s (2016b) methodological approach consists of calculating an updated baseline of infrastructure spending in 2015 for major countries, and projecting investment requirements on assumptions of growth and investment rates (which are in turn based on assessments of investment plans and identified gaps across major economies and regions).
3. Details of the assumptions on costs are available in IEA (2017).
4. Pre-construction includes power plants announced, in pre-permit development and permitted. "On hold" includes plants announced as being on hold. In the absence of an announcement that the sponsor is putting its plans on hold, a project is considered "shelved" if there are no reports of activity over a period of two years. At the global level, coal power plants in pre-construction development and "on hold" amount to 570 GW and 607 GW respectively.
5. See Iyer et al. (2015); Rozenberg, Vogt-Schilb and Hallegatte (2014); Johnson et al. (2015); Fay et al. (2015).
6. Although there is value in assessing the cost of shifting the building stock to meet the energy requirements of a low-carbon transition, retrofitting and renovation would add value to buildings, which is not the case of stranded assets in the energy sector. IRENA estimates stranded assets in the buildings sector to amount to USD 12.5 trillion in its Delayed Policy Action case and USD 5 trillion in the REmap reference case; computed as "the difference between cost of deep retrofit and the additional cost to build a new fossil-free building" (IRENA, 2017a).
7. It also assumes oil demand would be at 45% (IRENA) and 41% (IEA) of today's level by 2050. Other methodological differences include that IRENA estimates the impact on the oil and gas sector through the capital value of registered companies, then extrapolates to global oil and gas production. For power and industry, it calculates stranded assets based on the nominal value of a plant shutting down before the end of its economic lifetime.

References

Agrawala, S. and S. Fankhauser (2008), *Economic Aspects of Adaptation to Climate Change: Costs, Benefits and Policy Instruments, OECD Publishing, Paris,* http://doi.org/10.1787/9789264046214-en.

Alexandratos, N. and J. Bruinsma (2012), "World Agriculture Towards 2030/2050", Food and Agriculture Organization of the United Nations, Rome, www.fao.org/docrep/016/ap106e/ap106e.pdf.

Ang, G. and V. Marchal (2013), "Mobilising Private Investment in Sustainable Transport: The Case of Land-Based Passenger Transport Infrastructure", OECD Publishing, Paris, http://doi.org/10.1787/5k46hjm8jpmv-en.

Ansar, A. et al. (2016), "Does infrastructure investment lead to economic growth or economic fragility? Evidence from China", *Oxford Review of Economic Policy*, Vol. 32/3, Oxford University Press, Oxford, pp. 360-390, https://doi.org/10.1093/oxrep/grw022.

Armstrong, F. (2012), *Our Uncashed Dividend: The Health Benefits of Climate Action*, Climate and Health Alliance and the Climate Institute, Sydney, www.climateinstitute.org.au/verve/_resources/OurUncashedDividend_CAHAandTCI_August2012.pdf.

AsDB (2017), *Meeting Asia's Infrastructure Needs*, Asian Development Bank, Manila, http://dx.doi.org/10.22617/FLS168388-2.

ATAG (2014), *Aviaton: Benefits beyond borders, Air Transport Action Group*, Geneva, http://aviationbenefits.org/.

Berg, A. et al. (2012), *Public Investment, Growth, and Debt Sustainability: Putting Together the Pieces, International Monetary Fund, Washington DC,* www.imf.org/external/pubs/cat/longres.aspx?sk=25962.0.

Bhattacharya A., et al. (2016a), *Framework for Assessing the Role of Sustainable Infrastructure*, Brookings Institution, Washington DC.

Bhattacharya, A., et al. (2016b), *Delivering on Sustainable Infrastructure for Better Development and Better Climate, Brookings Institution, Washington DC,* www.brookings.edu/research/delivering-on-sustainable-infrastructure-for-better-development-and-better-climate/.

Booz Allen Hamilton (2007), *The global infrastructure investment deficit, Booz Allen Hamilton, McLean, Virginia.*

Brandon, N.P. et al. (2016), *UK Research Needs in Grid Scale Energy Storage Technologies*, Energy Superstore, Oxford and London, https://ora.ox.ac.uk/objects/uuid:c0a1d73e-3783-4298-b16b-d6108c241151.

Budischak, C. et al. (2013), "Cost-minimized combinations of wind power, solar power and electrochemical storage, powering the grid up to 99.9% of the time", *Journal of Power Sources*, Vol. 225, Elsevier, Amsterdam, pp. 60-74, http://dx.doi.org/10.1016/j.jpowsour.2012.09.054.

Caldecott, B. et al. (2017), "Stranded Assets and the Environment", *Journal of Sustainable Finance and Investment, Vol. 7/1, Taylor and Francis, Abingdon, United Kingdom, pp.* 1-138. www.tandfonline.com/toc/tsfi20/7/1?nav=tocList.

Calderon, C. and L. Serven (2014), "Infrastructure, Growth and Inequality: An Overview", World Bank, Washington DC, http://hdl.handle.net/10986/20365.

Carbon Tracker Initiative (2013), *Unburnable Carbon 2013: Wasted Capital and Stranded Assets, Carbon Tracker Initiative, London,* http://carbontracker.live.kiln.digital/Unburnable-Carbon-2-Web-Version.pdf.

Carney, M. (2015), *Breaking the tragedy of the horizon – climate change and financial stability, speech, Bank of England, London,* www.bankofengland.co.uk/publications/Pages/speeches/2015/844.aspx.

Carrington, D. (2016), "EasyJet's plans for a hybrid fleet, using hydrogen fuel cells", CNN, www.cnn.com/2016/02/02/aviation/easyjet-hybrid-plane-hydrogen-fuel-cell/index.html.

Climate Policy Initiative (2013), *Buildings Energy Efficiency in China, Germany and the United States*, Climate Policy Initiative, San Francisco, https://climatepolicyinitiative.org/wp-content/uploads/2013/04/Buildings-Energy-Efficiency-in-China-Germany-and-the-United-States.pdf.

Corfee-Morlot, J. et al. (2012), "Towards a Green Investment Policy Framework: The Case of Low-Carbon, Climate-Resilient Infrastructure", OECD Publishing, Paris, http://dx.doi.org/10.1787/5k8zth7s6s6d-en.

Darling, R.M. et al. (2014), "Pathways to low-cost electrochemical energy storage: a comparison of aqueous and nonaqueous flow batteries", *Energy and Environmental Science, Vol. 7, Royal Society of Chemistry, London, pp. 3459–3477,* http://dx.doi.org/10.1039/C4EE02158D.

Delgado, C., et al. (2015), *Restoring and protecting agricultural and forest landscapes and increasing agricultural productivity, New Climate Economy,* http://newclimateeconomy.report/2015/wp-content/uploads/sites/3/2015/12/NCE-restoring-protecting-ag-forest-landscapes-increase-ag.pdf.

Development Bank of Southern Africa (2012), *The State of South Africa's Economic Infrastructure: Opportunities and Challenges 2012*, Development Planning Division, Development Bank of Southern Africa, Johannesburg, www.gov.za/sites/www.gov.za/files/State%20of%20SAs%20Economic%20Infrastructure%20Report%202012a.pdf.

DfT (2007), *Low Carbon Transport Innovation Strategy*, UK Department for Transport, London, http://webarchive.nationalarchives.gov.uk/20081022212629/http:/www.dft.gov.uk/pgr/scienceresearch/technology/lctis/lctisdocpdf.

Dietz, S. et al. (2016), "Climate value at risk of global financial assets", *Nature Climate Change, Vol. 6, Springer Nature, London, pp. 676-679,* http://dx.doi.org/10.1038/nclimate2972.

Elgowainy, A. et al. (2012), "Life Cycle Analysis of Alternative Aviation Fuels in GREET", Argonne National Laboratory, Lemont, Illinois, https://greet.es.anl.gov/publication-aviation-lca.

Epstein, A.H. (2014), "Aeropropulsion for Commercial Aviation in the Twenty-First Century and Research Directions Needed", *AIAA Journal*, Vol. 52, American Institute of Aeronautics and Astronautics, Reston, Virginia, pp. 901–911, http://dx.doi.org/10.2514/1.J052713.

Erickson, E.M. et al. (2015), "Review—Development of Advanced Rechargeable Batteries: A Continuous Challenge in the Choice of Suitable Electrolyte Solutions", *Journal of the Electrochemical Society*, Vol. 162/14, Electrochemical Society, Pennington, New Jersey, pp. A2424–A2438, http://dx.doi.org/10.1149/2.0051514jes.

FAO (n.d.), *FAOSTAT*, Statistics Division, Food and Agriculture Organization of the United Nations, Rome Italy (accessed December 2016), www.fao.org/faostat/en/.

Fay, M., et al. (2015), *Decarbonizing Development: Three Steps to a Zero-Carbon Future, World Bank, Washington DC,* http://hdl.handle.net/10986/21842.

Fennell, P.S. et al. (2012), "CCS from industrial sources", *Sustainable Technologies, Systems and Policies*, Vol. 17, Hamad bin Khalifa University Press, Doha, http://dx.doi.org/10.5339/stsp.2012.ccs.17.

Florin, N. and P.S. Fennell (2010), *Carbon capture technology: future fossil fuel use and mitigating climate change, Grantham Institute, Imperial College, London,* www.imperial.ac.uk/grantham/publications/briefing-papers/carbon-capture-technology-future-fossil-fuel-use-and-mitigating-climate-change---grantham-briefing-paper-3.php.

FMESDE (n.d.), "Strategie National Bas-carbone", French Ministry of Ecology, Sustainable Development and Energy, Paris, http://unfccc.int/files/mfc2013/application/pdf/fr_snbc_strategy.pdf.

Fotouhi, A. et al. (2016), "A review on electric vehicle battery modelling: From Lithium-ion toward Lithium–Sulphur", *Renewable and Sustainable Energy Reviews*, Vol. 56, Elsevier, Amsterdam, pp. 1008–1021, http://dx.doi.org/10.1016/j.rser.2015.12.009.

GFMoENBN (2016), *Climate Action Plan 2050*, German Federal Ministry for the Environment, Nature Conservation, Building and Nuclear Safety, Berlin, www.bmub.bund.de/fileadmin/Daten_BMU/Download_PDF/Klimaschutz/klimaschutzplan_2050_kurzf_en_bf.pdf.

GI Hub (n.d.), "The Pipeline", Global Investment Hub, https://pipeline.gihub.org/, (accessed on 15 February 2017).

Government and Provinces of Canada (2016), *Pan-Canadian Framework on Clean Growth and ClimateChange*, Government of Canada, www.canada.ca/content/dam/themes/environment/documents/weather1/20161209-1-en.pdf.

Global Coal Plant Tracker (2017), *Global coal plant tracker*, http://endcoal.org/tracker/, (accessed on 28 February 2017).

Government of Canada (2016), *Canada's Mid-Century Long-Term Low-Greenhouse Gas Development Strategy*, Government of Canada Publications, http://publications.gc.ca/pub?id=9.825953&sl=0.

Grande, L. et al. (2015), "The Lithium/Air Battery: Still an Emerging System or a Practical Reality?", *Advanced Materials*, Vol. 27/5, Wiley, Hoboken, pp. 784–800, http://dx.doi.org/10.1002/adma.201403064.

Granoff, I. et al. (2015), "Zero Poverty, Zero Emissions: Eradicating Extreme Poverty in the Climate Crisis", Overseas Development Institute, London, www.odi.org/sites/odi.org.uk/files/odi-assets/publications-opinion-files/9844.pdf.

Gurwick, N.P. et al. (2013), "A Systematic Review of Biochar Research, with a Focus on Its Stability in situ and Its Promise as a Climate Mitigation Strategy", *PLOS ONE, Vol. 8/9, PLOS, San Francisco,* http://dx.doi.org/10.1371/journal.pone.0075932.

Gutman, J., A. Sy and S. Chattopadhyay (2015), *Financing African infrastructure: Can the world deliver?* Brookings Institution, Washington DC, www.brookings.edu/research/financing-african-infrastructure-can-the-world-deliver/.

Hisatomi, T. et al. (2014), "Recent advances in semiconductors for photocatalytic and photoelectrochemical water splitting", *Chemical Society Reviews*, Vol. 43, Royal Society of Chemistry, London, pp. 7520–7535, http://dx.doi.org/10.1039/C3CS60378D.

Hughes, G., P. Chinowsky and K. Strzepek (2010), *The Costs of Adapting to Climate Change for Infrastructure*, World Bank, Washington DC, http://siteresources.worldbank.org/EXTCC/Resources/407863-1229101582229/DCCDP_2Infrastructure.pdf.

Ignaciuk, A. (2015), "Adapting Agriculture to Climate Change: A Role for Public Policies", OECD Publishing, Paris, http://dx.doi.org/10.1787/5js08hwvfnr4-en.

Ignaciuk, A. and D. Mason-D'Croz (2014), "Modelling Adaptation to Climate Change in Agriculture", OECD Publishing, Paris, http://dx.doi.org/10.1787/5jxrclljnbxq-en.

IPCC (2014), *Climate Change 2014: Mitigation of Climate Change. Contribution of Working Group III to the Fifth Assessment Report of the Intergovernmental Panel on Climate Change, Cambridge University Press, Cambridge and New York*. http://www.ipcc.ch/report/ar5/wg3/.

IAEA (2016), *Power Reactor Information System, International Atomic Energy Agency, Vienna*, https://www.iaea.org/pris/ (accessed on 30 November 2016).

IEA (2017), *Chapters 1 and 2 in Perspectives for the Energy Transition: Investment Needs for a Low-Carbon Energy System, International Energy Agency/OECD Publishing, Paris*, www.energiewende2017.com/wp-content/uploads/2017/03/Perspectives-for-the-Energy-Transition_WEB.pdf.

IEA (2016a), *World Energy Outlook 2016, IEA/OECD Publishing, Paris*, www.worldenergyoutlook.org/publications/weo-2016/.

IEA (2016b), *World Energy Investment 2016*, IEA/OECD Publishing, Paris, http://dx.doi.org/10.1787/9789264262836-en.

IEA (2016c), *IEA Medium-Term Renewable Energy Market Report 2016: Market Analysis and Forecasts to 2021*, IEA/OECD Publishing, Paris, http://dx.doi.org/10.1787/renewmar-2016-en.

IEA (2016d), *Energy Technology Perspectives 2016*, IEA/OECD Publishing, Paris, http://dx.doi.org/10.1787/energy_tech-2016-en.

IEA (2012), *Energy Technology Perspectives 2012: Pathways to a Clean Energy System*, IEA/OECD Publishing, Paris, http://dx.doi.org/10.1787/energy_tech-2012-en.

IEA (2010), *Energy Technology Perspectives 2010: Scenarios and Strategies to 2050*, IEA/OECD Publishing, Paris, http://dx.doi.org/10.1787/energy_tech-2010-en.

ILO and IILS (2011), "A Review of Global Fiscal Stimulus", International Labour Organisation and International Institute for Labour Studies, Geneva, www.ilo.org/inst/research/addressing-crisis-challenges/discussion-paper-series/WCMS_194175/lang--en/index.htm.

IMF (2014), "Is It Time for an Infrastructure Push? The Macroeconomic Effects of Public Investment", in *World Economic Outlook 2014: Legacies, Clouds, Uncertainties*, International Monetary Fund, Washington DC, www.imf.org/external/pubs/ft/weo/2014/02/.

IRENA (2017a), *Chapter 3 in Perspectives for the Energy Transition: Investment Needs for a Low-Carbon Energy System, International Renewable Energy Agency, Abu Dhabi*, www.energiewende2017.com/wp-content/uploads/2017/03/Perspectives-for-the-Energy-Transition_WEB.pdf.

IRENA (2017b), "Rethinking Energy: Accelerating the global energy transformation", International Renewable Energy Agency, Abu Dhabi. www.irena.org/DocumentDownloads/Publications/IRENA_REthinking_Energy_2017.pdf.

Iyer, G. et al. (2015), "The Contribution of Paris to Limit Global Warming to 2°C", *Environmental Research Letters, Vol. 10/12, IOP Publishing, Bristol*, http://dx.doi.org/10.1088/1748-9326/10/12/125002.

Johnson, N. et al. (2015), "Stranded on a low-carbon planet: Implications of climate policy for the phase-out of coal-based power plants", *Technological Forecasting and Social Change, Vol. 90/A, Elsevier, Amsterdam*, http://dx.doi.org/10.1016/j.techfore.2014.02.028.

Kennedy, C. and J. Corfee-Morlot (2012), "Mobilising Investment in Low Carbon, Climate Resilient Infrastructure", OECD Environment Working Papers, No. 46, OECD Publishing, http://dx.doi.org/10.1787/5k8zm3gxxmnq-en.

LCICG (2014), *Coordinating Low Carbon Technology Innovation Support: The LCICG's Strategic Framework*, Low Carbon Innovation Co-ordination Group, London, www.lowcarboninnovation.co.uk/working_together/strategic_framework/overview/.

Magnuson, A. et al. (2009), "Biomimetic and Microbial Approaches to Solar Fuel Generation, *Accounts of Chemical Research*, Vol. 42/12, American Chemical Society, Washington DC, pp. 1899–1909, http://dx.doi.org/10.1021/ar900127h.

Mercer and IDB (2016), *Building a bridge to sustainable infrastructure: Mapping the global initiatives that are paving the way*, Mercer and Inter-American Development Bank, Washington, www.mercer.com/our-thinking/building-a-bridge-to-sustainable-infrastructure.html.

Moniz, S.J.A. et al. (2015), "Visible-light driven heterojunction photocatalysts for water splitting – a critical review", *Energy and Environmental Science, Vol. 8, Royal Society of Chemistry, London, pp. 731–759*, http://dx.doi.org/10.1039/C4EE03271C.

Morel, A. et al. (2016), *Stranded Assets in Palm Oil Production: A Case Study of Indonesia*, Smith School of Enterprise and the Environment, University of Oxford, Oxford, www.smithschool.ox.ac.uk/research-programmes/stranded-assets/Stranded_Assets_in_Palm_Oil_Production.pdf.

Napp, T. (forthcoming), *A survey of key technological innovations for the low-carbon economy, Grantham Institute, Imperial College, London.*

National Economic Council and the President's Council of Economic Advisers (2014), *An Economic Analysis of Transportation Infrastructure Investment*, White House, Washington DC, https://obamawhitehouse.archives.gov/sites/default/files/docs/economic_analysis_of_transportation_investments.pdf.

NCE (2016), *The Sustainable Infrastructure Imperative: Financing for Better Growth and Development, New Climate Economy, Washington DC and London,* http://newclimateeconomy.report/2016/wp-content/uploads/sites/4/2014/08/NCE_2016Report.pdf.

NCE (2014), *Better Growth, Better Climate*, New Climate Economy, Washington DC and London, http://newclimateeconomy.report/2014/.

Nykvist, B. and M. Nilsson (2015), "Rapidly falling costs of battery packs for electric vehicles", *Nature Climate Change*, Vol. 5, Springer Nature, London, pp. 329–332. http://dx.doi.org/10.1038/nclimate2564.

OECD (2016a), *The Economic Consequences of Outdoor Air Pollution*, OECD Publishing, Paris, http://dx.doi.org/10.1787/9789264257474-en.

OECD (2016b), "Producer and Consumer Support Estimates", database, www.oecd.org/tad/agricultural-policies/producerandconsumersupportestimatesdatabase.htm (accessed on 10 April 2017).

OECD (2015a), *Aligning Policies for a Low-carbon Economy*, OECD Publishing, Paris, http://dx.doi.org/10.1787/9789264233294-en.

OECD (2015b), *Governing the City*, OECD Publishing, Paris, http://dx.doi.org/10.1787/9789264226500-en.

OECD (2014), *Green Growth Indicators for Agriculture: A Preliminary Assessment*, OECD Publishing, Paris, http://dx.doi.org/10.1787/9789264223202-en.

OECD (2012), *Strategic Transport Infrastructure Needs to 2030*, OECD Publishing, Paris, http://dx.doi.org/10.1787/9789264114425-en.

OECD (2006), *Infrastructure to 2030: Telecom, Land Transport, Water and Electricity*, OECD Publishing, Paris, http://dx.doi.org/10.1787/9789264023994-en.

OECD/ITF (2017), *ITF Transport Outlook 2017*, OECD Publishing, Paris, http://dx.doi.org/10.1787/9789282108000-en.

Pan, Y. (2016), "China's Urban Infrastructure Challenges", *Engineering*, Vol. 2/1, Elsevier, Amsterdam, pp. 29-32, http://doi.org/10.1016/J.ENG.2016.01.007.

Pardee Center (2014), "Patterns of potential human progress", Pardee Center, University of Denver, Denver, http://pardee.du.edu/pphp-4-building-global-infrastructure.

Pardee Center (n.d.), "Access Ifs", *International Futures Model*, Pardee Center, University of Denver, Denver, http://pardee.du.edu/access-ifs.

Parry, M. et al. (2009), *Assessing the costs of adaptation to climate change: A review of the UNFCCC and other recent estimates*, International Institute for Environment and Development and Grantham Institute, Imperial College, London, http://pubs.iied.org/pdfs/11501IIED.pdf.

Platts WEPP (2017), *World Electric Power Plants Database*, Platts, London, www.platts.com/products/world-electric-power-plants-database (accessed on 31 March 2017).

Pultarova, T. (2016), "Hydrogen-powered aircraft takes to sky in Germany", *Engineering and Technology*, Institute of Engineering and Technology, Stevenage, United Kingdom, https://eandt.theiet.org/content/articles/2016/09/hydrogen-powered-aircraft-takes-to-sky-in-germany/.

Rafaj, P. et al. (2012), "Co-benefits of post-2012 global climate mitigation policies", International Institute for Applied Systems Analysis, Laxenburg, Austria, http://pure.iiasa.ac.at/10378/.

Rao, S. et al. (2016), "A multi-model assessment of the co-benefits of climate mitigation for global air quality", *Environmental Research Letters, vol. 11/124013, IOP Publishing, Bristol,* http://iopscience.iop.org/article/10.1088/1748-9326/11/12/124013/pdf.

Rautner, M., et al. (2016), *Managing the Risk of Stranded Assets in Agriculture and Forestry*, Chatham House, London, www.chathamhouse.org/publication/managing-risk-stranded-assets-forestry-and-agriculture.

Rozenberg, J., Vogt-Schilb, A. and Hallegatte, S. (2014), *Transition to clean capital, irreversible investment and stranded assets, World Bank, Washington DC,* http://documents.worldbank.org/curated/en/768841468171252748/Transition-to-clean-capital-irreversible-investment-and-stranded-assets.

Shearer, C. et al. (2017), *Boom and Bust 2017: Tracking the global coal plant pipeline, Global Coal Plant Tracker,* http://endcoal.org/wp-content/uploads/2017/03/BoomBust2017-English-Final.pdf.

Stotsky, J. and M. Ghazanchyan (2013), *Drivers of Growth: Evidence from Sub-Saharan Africa, International Monetary Fund, Washington DC,* www.imf.org/en/Publications/WP/Issues/2016/12/31/Drivers-of-Growth-Evidence-from-Sub-Saharan-African-Countries-41064.

UKCCSRC (2015), *RAPID Research and Pathways to Impact Development: Handbook 1.1b – Research Area Summaries and Needs*, United Kingdom Carbon Capture and Storage Research Centre, Sheffield and Edinburgh, https://ukccsrc.ac.uk/sites/default/files/documents/content_page/RAPID%201_1b.pdf.

UN (n.d.), *UN Comtrade Database, United Nations,* https://comtrade.un.org/ (accessed 30 April 2017).

UN (2016), *The Sustainable Development Goals Report 2016, United Nations,* https://unstats.un.org/sdgs/report/2016/The%20Sustainable%20Development%20Goals%20Report%202016.pdf.

UNCTAD (2016), *Review of Maritime Transport 2016: The long-term growth prospects for seaborne trade and maritime businesses, United Nations Conference on Trade and Development, Geneva,* http://unctad.org/en/pages/PublicationWebflyer.aspx?publicationid=1650.

UNFCCC (2015), "Intended Nationally Determined Contribution of Mexico", UNFCCC, Bonn, www4.unfccc.int/submissions/INDC/Published%20Documents/Mexico/1/MEXICO%20INDC%2003.30.2015.pdf.

UNFCCC (2007), *Background Paper on Analysis of Existing and Planned Investment and Financial Flows Relevant to the Development of Effective and Appropriate International Response to Climate Change*, UNFCCC, Bonn, http://unfccc.int/files/cooperation_and_support/financial_mechanism/application/pdf/background_paper.pdf.

Vogt-Schilb, A. & S. Hallegatte (2014), *Marginal abatement cost curves and the optimal timing of mitigation measures*. Energy Policy 66, 645–653. 2.

White House (2016), "United States Mid-Century Strategy for Deep Decarbonisation", White House, Washington DC, http://obamawhitehouse.archives.gov/sites/default/files/docs/mid_century_strategy_report-final.pdf.

Woetzel, J. et al. (2016), *Bridging Global Infrastructure Gaps, McKinsey Global Institute,* www.mckinsey.com/industries/capital-projects-and-infrastructure/our-insights/bridging-global-infrastructure-gaps.

Workman, M. et al. (2011), "An assessment of options for CO_2 removal from the atmosphere", *Energy Procedia*, Vol. 4, Elsevier, Amsterdam, pp. 2877–2884, http://dx.doi.org/10.1016/j.egypro.2011.02.194.

World Bank (2010, *The Costs of Adapting to Climate Change for Infrastructure*, World Bank, Washington DC, http://siteresources.worldbank.org/EXTCC/Resources/407863-1229101582229/DCCDP_2Infrastructure.pdf.

World Bank (n.d.a), *World Development Indicators*, World Bank, Washington DC, http://data.worldbank.org/data-catalog/world-development-indicators (accessed on 15 January 2017).

World Bank (n.d.b), "PPPIRC", World Bank, Washington DC, https://ppp.worldbank.org/public-private-partnership/ (accessed on 15 February 2017).

World Bank (n.d.c), "PPI Resources", World Bank, Washington DC, http://ppi.worldbank.org/resources/ppi-resources (accessed on 15 February 2017).

World Bank (n.d.d), "Public-Private-Partnership in Infrastructure Resource Center", World Bank, Washington DC, https://ppp.worldbank.org/public-private-partnership/ (accessed on 15 February 2017).

WEF (2015), *The Global Competitiveness Report 2015-2016*, World Economic Forum, Geneva, www3.weforum.org/docs/gcr/2015-2016/Global_Competitiveness_Report_2015-2016.pdf.

Zhang, S.S. (2013), "Liquid electrolyte lithium/sulfur battery: Fundamental chemistry, problems, and solutions", *Journal of Power Sources, Vol. 231, Elsevier, Amsterdam, pp. 153–162,* http://dx.doi.org/10.1016/j.jpowsour.2012.12.102.

Chapter 4

Growth implications of climate action

The current global economic environment provides governments with an opportunity to boost economic growth while also addressing the challenges of climate change. Ensuring that growth is low-emission, resilient and inclusive can help to meet the Paris Agreement goals while also delivering on the Sustainable Development Goals. While the synergies between climate and growth policies are substantial, capitalising on them requires fiscal initiatives to scale up public and private investment in the right technologies and infrastructure, combined with a well-designed structural reform package. This chapter shows how these pro-growth reform policies can support ambitious policy action on climate change to create a "decisive transition" to a low-emission, high-growth future. It presents model simulations that combine climate action with pro-growth policies including the impacts of delaying action.

The current global macroeconomic environment, including low interest rates in most countries, provides governments with an opportunity to create conditions for high-quality economic growth that is low-emission, resilient and inclusive. The synergies between climate and growth policies are substantial, but capitalising on them requires scaling up public and private investment in the right technologies and infrastructure, combined with an effective structural reform package. Growth and climate agendas can be integrated as their effectiveness depends partly on the same factors: developing and diffusing new technologies to attract investment, and reallocating resources towards high-productivity economic activities.

This chapter shows how these pro-growth reform policies can support ambitious policy action on climate change to create a "decisive transition" to a low-emission, high-growth future. Based on macro-economic model simulations, the chapter explores the potential impacts on growth and employment of scenarios that combine climate action with pro-growth policies. The chapter starts by providing context on the current global macro-economic conditions and the potential for fiscal and structural policy levers to promote growth. The following sections present the results of the model simulations, and examine the implications of a delayed action scenario and the consequences of a lack of co-ordinated action. The chapter concludes by shedding light on the structural and employment changes that economies face as they move to low-emission pathways.

The macro-economic context

Many G20 countries are in a low-growth trap

Global economic growth has hovered around 3% in the past five years, below the level needed to sustain the aspirations of citizens. In many countries, private and public investment has been weak, slowing growth in labour productivity and total factor productivity (OECD, 2015a, 2016a). In most high-income G20 countries, government, businesses and households have been investing substantially less than in the pre-crisis years (Figure 4.1).

Although overall investment as a share of GDP has increased in emerging G20 economies, its level varies significantly from country to country, and remains low in several. Yet investment is a key driver of growth, and a decisive factor enabling emerging economies to reach the levels of economic development of high-income countries. In response to these bleak economic conditions, expectations for GDP growth and investment for the next decade have been revised downwards in both advanced and major emerging economies (Figure 4.2).

As G20 governments seek to revive economic growth, the quality and inclusiveness of growth also matters. To continue to improve well-being beyond the short term, the sources of growth need to be sustainable economically, socially and environmentally. Over the longer term, the fundamentals of continued economic growth are at great risk due to the scale of potential damage from climate change described in Chapter 2. Climate change poses a major systemic risk to all economies, but particularly for societies in less-developed, less-resilient countries. Delaying action on climate change is likely to result in a more disruptive, substantially costlier transition, as high-carbon infrastructure and other assets will be made obsolete.

As well as supporting low-emission, climate-resilient development, new growth also needs to be inclusive. Widening inequalities and an increasing realisation that recent growth has benefited only parts of the population have made enhanced inclusiveness a key priority for governments. The benefits of low-emission, climate-resilient growth, including new economic opportunities, need to be equitably distributed across society, reducing potential opposition to climate change policies and helping to ensure existing inequalities are not compounded in the transition.

Figure 4.1. Investment as a share of GDP in 2015 relative to the average in the pre-crisis decade*

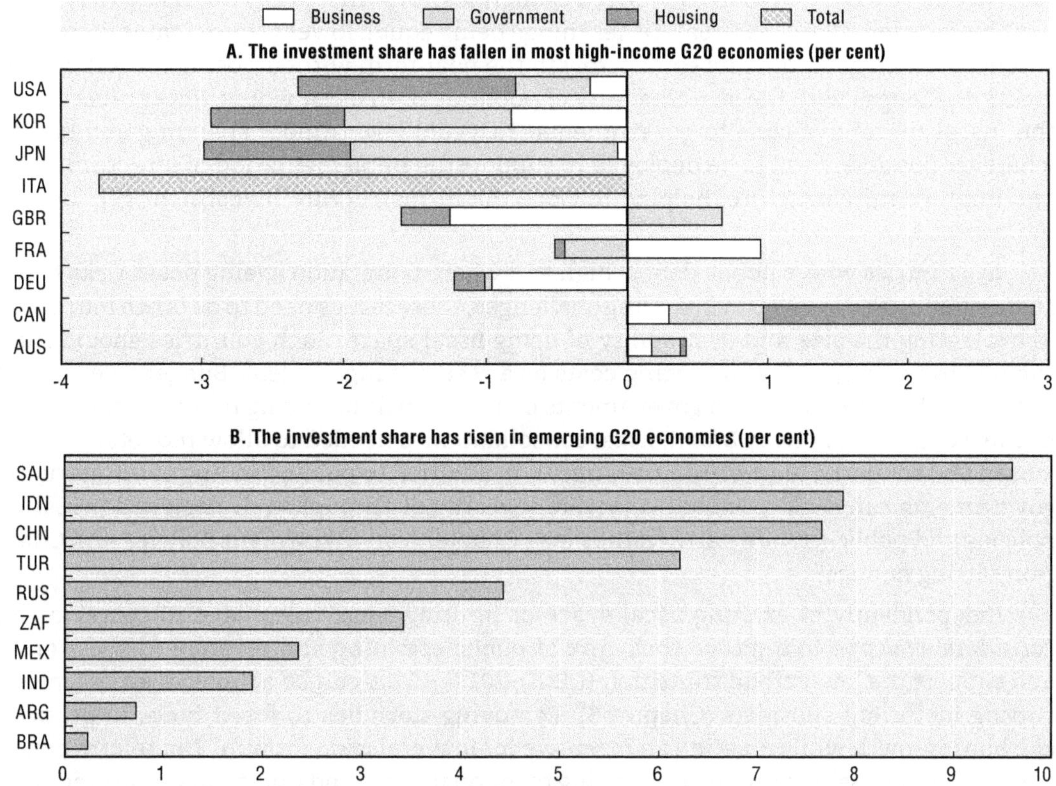

Note: *Average of investment shares in GDP from 1996 to 2007. No breakdown available for Italy and the emerging economies

Source: OECD (2016a), Economic Outlook: Statistics and Projections (database).

Figure 4.2. Long-term growth expectations have declined

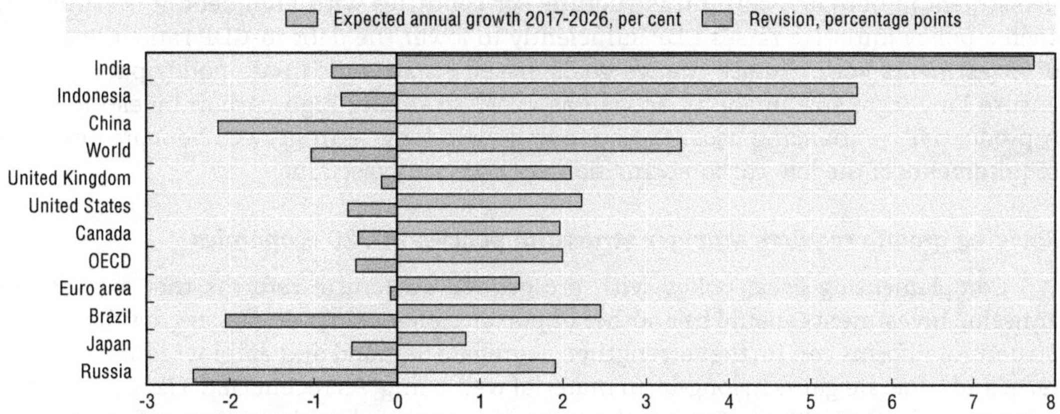

Note: The revision is the difference between April 2011 projections of average annual GDP growth over 2012-21 and April 2016 projections of average annual GDP growth over 2017-26. OECD and World estimates based on weighted average of available countries, using 2015 PPP shares.

Source: Consensus Forecasts; and OECD calculations.

A window of opportunity to escape the low-growth trap

To get out of the low-growth trap, collective, well-designed policy decisions are needed to support aggregate demand in the short term and provide an impulse to longer-term growth. There is little scope left for monetary policy to provide such a stimulus in most countries, so a proactive fiscal policy response is required. Such a response is feasible because interest rates are close to zero in many advanced economies, and where they are rising they are still near historic lows (OECD, 2016a). Other things being equal, lower interest rates increase the extent to which governments can borrow in the near term without losing market access or facing challenges with the sustainability of public debt. In other words, lower rates increase the "fiscal space" available to governments. Targeted government spending and taxation, wherever possible, need to be deployed to support the implementation of structural reform and improve infrastructure, helping to close the infrastructure investment gap identified in Chapter 3.

In countries where public debt is high and where population ageing poses risks to fiscal sustainability and long-term financing challenges, these issues need to be taken into account in evaluating the size and desirability of using fiscal space. Such countries should indeed avoid substantially higher financing costs as a result of higher debt. Budget rules can also constrain the extent to which governments can use deficit-financing to fund infrastructure. In a few countries, notably Brazil, fiscal consolidation is needed to allow monetary policy to loosen and support a recovery of investment. In addition to public funding of infrastructure, governments can seek to mobilise private investment through well-designed investment policies and public-private partnerships (see Chapter 5 on investment policies and Chapter 7 on raising finance).

Independently of existing fiscal space or limiting fiscal rules, all G20 countries have considerable scope to improve their mix of public spending and revenue to boost growth and support the low-carbon transition (OECD, 2016a). This can be achieved, for example, by cutting inefficient subsidies (Chapter 5). Removing subsidies to fossil fuels, in particular, can boost growth while creating an incentive for low-emission growth. Tax reform can also support low-emission growth, by reducing taxes on income and raising taxes on greenhouse gas (GHG)-emitting and other polluting activities or immovable property (Johansson et al, 2008).

The fiscal space should be used wisely to boost production capacity with appropriate investment in hard and soft infrastructure. For countries with high debt, it is critical that such a policy initiative raises GDP sufficiently to lower the debt-to-GDP ratio. Consistency of investments with climate change goals, based on strong climate policy packages, will ensure long-term sustainability as well as avoid stranding high-carbon investment later. A productivity-enhancing fiscal initiative will yield long-term growth benefits only if the requirements of the low-carbon transition are taken into account.

Reviving growth requires stronger structural policies in G20 economies

Complementing fiscal policy with pro-growth structural reforms that support low-emission investments should be another important pillar of low-emission growth packages. Structural reforms can further strengthen aggregate demand and employment in the short term and generate gains in long-term material well-being. As recent OECD *Economic Outlooks* have argued (OECD, 2016a, 2015b), the pace of growth-enhancing reforms has slowed in high-income and emerging economies, particularly in cross-cutting policy areas with a strong influence on labour productivity, such as education and innovation (OECD, 2017). Governments have tended to concentrate reform efforts in specific policy areas, which suggests that potential gains from policy synergies and reform complementarities are being missed (OECD, 2017). Enhanced education and innovation policy is vital not only to address the persistent and widespread decline in productivity growth but also to manage the low-carbon transition successfully and to make economic growth more inclusive.

To strengthen economic growth, renewed efforts are needed across a wide range of reform areas in both advanced and emerging economies. Possible reform packages include measures to enhance entry of new firms and product market competition, particularly in services sectors with pent-up demand (Gal and Hijzen, 2016). Firm-level evidence suggests that reforms to strengthen competition, market entry and entrepreneurship can boost investment by around 4% after two years in high-income economies (Gal and Hijzen, 2016). As shown below, they would improve the response of firms to increases in energy prices and tighter environmental regulation, boosting investment, innovation and productivity.

Such reforms encourage the take-up of new technologies and more efficient use of resources; they can also hasten the development of low-carbon business models, such as new transport solutions through, for example, the development of start-ups. Reallocation-friendly banking sectors and insolvency regime reforms could ease the exit of failing firms, thereby facilitating the reallocation of resources to more productive and innovative activities, including low-carbon activities (Adalet McGowan et al., 2017). Such policies would also boost investment in knowledge-based capital, such as high-productivity technologies, research and development, management skills and worker qualifications across countries, businesses and households – for example, through education and trade – also increasing diffusion of new, lower-carbon technologies.

Steps to better match skills to jobs and to ensure that skills are used fully could also boost productivity by enabling firms and workers to adopt and use innovation and new technologies (OECD, 2016a). Reforms to housing policies and active labour market policies that combine benefits with retraining and upskilling can lower unemployment, facilitate geographic mobility and improve the matching of skills and jobs. Such policies can help workers in declining fossil fuel-intensive production find new jobs in low-carbon sectors while encouraging upward social mobility – part of ensuring a "just transition" for workers (see Chapter 6).

Reform efforts will only work if they are coherent. Regulatory policies need to encourage the emergence of new business models, especially in low-emission activities. Complementarities also need to be exploited to make the most of growth impacts of policy reforms. For example, relaxing labour regulations in an environment of rigid product markets may reduce employment and wages. In contrast, deregulating the business environment at the same time enhances the likelihood that businesses will compete for workers. Overall, integrating climate policies with growth policies is a policy challenge with substantial benefits. The specifics of how structural reform policies can support climate change mitigation strategies are covered in Chapter 5.

Combining economic and climate policies could both achieve the Paris climate objectives and spur economic growth

Governments have at their disposal a range of policy options to both generate economic growth and to combat climate change. How these policies are combined will influence differently both economic growth and the extent to which countries move towards low-emission and resilient pathways. The policy combinations can be stylised as choices along two dimensions: between current economic policies and pursuing a high-growth policy package on one dimension, and between no climate action and action to pursue the Paris

Agreement goals on the other dimension (Figure 4.3). If chosen appropriately, a combination of climate policy instruments, and well-aligned fiscal initiatives and structural reforms would allow G20 countries to both achieve climate goals and escape the low-growth trap.

Figure 4.3. **Identifying a pathway for the "decisive transition"**

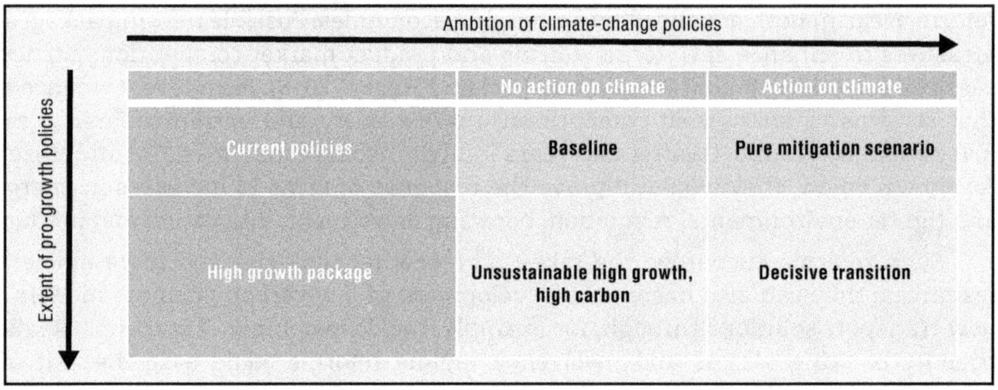

The *baseline* scenario used in the model simulations assumes no climate policy change from the current situation, and no new policy to support growth beyond what is currently planned (top-left quadrant of Figure 4.3). In such a situation, the world would remain in a low-growth trap and miss the Paris Agreement goals.

Governments may be tempted to pursue fiscal and structural policies to provide an impulse to economic growth "at all costs", without ensuring that new investments support the low-emission transition (bottom-left quadrant). Reasons for pursuing this path could include low prices for fossil fuels, the strength of incumbent fossil fuel technologies (see Chapter 3), inaccurate or non-existent pricing of GHG emissions and other pollution externalities, and a disregard for the longer-term consequences of today's infrastructure investment decisions (see Chapter 5). There may also be a lack of "bankable" low-carbon and climate-resilience projects, due to policy misalignments and the incumbency advantage of existing technologies and business models. This "unsustainable high-carbon" pathway is not examined in this report.

Alternatively, governments may pursue policies to decarbonise economies but without taking action to provide an impulse to economic growth: a "pure mitigation" scenario (top-right quadrant). As argued below, however, this would result in higher adjustment costs, less take-up of business opportunities in the context of the low-carbon transition, and lower material well-being and employment, all of which would make the transition more difficult politically.

The *"decisive transition"* scenario is a high-investment, high-innovation and low-carbon transition path, combining pro-growth policies with more ambitious climate policies (bottom-right quadrant). In this scenario, countries implement a policy package that spurs growth while accelerating the transition towards long-term climate change objectives. This package comprises a fiscal initiative in support of climate objectives – for example, additional investment in infrastructure, education and R&D – and structural reforms that have been found to boost long-term growth and can be made coherent with the low-carbon transition (OECD, 2017). In the model simulation, a "typical" package is considered to combine an increase in public investment with a cut in the stringency of product-market regulations and an increase in R&D spending. In reality, the composition of this package would be country-specific, reflecting existing institutions, regulatory frameworks and preference for equity. Box 4.1 provides an overview of the underlying model assumptions while Annex 4.A2 provides further details on the models' structure and parameters.

The macro-economic impacts of the different scenarios are expected to vary over time. In particular, while decarbonisation policies are likely to create adjustment costs in the short to medium term (5 to 20 years depending on policy stringency), high-carbon pathways would mean, over the longer term, increasing damage from climate change. This would weigh negatively on output and on well-being more generally. The high-carbon pathways would also entail increasing tail risks (such as rapid sea-level rise from melting of ice sheets, and systemic effects of repeated extreme weather events) and well-being costs (such as increased mortality from air pollution). Ambitious climate mitigation action offers the benefit of decisively lower long-term costs from climate change and provides decisively more insurance against the risk of destructive extreme weather events. Conversely, it requires more investment and more stringent climate mitigation policies in the short term. Undertaking structural reforms so these adjustments occur in the context of high, inclusive growth ought to be seen as an integral part of making the economic case for climate policy action.

The macro-economic impacts of the different scenarios will also substantially differ across countries, depending on their sectoral structure and energy consumption. For example, fossil-fuel exporting countries are usually seen as incurring the highest costs in the transition. In reality the situation may be more complex, as demand for fossil fuels will continue for some time, and relative costs of extraction will determine market shares and revenues in a scenario with lower demand. Similarly, countries whose public investment is low could benefit the most from the additional increase in investment (Fournier, 2016). More flexible product and labour markets would also facilitate the transition toward a decarbonised economy (see below).

Given the importance of path dependence, the implications of delayed action are also examined, building on the IEA disjointed scenario (IEA, 2016), whereby investment to meet the goal of limiting global warming to 2°C is delayed to 2025. This implies a more abrupt path to decarbonisation from 2025 on, combining higher carbon taxes, more support to low-carbon technologies in general, and significantly larger stranding of fossil-fuel-based assets.

The model simulations in this chapter explore the implications of a decisive transition for the main macroeconomic aggregates, including GDP, employment, business investment and the ratio of public debt-to-GDP. The decisive transition scenario is compared with a baseline scenario that assumes no change from current climate policy and no new policy to support growth. The transition would involve not only undertaking mitigation policies to reduce emissions and achieve a 2°C path with a 50% probability, but also complementing these policies with a fiscal initiative (e.g. additional investment in hard and soft infrastructure) and structural reforms that would support long-term growth and reduce adjustment costs. A simulation is also run for a more ambitious climate scenario, assuming a 66% probability of keeping temperature below 2°C. The simulations build on the results from a parallel report for the German G20 Presidency on the scale and scope of energy sector investments needed to increase the chances of reaching this goal (IEA, 2017).

Box 4.1. Key modelling assumptions

Several assumptions underpin the decisive transition scenario:

- Governments absorb some of the incremental costs of low-carbon and climate-resilient infrastructure investments (e.g. via public investment programmes or procurement of innovative solutions). This is justified by market failures that may prevent firms from responding to more direct price-based instruments (see Chapter 5). Such public investments could trigger higher growth in business investment, at a time when it remains modest. As many climate-friendly investments involve new technology, they should also open markets for innovative firms. One important assumption underlying the simulations is that the investment undertaken is of good quality, and there is relatively strong institutions and effective public governance in place.

Box 4.1. **Key modelling assumptions** *(cont.)*

- All G20 countries take action on climate. If a leadership group of countries were to act alone, free riders would benefit from lower fossil fuel prices and substitution of carbon-intensive economic activity away from this group of countries, undermining global emission reductions (see below and Annex 4.A1 for a discussion).
- The boost in investment is assumed to be budget-neutral in the medium term and financed by better reallocating tax and spending, which would leave the public deficit unchanged. From 2017 to 2020, the measure would be financed through a higher public deficit in all the G20 countries covered in the analysis except Brazil, Japan, India and South Africa. In these four countries, assuming no change in policies, fiscal space is limited, and the initiative is expected to be budget neutral from 2017 onwards.
- In most countries, revenues from the taxation of carbon emissions are used to pay down public debt. In countries where the ratio of public debt-to-GDP is low, however, it is assumed that, in the medium term, those revenues are used to support a further increase in public consumption. A detailed discussion of the potential uses of carbon tax revenue follows in Chapter 5. Other recycling options could lead to higher growth outcomes depending on pre-existing tax levels and their distortionary effects in various economies. In most cases, the effect of the recycling is expected to vanish over time, except if revenues are used to increase good-quality productive investment.
- Interest rates set by central banks are assumed to remain at their current level for three years in the euro area and Japan, and to follow inflation and growth developments elsewhere.

Simulations were performed using Yoda, an OECD in-house semi-structural model for selected G20 economies. This model encompasses international spill-overs and delayed labour-market response to policy (hysteresis effects). To assess the robustness of simulation outcomes and complement the analysis, some simulations were also performed using the macro-economic Oxford model which has detailed trade and financial inter-linkages, but can be simulated only up to 2045 (see Annex 4.A2). Like all empirical tools, these models have several limitations. In particular, they are a stylised representation of the economy. Political decisions, social acceptance and institutional factors, for example will also play a major role in the real world, but are not taken into account in the simulations.

The quantitative analysis covers most G20 countries (representing 88% of the total of G20 economies excluding the European Union), based on data availability and the geographic scope of modelling tools.[1] With a view to identifying categories of countries that would respond differently to mitigation and pro-growth policies, four types of stylised economies are presented depending on their reliance on fossil fuels (net importers or exporters) and their level of development (advanced or emerging economies).

A "pure mitigation" scenario, without supporting growth policies, would have overall limited growth effects

Achieving the goal of limiting global warming to 2°C with 50% probability will clearly require ambitious climate policies. This includes strengthening the use of carbon pricing instruments in order to direct private investment and technological change into low-emission activities in a cost-effective way. In the model simulations, the move is achieved through higher carbon taxes and a range of energy efficiency and technology support policies (IEA, 2016, 2017). Without specific additional measures to boost growth, the impact of decarbonisation is estimated to be small in the medium term on average for G20 countries. For each country, the effects of such a "pure mitigation" scenario would depend on whether it is a net exporter or importer of fossil fuels, with a more pronounced negative effect on net fossil-fuel exporters.

Long-term effects of mitigation policies on output in a pure mitigation scenario would be more pronounced, though still relatively small given the time horizon involved. Net fossil-fuel exporters in OECD member countries and emerging economies would experience significant losses in the long term, reflecting to a large extent massive disinvestment in high-carbon industries and lower fossil-fuel export prices. In countries where the level of public debt is low, however, recycling of carbon-tax revenues through additional spending could mitigate those losses. By contrast, net fossil-fuel importers would benefit in the long term from the increase in net investment and, to a lesser extent, the decline in international commodity prices from lower global demand. Overall, the GDP impact of mitigation policies would be small in net fossil-fuel importer economies in the long term.

Examining these results in more detail, the move to a low-emission pathway implies reducing high-carbon investment, and using the freed resources to fund part of the increase in low-carbon investment spurred by mitigation policies. According to IEA (2016, 2017) estimates, the energy sector requires net investment of around 0.2% of GDP in order to reduce emissions to a level that keeps warming at or below 2°C (Figure 4.4). This includes investment in renewable energy technologies, nuclear power and energy efficiency, and the lower investment in fossil-fuel supply and power transmission and distribution, due to lower electricity demand. This estimate is based on the assumption that world GDP would grow around 2% between now and 2045-50. This scenario assumes an orderly transition to a low-carbon economy and is contingent on the deployment of energy technologies that deliver net negative emissions.

The investment data used in the scenario is restricted to the energy sector. Agriculture, forestry and land use are likely to play an important role in achieving a 2°C objective, especially in countries such as Brazil and Indonesia, but the majority of infrastructure investment needs relates to energy supply and use (see Chapters 2 and 3) (OECD, 2015b; OECD, 2016b). In addition, there are limited up-to-date data for land-use investment, making their inclusion in macro-economic estimates difficult.

Further, investment in total urban and non-urban transportation infrastructure could add the equivalent of 0.2-0.3% of world GDP on average per year, according to estimates from the IEA Energy Technology Perspectives and data from the International Transport Forum. This information could not be included in the simulation exercise, however, because detailed data are not available for all scenarios and individual countries. Adding this infrastructure investment would have increased the positive GDP impact of investment in this scenario, though to a limited extent.[3]

Figure 4.4. **World net additional energy-related investments implied by a transition from current policies to a 50% 2˚C probability scenario**

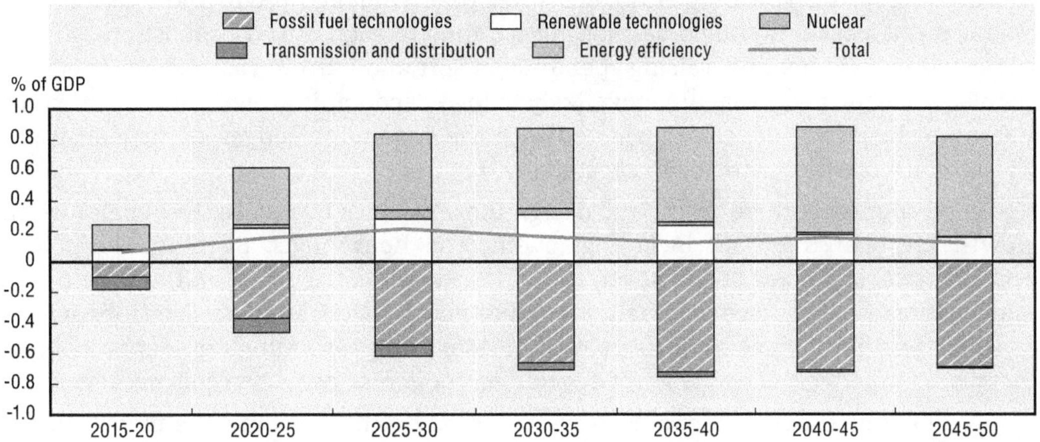

Note: based on Current Policy Scenario and 50% 2°C Scenario.
Source: IEA 2016.

Studies based on integrated assessment models have concluded that mitigation policies will have a negative impact in the long term (IPCC, 2014). Once economies adjust to new policies, however, the cost in terms of lower material living standards is low (OECD, 2015c). Other studies have suggested that limiting global warming to 2°C with 50% probability would entail global consumption losses of 2-6% by 2050; this would amount to a small fraction of the consumption gains in the context of continued economic growth (IPCC, 2014). The magnitude of the impact depends crucially on assumptions about availability and costs of low-emission technologies, the degree of market flexibility and the options for recycling carbon tax revenues.

Box 4.2. Modelling exercises for growth and climate policy analysis

A range of modelling approaches has been used to assess the effects of climate mitigation policies on economic growth (see Clarke et al., 2014). Differences in results arise, among other things, from the choice of modelling approach. Computable general equilibrium models such as ENV-Linkages have been used for instance in OECD (2009). Other studies have used macro-economic models accounting for short-term market failures, such as E3me used in IRENA (2017) to estimate GDP and employment effects of a low-carbon scenario based on extensive deployment of renewable energy and energy efficiency. Other differences arise from key parameter assumptions (e.g. the cost of low-carbon technologies, endogenous or exogenous technical change, the degree of crowding out of investment); and the choice of policy instruments (carbon pricing alone or in combination with low-carbon technology support, energy efficiency measures, etc.).

Most economic analyses to date have focused primarily on climate policy. The "decisive transition" scenario in this report is driven by the need to consider climate policy in the current macro-economic context of low-growth, low productivity growth, under-investment and low interest rates. It also broadens the policy tool-kit to include dedicated fiscal initiatives (beyond the carbon tax, used for mitigation purposes) and structural reforms that are aligned with both the growth imperative and the requirements of the transition to low-emission, climate-resilient economies.

A decisive transition would spur growth while limiting climate change

Complementing climate change policies with a combination of a fiscal initiative and structural reforms would help achieve both climate and growth objectives. The fiscal initiative comprises spending or tax measures that will foster productivity in the medium to long term. Measures should be chosen according to each country's most pressing needs and could include not only raising spending on soft and hard infrastructure or education, but also reducing taxes that are most likely to lower economic growth, such as income tax. Measures should also be closely aligned with the general objectives of the transition to a low-carbon, climate-resilient global economy.

In all countries, there is scope to design these policies to ensure the benefits of higher growth are shared by all, including low-income households. Such measures include decreasing labour taxes at the lower end of the wage distribution, and improving access to education, health, low-cost quality housing and public transport. Priorities will differ depending on individual countries (OECD, 2016a, 2017). Spending on adaptation policies that would improve the resilience of economies could also be part of this package.

To illustrate the impact of a decisive transition on the economy, the model simulation presented below assumes that in addition to mitigation policies, countries put in place a pro-environment, pro-growth policy package that combines fiscal and structural measures.

This approach differs from previous work on the issue in that it places climate policy actions within the current macro-economic context of low-growth, low productivity growth, under-investment and low interest rates, and broadens the policy tool-kit to include dedicated fiscal initiatives (beyond the carbon tax, used for mitigation purposes) and structural reforms that can support the low-carbon transition. The policy package used in the simulations includes:

- A fiscal initiative that corresponds to an increase in public investment of 0.5% of GDP. This represents a larger increase than in the pure mitigation scenario in countries that are net fossil fuel importers, as there is a need to invest more to compensate for the disinvestment triggered by decarbonisation policies. In many countries, such a package could be deficit-financed for a few years, before turning budget-neutral. OECD analysis suggests that thanks to low real interest rates, OECD member countries could afford to finance a fiscal initiative equivalent to 0.5% of GDP per year for about three to four years, on average (Mourougane et al., 2016). After this period, reallocating spending and taxation to the most growth-friendly and equity-enhancing measures would help to free up resources (see Fournier and Johansson (2016) for examples). Assuming it takes the form of an increase of good-quality public investment, such an initiative would leave the public debt-to-GDP ratio unchanged in the long term. These measures are also to be aligned with low-emission and climate-resilience objectives.
- Changes in R&D spending that would be needed at the world level to achieve a 50% 2˚C scenario. Estimates have been derived from Marangoni and Tavoni (2014), assuming all countries act collectively. The impact of R&D spending on total factor productivity draws on recent OECD analysis (Egert and Gal, 2016).
- Reforms to make product-market regulation more conducive to competition and market entry, essential to facilitate the transition. The impact of product-market reforms on long-term output is based on new OECD analysis on the impact of selected structural reforms for both OECD member countries and emerging economies (Egert, forthcoming ; Egert and Gal, 2016). The measure is calibrated using past observations of reform changes. In practice, the reform is assumed to be more ambitious in emerging economies than in OECD member countries, explaining why the resulting output impact is larger for emerging economies. A more flexible regulatory environment, for example as measured by the OECD product-market regulation (PMR) indicators, reduces the cost of the transition to a low-carbon economy. In particular, new results show it can reduce the negative effects from higher fossil fuel prices on business investment. In countries with most flexible product markets, the effect of higher end-user fossil-fuel prices on investment seems neutral or could even be positive (Box 4.3).

The simulations indicate that selected pro-growth policies can offset the negative impact of mitigation policies aimed at limiting global warming to 2°C with 50% probability, showing that combined climate and growth policies can be, on average across countries, good for growth, in both the long term and the short term. After five years, average gains in output for G20 economies would amount to around 1%, thanks to well-aligned pro-growth policies (Figures 4.5 and 4.6). Estimates suggest that those effects could raise long-run output by up to 1-4% in most of the large advanced economies and emerging economies by 2050. Those gains would be just below 3% by 2050 for G20 countries. The detailed results for the country types are provided in Table 4.1.

An important part of the output effect is an overall boost in investment, including in low-emission infrastructure, by 0.5% of GDP, increasing long-term output by up to 2% in advanced economies and emerging economies by 2050. Medium-term gains would amount to 0.2-0.5 % in most countries.

The countries that would experience the highest gains are those where the initial stock of public capital is lowest, and those where long-term unemployment is high. Net fossil-fuel exporters would benefit the most from the initiative, which is supposed to offset the negative impact of the disinvestment caused by mitigation policies. In a few economies that are heavily reliant on fossil fuels, a typical pro-growth package may not be sufficient to fully compensate for mitigation cost in the short term. Designing the policy package that would best fit each country's needs, however, would have a stronger counterbalancing effect.

Adding reforms that favour innovation and growth (such as basic public R&D and structural reforms) to the fiscal initiative would increase the output gains. Indeed, implementing structural reforms can enhance output and lower the public-finance impact of an increase in public investment, through their gradual effect on total factor productivity and potential output. In particular, reforms aimed at removing barriers that hold back demand for investment, such as improving the design of regulations to reduce unnecessary burdens on entrepreneurship, can markedly boost output in the long term through their effect on total factor productivity.

Figure 4.5. **Contribution of selected pro-growth and mitigation policies in the G20**
(50% probability of achieving 2°C)

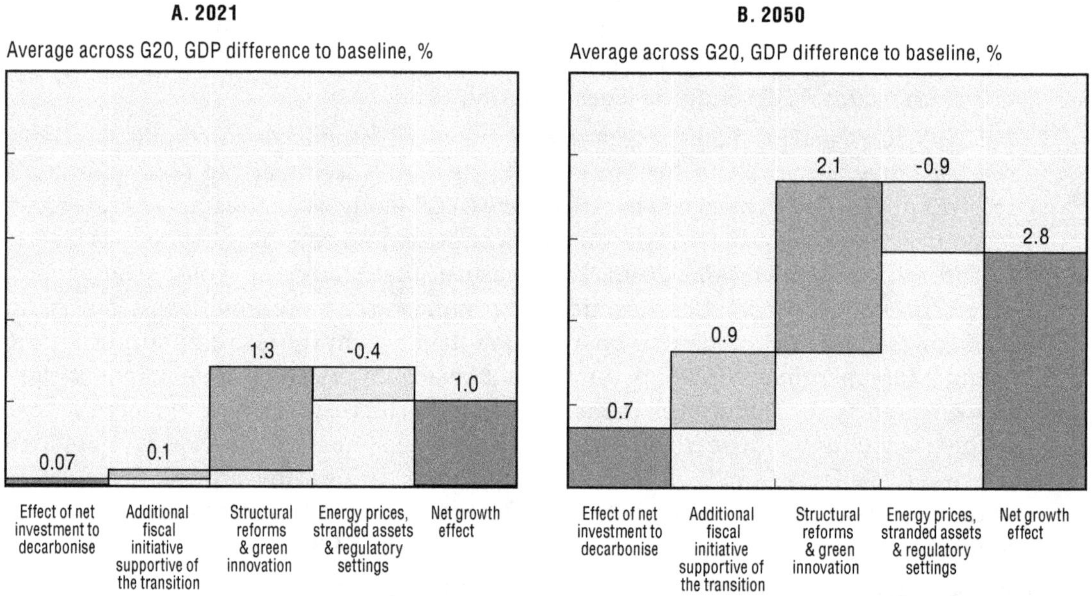

Note: The average G20 is a weighted average of selected G20 economies, representing 88% of the G20 countries (i.e. excluding the European Union). Energy prices and stranded assets are based on the IEA scenario and correspond to a move to a 2°C climate objective with a 50% probability. Regulatory setting captures the reduced costs of the transition in a more flexible regulatory environment, based on firm level investment regression that interacts energy price inflation with product-market regulation settings. Fiscal initiative corresponds to an increase in public investment that complements the net investment from decarbonisation so that in total, investment would increase by 0.5% of GDP. This means that net fossil fuel exporters who experience disinvestment from mitigation policies are assumed to invest more to compensate for this disinvestment. The structural reform considered here is a lowering in barriers to investment by 0.35 point for the OECD member countries and by 0.85 point for emerging economies, which correspond to the average change in this measure in the past. The impact has been calculated using estimation of business regulation on income per capita by Égert (forthcoming) and Égert and Gal (2016). Innovation corresponds to the increase in R&D spending necessary to reach a 2°C scenario using estimates from Marangoni and Tavoni (2014). It is assumed that the stylised fossil fuel exporters recycle their carbon tax revenues into higher public consumption in the medium term, given their initial low level of public debt. No recycling is assumed for the net importers.

Figure 4.6. Net growth effect of selected pro-growth and mitigation policies in stylised economies

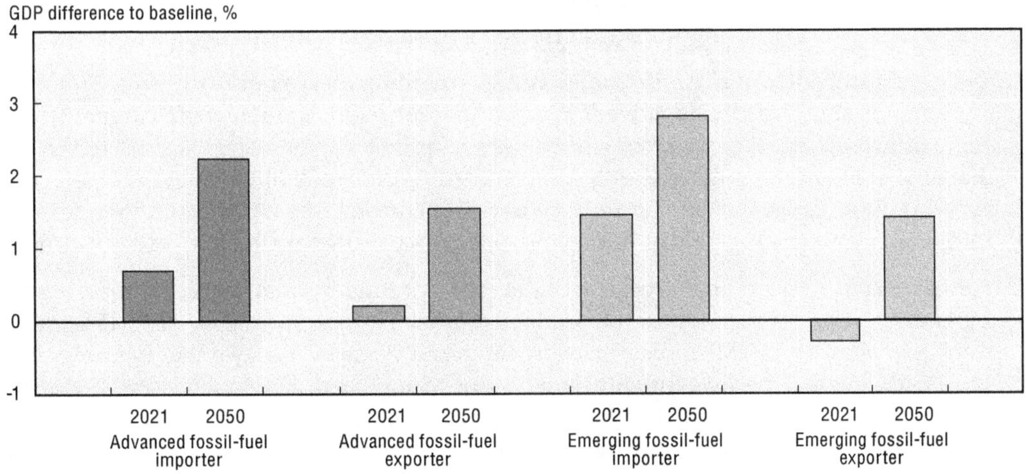

Note: See the Note for Figure 4.5.

Table 4.1. Contribution of selected pro-growth and mitigation policies in stylised economies and the G20

GDP difference to baseline, %

Policy component	Advanced fossil-fuel importer		Advanced fossil-fuel exporter		Emerging fossil-fuel importer		Emerging fossil-fuel exporter		G20 average	
	2021	**2050**	**2021**	**2050**	**2021**	**2050**	**2021**	**2050**	**2021**	**2050**
Effect of net investment to decarbonise & additional fiscal initiative supportive of the transition	0.4	1.5	0.3	1.1	0.0	0.7	0.2	1.6	0.2	1.6
Structural reforms & green innovation	0.7	1.3	0.6	1.4	1.7	2.6	1.9	5.1	1.3	2.1
Energy prices, stranded assets & regulatory settings	-0.3	-0.6	-0.7	-0.9	-0.3	-0.5	-2.4	-5.3	-0.4	-0.9
Net growth effect	**0.7**	**2.2**	**0.2**	**1.6**	**1.5**	**2.8**	**-0.3**	**1.4**	**1.0**	**2.8**

Note: See the note for Figure 4.5

Box 4.3. A regulatory environment that encourages competition and firm entry improves firms' investment and innovation response to climate change mitigation action

Climate change mitigation action requires flexibility: old technologies and infrastructure need to be replaced by new ones. New OECD research undertaken for this project shows that governments need to provide a flexible regulatory environment – which does not restrict firm creation, market entry or competition – to encourage private investment and innovation and thus make the most out of the low-carbon transition (Annex 4.A4). Previous research has already shown that such a flexible regulatory environment can boost productivity, investment and employment across the income spectrum.

The new OECD research undertaken for this project has econometrically estimated the response of firms' investment to higher energy prices, taking into account the regulatory environment in which firms operate, as measured by the OECD's Product Market Regulation (PMR) indicator. Pricing carbon emissions in energy use is a key element of policies to lower CO_2 emissions. Carbon pricing results in higher energy prices. The new research shows that the effect of higher energy prices on manufacturing investment tends to be neutral

Box 4.3. A regulatory environment that encourages competition and firm entry improves firms' investment and innovation response to climate change mitigation action *(cont.)*

or even significantly positive if product market regulations are not restrictive. By contrast, the effect is significantly negative if product market regulations restrict competition and entry of new firms. The research can be seen as providing tentative support to the claim that environmental regulations may induce firms to innovate and improve efficiency, thus boosting productivity – the "Porter hypothesis" (Porter, 1991; Porter and van der Linde, 1995) – provided the regulatory framework encourages competition and entry of firms.

The estimated effect of an increase in the energy price index on manufacturing investment depends on the restrictiveness of product market regulation (Figure 4.7). If regulatory restrictiveness is low (PMR indicator below about 1.5), a rise in energy prices has a significant and positive effect on investment: firms adapt to higher prices by boosting investment. By contrast, if regulatory restrictiveness is high (PMR above about 2.3), a rise in energy prices has a significant and negative effect on investment. For example, in a country with restrictive regulation (PMR of 2.5), a typical firm's investment would diminish by about 1% in response to an increase of energy prices of 10%. In a country with competition-friendly regulation (PMR of 1), a typical firm's investment would rise by about 1%. The OECD has recorded PMR values of 2.5 or higher for several emerging economies among the G20. A PMR value of about 1 has been recorded for the United Kingdom.

Figure 4.7. The effect of energy price inflation on investment depends on product market regulations

Effect of a one per cent rise in energy price inflation on investment to capital ratio, in per cent

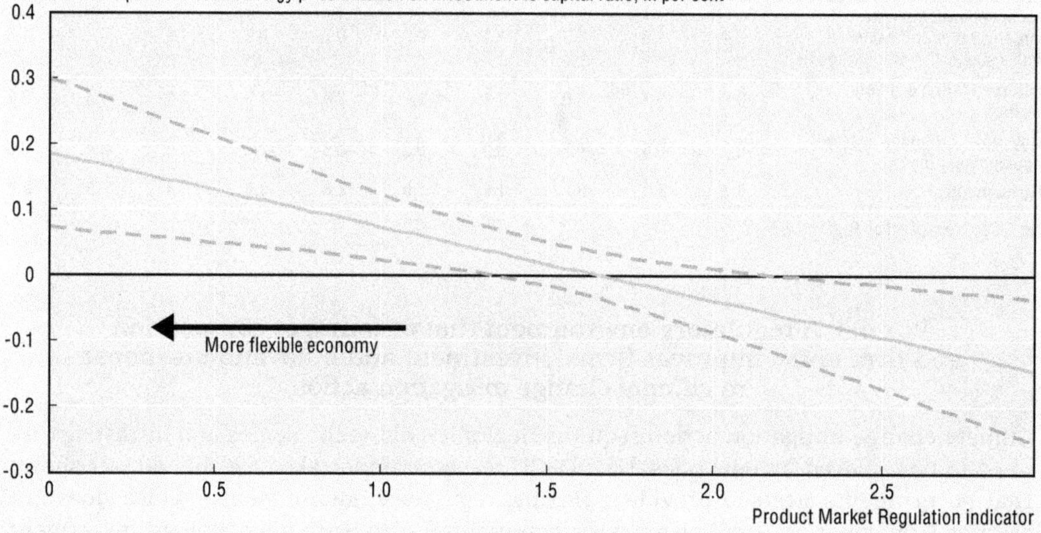

Note: The econometric model described in Annex 4.A4 is built on a baseline used in Dlugosch and Koźluk (2017). It includes country-year, industry-year and firm-fixed effects, a measure for stringency in labour market index, a measure of access to finance, sales over total capital, lagged out gap and lagged real interest rates as further controls. Energy price inflation is the three-year moving average of changes in the energy price index. Time sample: 1999-2011. The dashed line indicates the 95% confidence interval, using firm clustered standard errors.

The fiscal initiative in the decisive transition would also trigger an increase in business investment, especially in countries where investment needs are the highest (Figure 4.8). Business investment in the average of selected G20 economies could rise by almost 4% by 2050, according to the Yoda model. Simulations from the Oxford model would point to smaller increases on average in G20 economies. One condition underlying those positive outcomes is that governance and framework conditions are good enough to mobilise business investment. Policies to achieve this outcome are discussed in Chapter 5.

Figure 4.8. **Business investment impact of a decisive transition to decarbonisation**

Difference to baseline, per cent

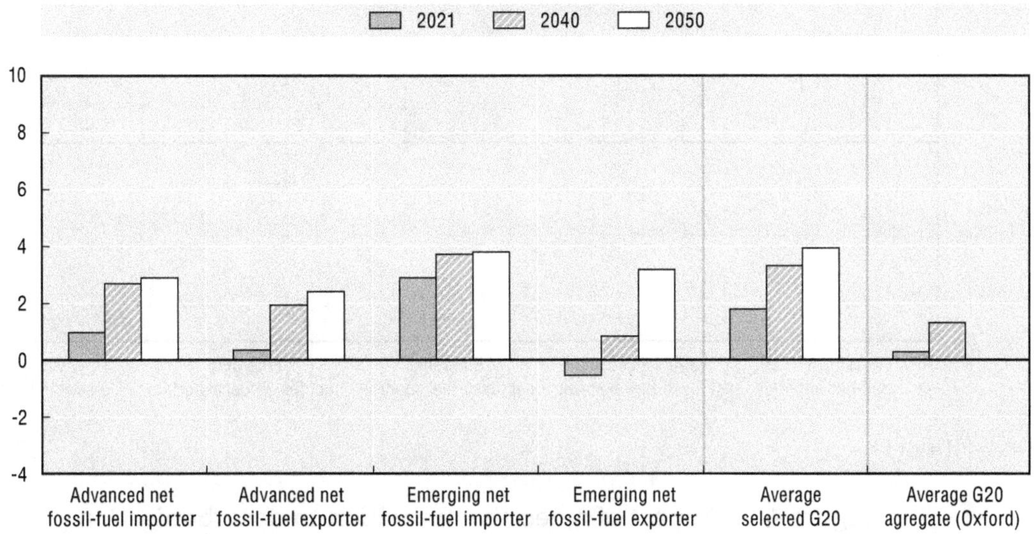

Note: The average G20 is a weighted average of the G20 economies covered in the analysis, which represent 88% of the G20 countries (i.e. excluding the EU).

Consistent with growth and real wages developments, countries are expected to experience sizeable gains in employment (Figure 4.9). Gains would amount to 0.2% in the average of G20 countries after five years. Those gains would mostly come from the additional investment and structural reforms. Long-term developments in employment and gross employment reallocation effects from those policies are examined in more detail later in this chapter.

Finally, the likely impact of a decisive transition on public finances is expected to be small, assuming investment needs will be financed by the deficit for three years and be budget-neutral thereafter (Figure 4.10). Over the medium term, the ratio of public debt-to-GDP would fall compared with a no-policy-change scenario in net fossil-fuel importers, reflecting mostly gains in carbon tax revenues. In some countries, output gains more than finance the initial fiscal impulse. The ratio of public debt-to-GDP could fall by 5-7 percentage points compared with the baseline scenario in the average of selected G20 economies after five years, and by up to 20 percentage points by 2040.

Overall, these estimates rely on very specific assumptions and should be interpreted with care. A crucial assumption is that governments invest in good-quality projects and that the fundamental framework conditions are in place to get the most out of these investments (see Chapter 5 for a discussion of policies). Furthermore, a typical policy package has been simulated in all countries, for practical reasons. On the one hand, choosing the composition of this package in light of each country's institutional and regulatory frameworks, as well as its social preferences, would certainly be the most effective way of maximising the impact on output. On the other hand, poor choice of policy settings and ineffective implementation and governance of reforms would lower the output impact of the policy package. In addition, the simulations presented in the chapter do not account for the political economy of reforms and any difficulties in ensuring reform acceptance (see Chapter 6).

Figure 4.9. **Impact of a decisive transition on employment**

Difference to baseline, per cent, 2021

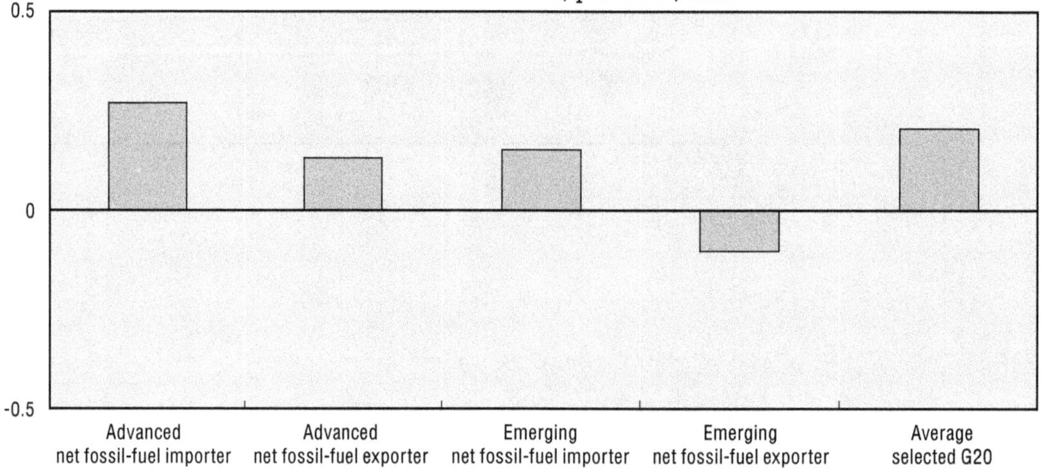

Note: see Figure 4.8.

Figure 4.10. **Impact of a decisive transition to decarbonisation on the debt-to-GDP ratio**

Difference to baseline, percentage points

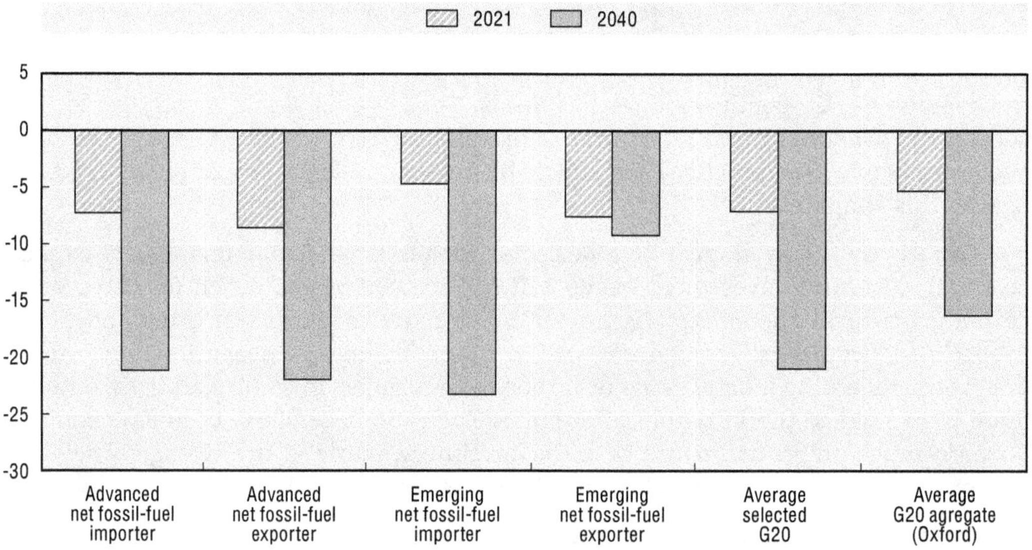

Note: The average G20 is a weighted average of the G20 economies covered in the analysis, which represent 88% of the G20 countries (i.e. excluding the EU).The net fossil fuel exporter takes the example of a country where public debt is low and carbon tax revenues are recycled into additional public consumption in the medium term. No recycling is assumed in the other stylised economies.

Avoided climate change damages bring additional economic gains

The simulations presented above do not account for the costs associated with climate change damages, which would weigh on long-term economic growth, through lost output as well as reduced well-being. The mitigation effort in the decisive transition would significantly reduce these long-term risks. Accounting for avoided climate damages enhances the output impact of the decisive transition, as damages affect the long-term capacity of the economy (Figure 4.11). Those damages depend heavily on countries' locations (damages are usually larger in countries close to the Equator), geographies and economic and social structures. The effects of climate damages on output are estimated to be small in the short to medium term but more marked in the long term and in some emerging economies. Evidence from the literature suggests that climate change damages could have a disastrous effect after 2050, beyond the scope of the analysis considered here (IPCC WGII, 2014; OECD, 2015d).

Figure 4.11. **Effect of including avoided damages in a decisive transition scenario in 2050 (50% probability of meeting 2°C)**

Average of selected G20 economies

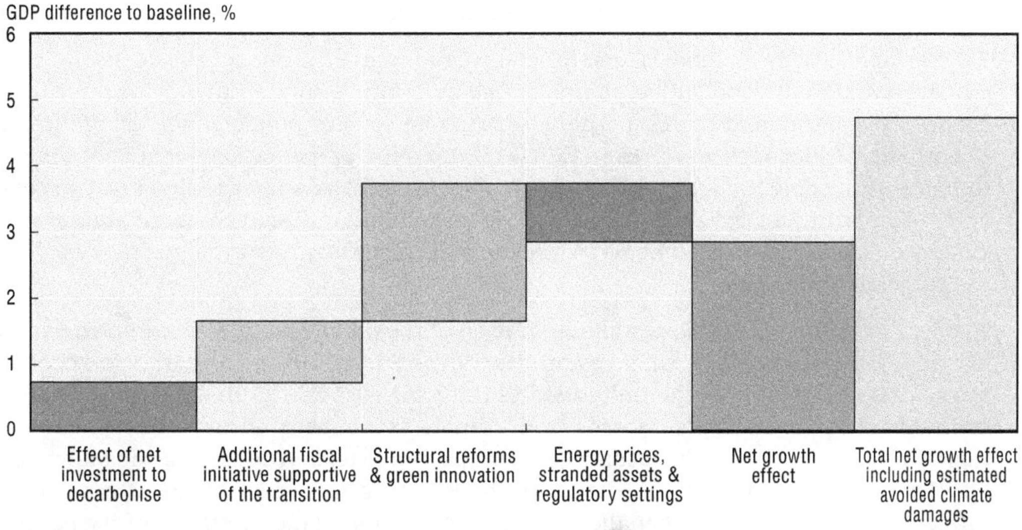

Note: See Figure 4.5.

It should be noted that the scale of climate damages is very hard to gauge with standard modelling tools (Box 4.4). It has been computed in this report using simple rules for the baseline and decisive transition scenarios. The emissions profile that is consistent with the Yoda model is used to calculate the average expected global temperature increase by 2100, using the MAGICC model (Meinshausen, Raper and Wigley, 2011). The resulting average global temperature increases from pre-industrial levels are 4.3°C in the baseline scenario and 1.6°C (with a 50% chance of remaining below 2°C) in the decisive transition scenario. These estimates correspond to the upper range of the global temperature sensitivity as used in Clarke et al. (2014). Global damages associated with these temperature increases have been calculated using the climate damage function in Nordhaus (2016) and correspond to a very small subset of likely climate-change damages to the economy, excluding in particular extreme events (Box 4.4). Those global estimates are then distributed across regions and countries following OECD (2015d).

Box 4.4. The challenge of estimating the consequences of climate impacts

Economic analyses do not yet adequately capture the full range of climate impacts and should not be seen as providing the complete picture of the costs of climate change impacts (for a discussion see e.g. Stern, 2013). Some of the main challenges include:

- *Uncertainty:* The impacts of climate change depend upon the interactions between natural systems, socio-economic changes and the severity of temperature changes, all of which are subject to considerable uncertainties in estimating future impacts. These interactions can be complex and non-linear.
- *Modelling assumptions:* Decisions about the extent of autonomous adaptation, macroeconomic impacts and the weight to place on future losses can all significantly affect the results produced.
- *Data gaps and modelling constraints:* Some impacts, such as changes in climate extremes, are subject to very limited available data, but could be a significant source of future losses from climate change. In addition, non-market impacts (such as impacts on biodiversity) are not well captured in existing models.

Furthermore, economic models tend to struggle to capture the impacts of localised extreme climate events on global value chains, combined with a limited ability to project their frequency, severity and location. The latest climate science projects increasing incidence of episodes of high temperatures (IPCC, 2013). Extreme precipitation events are projected to increase in some regions, while rising sea levels will also increase flood risk in coastal areas. The processes governing cyclones are particularly difficult to model, but available evidence suggests that cyclone wind speeds will increase while cyclone frequency stays the same or diminishes.

Historical experience provides some indication of the potential economic impact of future extreme weather events. A single event, Hurricane Sandy, led to 43 deaths and economic losses of USD 50 billion in the United States (City of New York, 2013). In general, wealthier countries tend to suffer larger losses from climate extremes in absolute terms, due to the higher value of assets at risk, but smaller in proportional terms (Cummins and Mahul, 2009; Bosello and Dasgupta, 2015). Evidence on the longer-term impacts of disasters is mixed, reflecting both measurement challenges and the counterbalancing effects of the economic stimulus from reconstruction activities. Lis and Nickel (2010) found that natural disasters lead to median GDP being 4% lower five years later in developing countries. Disasters do not appear to have an impact on measured growth in OECD countries. Meanwhile, Cavallo et al. (2013) found that even extremely large disasters do not display a significant long-term impact on economic growth, unless they are followed by a "radical political revolution".

The ambiguous evidence on GDP impacts should not hide the underlying issue that impacts on welfare are undoubtedly negative. First, reconstruction activities are recorded as additional value-added, although they may merely replace destroyed capital stock. Second, the poor tend to bear the brunt of climate-related disasters. Their economic losses are smaller in absolute terms but have a disproportionately negative impact on welfare (Hallegatte et al., 2017). Third, only a subset of the impacts from extreme events is included in GDP, with impacts such as deaths and injuries only being captured indirectly. These costs are predominantly borne by developing countries. For example, between 1970 and 2008, 95% of deaths from natural disasters occurred in developing countries (Handmer et al., 2012).

Avoided damages from decisive action would appear more markedly in the second half of the century, when increases in global temperature diverge between a business-as-usual scenario and the decisive transition scenario. With emissions reaching net-zero in the second half of the century, damages to GDP would hardly increase in the decisive transition scenario, at about 1% of GDP, while upper estimates without climate action show a rapid increase towards 10 to 12% annually on a global scale by 2100 (OECD; 2015d; Nordhaus, 2016; Weitzman, 2012), with much more pronounced impacts for the most vulnerable regions.

The results are subject to several caveats:

- The method does not include potential co-benefits such as reduced air pollution, which could alter the macroeconomic impact of a transition to a low-carbon path via their effect on health and productivity (see Chapter 3).
- Non-market damages are captured in a very crude manner, through a 25% increase in the estimated damages.
- Extreme events and their possible systemic effects are highly uncertain and difficult to quantify (Box 4.4).
- Uncertainties surrounding market damage estimates are large, reflecting uncertainties that occur in every stage of the process of calculating damages (Clarke et al., 2014; OECD, 2015). Results are highly dependent on assumptions (e.g. the rate of economic growth in different countries; or when certain technologies will come online). Alternative measures of climate damages, such as the social cost of carbon (the economic cost caused by an additional tonne of CO_2) are also subject to high uncertainties and depend in particular on the assumption made on the rate of time preference, as part of the discount rate (Nordhaus, 2016; Anthoff and Tol, 2013; Greenstone, Kopits and Wolverton, 2013).
- The macro-economic adjustment effects of climate change due to changing relative prices and marked differences in sectoral labour productivity are not covered. Should climate change induce structural shifts toward the less productive sectors, climate change damages would be exacerbated by the sectoral and temporal reallocation of factor inputs (Kalkuhl and Edenhofer, 2016).

Pursuing a more ambitious climate scenario: 66% probability of limiting global warming to 2°C

Limiting warming to 2°C is not enough to satisfy the objectives of the Paris Agreement. While it is difficult to precisely define what "well below 2°C" and "efforts to limit to 1.5°C" mean, a step towards a more ambitious scenario can be described in which more stringent action raises the probability of holding warming below 2°C from 50% to 66% (see IEA, 2017, for policy and technology details). This scenario will require more investment effort at the global level, in response to more ambitious GHG emission-reduction policies than in the 50% 2°C scenario. Simulations presented here suggest that this more ambitious scenario can still deliver positive output outcomes, provided that mitigation is accompanied by strong pro-growth reforms.

The net impact on growth will depend on the relative changes of a range of factors. More stringent environmental policies and higher stranded assets in this scenario than in the 50% 2°C path will have a stronger dampening effect on the productive capacity of economies. Estimates suggest that mitigation costs could be about three times higher in a 66% 2°C scenario than in a 50% 2°C scenario (Hof et al., 2017). The need for more investment in a 66% 2°C scenario will boost growth, however. Recycling of carbon revenues, which will be bigger in the 66% scenario, could also offset some of the mitigation costs. More importantly, the benefits from avoided damages from climate change would grow from the 50% 2°C scenario, although more markedly so in the second half of the century, and with the above mentioned uncertainties.

Assessing the impact of a transition to a 66% 2°C scenario is challenging, as it represents a non-linear step change from the pathway to reach 2°C with a 50% probability – energy-related CO_2 emissions in 2050 would need to be roughly halved from their level under the 50% 2°C projection (IEA, 2017).

- First, the marginal returns of the additional investment required in the energy sector will most likely be lower, as all the easily attainable benefits will already have been grasped in a 50% 2°C scenario.
- Second, reaching a 66% 2°C target will require substantially more action on emissions sources related to energy use (for instance, retrofitting existing coal or gas power plants with carbon capture and storage; significant structural changes to the transport sector). The 66% scenario will also require even more stringent action on non-energy-use emissions, including from land use, which also has implications for the energy sector because of constraints on biomass supply. As a result, the carbon taxes will need to rise more than proportionately throughout the economy, reflecting the high marginal costs of emissions reductions.
- Third, the level of stranded assets would be higher than in a 50% 2°C scenario, with a correspondingly higher impact on GDP outcomes. Although some global estimates exist for the different scenarios, reliable country estimates of stranded assets are not publicly available (see Chapter 3). The timing of the stranding of these assets is also uncertain.
- Finally, the availability, cost and future performance of key technologies has an important role in achieving ambitious climate targets. In many studies, remaining below the 2°C target requires net global emissions in 2100 to be zero or negative (Dessens, Anandarajah and Gambhir, 2014). This requires relying on "negative emissions" technologies such as biomass energy with carbon capture and storage (BECS) to counterbalance unavoidable GHG emissions. Research is under way to better understand these different factors (see Chapter 2).

There has been little analysis of the outcome of a 66% 2°C scenario and considerable uncertainties surround the estimates that are available. Most of the IPCC's Fifth Assessment Report (2014) scenarios for ambitious mitigation hover around 2°C. These show consumption losses in the long term, with higher mitigation costs for more stringent scenarios. The special IPCC report asked for by Parties to the UNFCCC as part of the Paris Agreement should provide more insights into pathways leading to a maximum temperature increase of 1.5°C.

In addition to the very large uncertainties around estimates of mitigation costs in a 66% 2°C scenario, it is extremely difficult to gauge the extent of the fiscal impetus that will be needed to offset mitigation costs in such a scenario. Such an impetus could amount to several percentage points of GDP in some economies that rely heavily on fossil fuels. There are also major uncertainties regarding the level of R&D spending that would be required to achieve this more ambitious climate target. Scenarios quantifying those needs are being developed, and will be available in the coming years.

With these caveats in mind, Figure 4.12 provides an illustration of the long-term output impact of a decisive transition to a 66% 2°C scenario. The net impact on output in 2050 is estimated to be 2.5%. The results suggest that a larger pro-growth policy package will be required to offset the additional mitigation costs of the more stringent climate policy required to meet the goal. Incorporating the output impacts from avoided climate damages provides an additional boost to growth in the long term, with a total increase in output of 4.6% above the baseline in 2050.

Figure 4.12. Illustration of the output impact of decisive transition to a 66% 2°C scenario in 2050, including avoided damages

Average of selected G20 economies

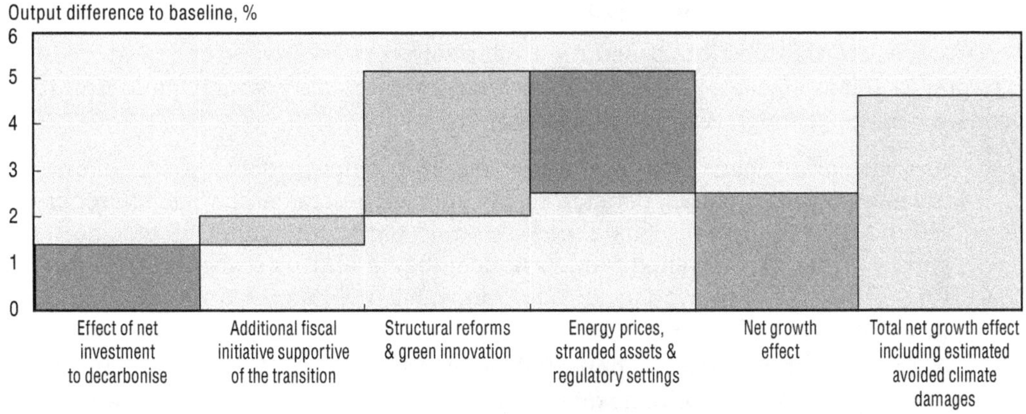

Note: The average G20 is a weighted average of selected G20 economies, representing 88% of the G20 countries (i.e. excluding the European Union). Energy prices and stranded assets are based on the IEA scenario and correspond to a move to a 2°C climate objective with a 66% probability. Regulatory setting captures the reduced costs of the transition in a more flexible regulatory environment, based on firm level investment regression that interacts energy price inflation with product-market regulation settings. Fiscal initiative corresponds to an increase in public investment that complements the net investment from decarbonisation so that in total, investment would increase by 0.5% of GDP. This means that net fossil fuel exporters who experience disinvestment from mitigation policies are assumed to invest more to compensate for this disinvestment. The structural reform considered here is a lowering in barriers to investment by 0.35 point for the OECD member countries and by 0.85 point for emerging economies, which correspond to the average change in this measure in the past. The impact has been calculated using estimation of business regulation on income per capita by Égert (forthcoming) and Égert and Gal (2016). Innovation corresponds to a 0.1% of GDP increase in R&D spending. It is assumed that the stylised fossil fuel exporters recycle their carbon tax revenues into higher public consumption in the medium term, given their initial low level of public debt. No recycling is assumed for the net importers. For damages, simulations presented here include only a subset of potential damages, excluding for instance damages from extreme climate events, due to difficulties in projecting their frequency, severity and location. The exercise models global damages associated with temperature increases, using the Nordhaus (2016) damage function.

Acting now and acting together: implications of a delayed transition and of actions limited to a leadership group

The costs of delayed action

The decisive transition scenario requires a rapid change in the direction of investment, based on a set of incentives that may be disruptive at some level in the near term, even if combined with a broader economic package for growth. Faced with potential short-term costs, countries may be tempted to delay action to ward off the long-term climate threat, despite the availability of policies to manage adverse impacts on industry, labour markets, households and communities (Chapters 5 and 6).

The scenario presented in this section assumes the decisive transition (i.e. action and investment to limit global warming to 2°C with a probability of 50%) is delayed to 2025, at which point a reassessment of climate risks triggers an abrupt transition toward decarbonisation. The delay in investing in decarbonisation increases costs, as more stringent climate policies have to be introduced more rapidly, leading to more stranded assets of emission-intensive activities that for which investment continued up to 2025 under a less ambitious climate policy outlook. The scenario relies on the IEA global estimates of around USD 310 billion of additional stranded assets in the upstream oil sector alone (IEA, 2016). This amount is distributed across countries, over the first 10 years following the delay, following the methodology outlined in Glades and Ekins (2014).

Delaying action would be costly for all countries (Figure 4.13). The average output loss for selected G20 economies would amount to 2%, with most of the loss incurred a year after the delayed transition starts. Losses would be particularly marked in net fossil fuel exporters, with a significant amount of additional stranded assets compared with a non-delayed scenario. Broadening the scope of stranded assets beyond the upstream oil industry would also result in higher losses – coal-based power generation in particular can be a significant share of stranded assets in mitigation scenarios (see Chapter 3 for IEA, 2017 and IRENA, 2017 estimates).[4]

In addition, this result is based on a relatively conservative scope of stranded assets (Chapter 3). Capital losses may also trigger financial instability, which could lead to further economic losses through two principal channels:

- A stock market channel: stock market instability through the exposure of institutional investors such as banks, pension funds and insurance companies to stocks of listed oil and gas companies. This exposure could harm economies as a tightened credit supply – due to weakened bank balance sheets and increased volatilities – would further hamper investment. In addition, household exposure to stocks of affected companies could lead to a decrease in savings with negative effects on consumption.
- A debt and loan channel: potential default by some affected companies on their fixed income and bank loans due to capital losses, with similar implications for bank balance sheets and credit supply as the stock market channel.

To illustrate the stock market channel, a full write-down of companies is simulated in the oil and gas sector on domestic equity markets. Although exposure of stranded assets of firms in climate-sensitive sectors is not likely to lead to a full write-down of all assets, a 100% shock to market capitalisation provides an upper bound estimate. This choice is consistent with Battiston et al. (2016) who stress test the EU banking system in order to identify systemic impacts of climate change policies.

The magnitude of additional economic losses triggered by financial instability from capital losses from stranded assets would depend on countries' reliance on fossil fuels and the extent of market capitalisation, with sizeable impacts limited to the short term.

Overall these results are in line with existing literature which also points to significant cost of delaying action and a trade-off between reduced short-term costs and higher economic adjustments caused by continued carbon lock-in. This implies the need to sharply increase carbon prices and to introduce more stringent regulations as governments seek to return to a 2°C-compatible pathway (Bosetti et al., 2009; Kriegler et al, 2015; Jakob et al., 2012). Inertia in the extraction sector is also found to be a factor affecting the benefit of climate change action and thus strengthens the case for early action (Bauer et al., 2016).

Figure 4.13. **Macro-economic implications of delaying action on climate (without growth-enhancing policies)**

GDP difference to 50% 2°C scenario

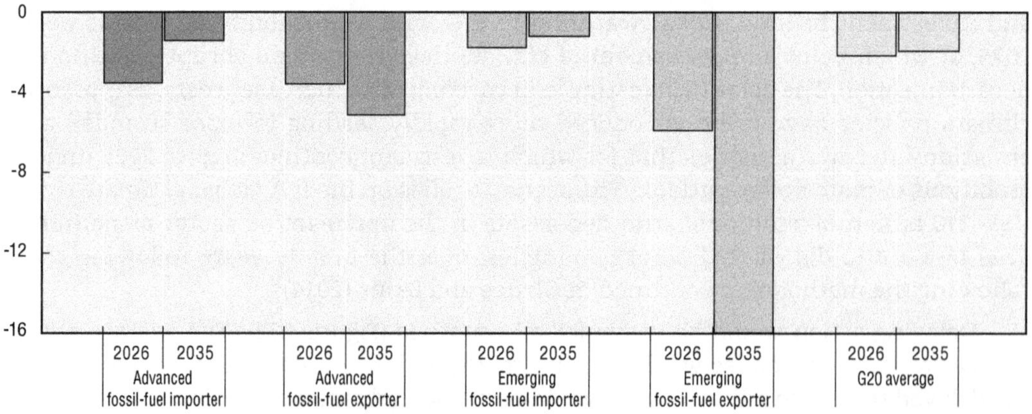

Decisive action taken by a leadership group of countries

Decisive action on climate mitigation, beyond the Nationally Determined Contributions submitted for the Paris Agreement, may also be taken by a group of leading countries, even if multilateral action is lacking. Such action would require bigger structural adjustments than the multilateral case, as emission-intensive activities in coalition countries may move to other countries. This would reinforce the case for accompanying structural reforms to boost growth and facilitate economic adjustment in the leadership coalition countries (see above and Chapter 5). Countries outside the leading group may face more stranded assets later, and the costs of delayed action. Achieving climate change mitigation objectives would be compromised, resulting in substantial long-term costs and risks. Such a leadership group scenario cannot be accurately modelled, but the following discussion analyses qualitatively the likely forces at play.

Collective action is needed to mitigate climate change. Countries may be tempted to "free ride", however, enjoying the benefit of climate change mitigation while limiting their own mitigation efforts to avoid the potential costs or loss of competitiveness. There is also an incentive to wait for low-carbon technologies to become less expensive. A group of countries may nevertheless choose to take the lead even if others do not: leader countries can gain from the co-benefits of climate change mitigation action, especially the positive impacts of lower air pollution for human health (see Chapter 3 and Tirole, 2012). Some large countries, in particular China, bear a significant share of the cost of global warming and may therefore have a stronger incentive for action. In addition, public opinion may put environmental concerns high enough on the political agenda to induce governments to take the lead. Leader countries may choose to act without an explicit agreement or in a coalition; there is a vast literature investigating the rationale for building environmental coalitions, their credibility and enforcement, and mechanisms to increase their breadth and stability (Barrett, 1994; Barrett, 2003; Nordhaus, 2015; Dellink, 2001; Finus et al., 2006). Annex 4.A1 looks in more detail at some of the mechanisms resulting from a group of countries taking the lead in terms of climate mitigation.

Evidence suggests that countries taking the lead on climate change may not suffer in aggregate macro-economic terms, particularly in the medium term (Dechezleprêtre and Sato, 2015). More stringent environmental policies do not seem to affect aggregate competitiveness, but do have small but significant effects on sectoral specialisation: countries with stringent environment policies (or high energy prices) tend to have an advantage in less pollution-intensive exports, after controlling for other factors such as capital intensity, endowments or geography (Koźluk and Timiliotis, 2016). Most technologically advanced firms are also more likely to further increase their productivity. Less advanced firms, especially those relying on carbon-intensive production, would either need to increase investment to reorient their production or exit the market (Albrizio et al., 2017). Caution is required when extrapolating these results to unprecedented increases in environmental stringency. However, a regulatory environment that is conducive to competition, entry of new firms and smooth reallocation of resources from exiting firms and activities makes it more likely that the economy will respond through higher innovation and productivity (see Box 4.3 above).

Conversely, if climate action is limited to a leadership group, high-emissions production in countries outside the group is likely to increase, as it would benefit from less stringent climate policies, lower associated costs (e.g. carbon taxes) and lower global fossil-fuel prices. Empirical evidence, as cited above, suggests these production cost differences may not be large. Nevertheless, the more easily production in the leadership coalition can be substituted by production in other countries (e.g. of energy-intensive manufacturing goods traded in world markets), the stronger the competitiveness gains of opt-out countries in carbon-intensive sectors. In the longer term, however, if the opt-out countries decide eventually to pursue mitigation, they would face the risk of substantially more stranded assets as well as higher adjustment costs. In the meantime, their populations would also suffer the health costs of higher local pollution and environmental degradation.

Attaining global climate goals in a leadership group scenario would require larger climate mitigation efforts within the leadership coalition than in the case of a global decisive action scenario, due to the lack of action in the opt-out countries. Hence, such a scenario can only be realistic if the coalition is sufficiently large. Moreover, the structural changes and associated costs would be more marked. Reforms to reduce the costs of structural change and boost technology diffusion would therefore be particularly relevant for the leading countries. Adjustment costs and redistributive effects across sectors would result in intensive lobbying against mitigation action. The leading countries could hence use carbon tax revenues for growth-enhancing tax reforms (e.g. to lower labour taxes) and to mitigate adjustment costs. Pollution haven effects could give rise to demands for tariffs (e.g. border carbon adjustments) or other restrictions on carbon-intensive imports – even if the welfare gains of freer trade are likely to outweigh the costs of abating the trade-induced leakage (Kuik and Gerlagh, 2003).

By pursuing climate mitigation policies, the leadership group would also accelerate the development of low-carbon technologies, reducing the cost of their deployment. In principle, they could gain "first-mover" advantages in these technologies, but these "learning by doing" benefits could spill over worldwide. This would have the benefit of reducing "carbon leakage" (Castelnuovo et al. 2005). This would result in lower emissions and reduce deployment costs for other countries through knowledge and technology. The positive technological spillover benefits from trade could even dominate the carbon leakage effect (Gerlagh and Kuik, 2014).

Structural and employment composition effects of low-emission pathways

The decisive transition requires advances in technology and policy measures to promote less polluting economic activity; these will inevitably result in structural change. The evidence on existing policies to foster decarbonisation, such as the Emissions Trading System in the European Union, suggests that the sectoral employment effects have been small (OECD, 2014a). The estimated competitiveness effects on employment through relocation of industry have also tended to be small. However, this may in part reflect the lack of ambition of policies and a tendency to grant concessions and exemptions to trade-exposed industry (OECD, 2014a).

Simulations with the OECD computable general equilibrium model ENV-Linkages (Chateau et al., 2014) consider the sectoral shifts under a global "mitigation-only" policy scenario to limit global warming to 2°C and find sectoral reallocation effects to be modest in terms of overall value-added and employment.[5] At the individual sectoral level, expansions and contractions are large for some sectors, however. Not surprisingly, fossil-fuel industries and energy-intensive manufacturing will experience the steepest declines and renewable-energy industries the sharpest increases. For example, by 2030, GHG mitigation policies are estimated to lower value-added in the coal-extraction sector by around 40% compared with a situation in which governments take no further action on climate. However, the impact of this reduction on aggregate global value-added would only be 0.1 percentage point. Conversely, value-added in solar and wind energy is estimated to increase by 40%.

Large impacts in individual sectors may not translate into a large overall reallocation of activity and jobs because the most carbon-intensive industries represent only a small share of total value-added and employment. Job reallocation as a result of climate action across sectors (summing up the creation and the shedding of jobs) is estimated at 1.5% of total employment by 2050. This adds a modest amount to the job reallocation to be expected on the basis of past experience. For example, between 1995 and 2005, the amount of sectoral job reallocation in OECD member countries amounted to 20% of employment (OECD, 2012).

In some countries where the scope for expansion of renewable energies (including hydro and geothermal) appears strong, manufacturing sectors may grow significantly as a result of competiveness gains, as countries collectively move to reduce GHG emissions from fossil fuels. This is the case in the United States and Brazil. By contrast, manufacturing is projected to shrink as a result of climate action in China and India, as their economies are more energy-intensive. Increases in non-transport services, including housing services, are estimated to be substantial in low-income countries because of their large weight in the economy (close to 2% of aggregate value-added in India, for example), whereas there may be reductions in high-income countries. In a few countries, including Brazil and Russia, downscaling of value added in land transport also has some weight relative to total economy value-added.

The model simulations also suggest that mitigation policies would result in modest additional employment reallocation world-wide by 2050, relative to total employment, as different sectors create and shed jobs (Figure 4.14). To focus on the reallocation effects across sectors, the model assumes that overall national employment levels remain unaffected by climate mitigation action in the long run. Job creation and job shedding balance, as wages adjust to changes in labour supply and demand across sectors. As argued above, higher GDP growth in the decisive action scenario – in which mitigation policies are implemented in an integrated way with growth-enhancing policies – is likely to result in higher employment in the short term. In the longer term, higher GDP growth would typically result in higher wage growth on aggregate, with employment effects largely depending on individual labour markets. Sectoral job creation and destruction are each estimated to amount to 0.7% of employment. Overall, job shedding in emissions-intensive energy sectors is only partly offset by job creation in low-emission energy sectors, reflecting the important role of improving energy efficiency. Individual country results would vary according to the type of renewables or fossil fuels they use. Where biofuels may expand, such as in Brazil, renewables may generate more employment, as this activity is labour intensive. Job creation may then exceed job loss in services and manufacturing sectors.

Job reallocation is estimated to be strongest in some of the emerging economies, notably in India, Indonesia and Russia, although these results need to be treated with caution in view of widespread labour-market informality in these countries (Figure 4.15). In India and Russia, climate change mitigation policies result in more substantial job losses in high-emissions energy production, whereas in Indonesia agriculture may shed jobs equivalent to about 1% of total employment. Non-transport services account for much of the job creation as a result of climate change action. In South American economies and South Africa, job reallocation is smaller. This is also true for all high-income countries, including Australia and Canada, where fossil-fuel extraction activities are relatively important.

Figure 4.14. **Sectoral composition of job reallocation in the world**

Deviation to baseline in 2050 in a 2°C scenario, percentage of total employment

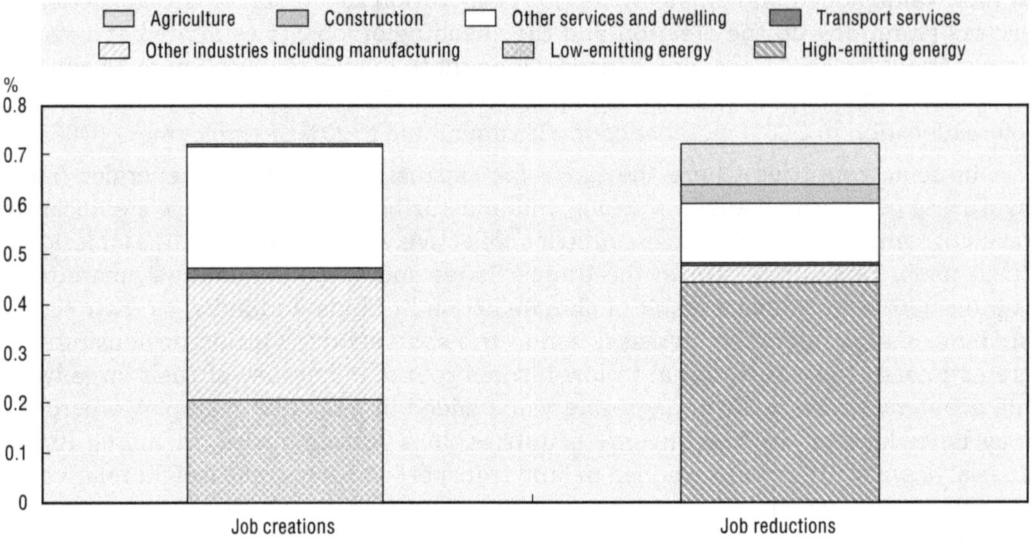

Figure 4.15. **Job reallocation is relatively strong in some emerging economies**

Deviation to baseline in 2050 in a 2°C scenario, percentage of total employment

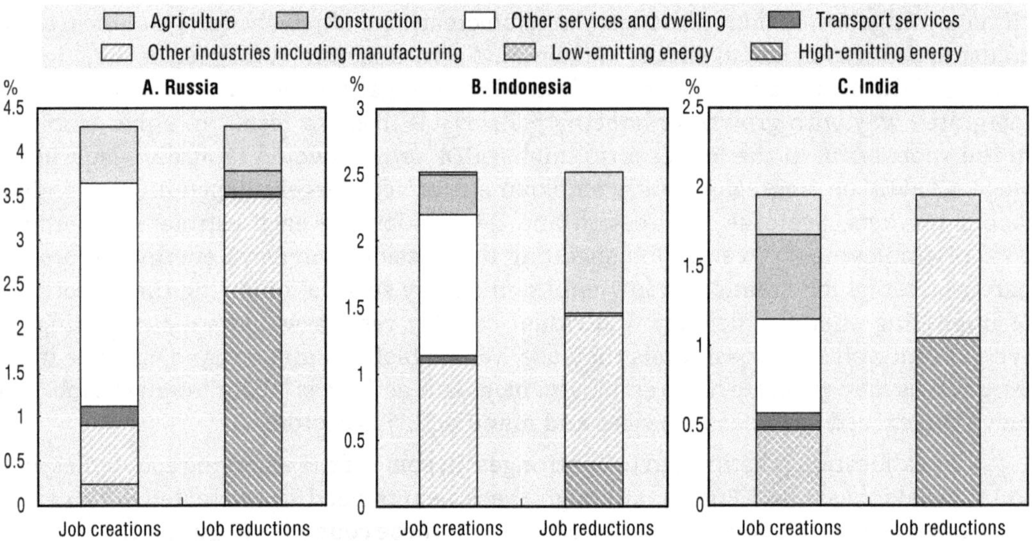

Model simulations suggest that the majority of job reallocations from climate change mitigation action affect low-skilled workers, because they constitute over half of total labour costs in energy-producing sectors (OECD, 2016c). This is especially the case in emerging economies. In OECD member countries, high-skill occupations such as managerial staff and technical experts are also expected to undergo significant reallocation.

In the long term, the scale of change implied by the transition to low-emission and climate-resilient economies will be much greater than modelled here because such economies will ultimately require greater reliance on knowledge-based outputs and human capital, and less reliance on the use of material resources and environmental services

(OECD, 2014a). More generally, in the long run, it has been argued that low-carbon economies require more reliance on new ideas – the "knowledge economy" or "weightless economy" – and less on raw materials and energy, boosting demand for highly educated workers and workers providing services that do not require a lot of material inputs (OECD, 2014a). These are likely to require resource reallocation not only between the broadly defined sectors shown above, but also within sectors.

For mitigation policies, as for any policy resulting in structural change, the functioning of the labour market will have a significant impact on aggregate employment outcomes. Rigidities in the labour market may hamper structural adjustments, with negative effects on employment and GDP (OECD, 2012). Wage rigidity in the formal sector is more of a problem in rapidly growing developing countries, such as in China and India, because of the larger sectoral reallocations of labour necessary (OECD, 2014a).

The decisive transition policy scenario shows how climate policies combined with well-aligned pro-growth reforms and fiscal initiative can put economies on track for low-emission, climate-resilient growth, while also facilitating the reallocation of jobs across sectors. It is also essential that sector-level changes be accompanied by proactive measures to plan and invest in sustainable jobs, as discussed in Chapter 6.

Notes

1. The simulations presented in the chapter cover the G7 countries, Australia, China, India, Indonesia, Brazil, South Africa, and the Russian Federation. The G7 countries and Australia are considered to be advanced economies and China, India, Indonesia, Brazil, South Africa, and the Russian Federation emerging economies.
2. Studies based on integrated assessment models also find a negative impact of mitigation policies in the short to medium term, although the magnitude of the impact is more pronounced than in the current analysis (IPCC, 2014). Differences reflect the nature of the modelling tools used in the two approaches and – more important – the fact that the present analysis accounts for the specificity of the current macroeconomic environment of low growth and low interest rates. Another difference between simulations presented in this chapter and those cited in IPCC (2014) is the focus on selected G20 economies, primarily due to data constraints.
3. The implementation of a carbon tax and other energy taxes can reduce growth prospects in the short term. In the IEA scenario, carbon and other energy taxes lead to a fall in fossil fuel demand, including through the change in the sectoral composition of the economy and gains in energy efficiency. In practice, however, the simulation outcomes presented in this chapter are primarily explained by the magnitude of the investment needs estimates and carbon price increases to achieve the climate objectives.
4. There is no universally agreed definition of stranded assets: see the discussion in Chapter 3.
5. This scenario focused on the effects of global climate mitigation based on a price applied globally to all greenhouse gas emissions, without fiscal initiative and pro-growth reforms assumed elsewhere in this chapter. Overall impacts on GDP would be more negative in the absence of these measures (Chateau, Dellink and Lanzi, 2014).

References

Adalet McGowan, M., D. Andrews and V. Millot (2017), "The Walking Dead?: Zombie Firms and Productivity Performance in OECD Countries", OECD Publishing, Paris, www.oecd.org/eco/The-Walking-Dead-Zombie-Firms-and-Productivity-Performance-in-OECD-Countries.pdf.

Albrizio, S., T. Koźluk and V. Zipperer (2017), "Environmental Policies and Productivity Growth: Evidence across Industries and Firms", *Journal of Environmental Economics and Management*, Vol. 81, Elsevier, Amsterdam, pp. 209-226, https://doi.org/10.1016/j.jeem.2016.06.002.

Anthoff, D. and R. Tol (2013), The uncertainty about the social cost of carbon: A decomposition analysis using FUND, *Climatic Change*, Vol. 117, Springer Nature, London, pp. 515-530, http://doi.org/10.1007/s10584-013-0706-7.

Barrett, S. (2003), Environment and Statecraft: *The Strategy of Environmental Treaty-Making, Oxford University Press, Oxford* and *New York*.

Barrett, S. (1994), "Self-Enforcing International Environmental Agreements", *Oxford Economic Papers*, *Vol.* 46, *Oxford University Press, Oxford*, pp. 878-894.

Bosello, F. and S. Dasgupta (2015), *Development, Climate Change Adaptation, and Maladaptation: Some Econometric Evidence*, Fondazione Eni Enrico Mattei, Milan.

Bosetti, V. et al. (2009), "Delayed action and uncertain stabilisation targets. How much will the delay cost?", *Climatic Change*, Vol. 96, Springer Nature, London, pp. 299-312, http://doi.org/10.1007/s10584-009-9630-2.

Castelnuovo et al. (2005), "Learning-by-Doing vs. Learning by Researching in a model of climate change policy analysis", *Ecological Economics*, Vol. 54/2-3, Elsevier, Amsterdam, pp. 261-276, https://doi.org/10.1016/j.ecolecon.2004.12.036.

Cavallo et al. (2013), "Catastrophic natural disasters and economic growth", *The Review of Economics and Statistics*, Vol. 95/5, MIT Press, Cambridge, Massachusetts, pp. 1549–1561, http://doi.org/10.1162/REST_a_00413.

City of New York (2013), "A stronger, more resilient New York", http://s-media.nyc.gov/agencies/sirr/SIRR_singles_Lo_res.pdf.

Chateau, J., R. Dellink and E. Lanzi (2014), "An Overview of the OECD ENV-Linkages Model: Version 3", *OECD Environment Working Papers, No. 65*, OECD Publishing, Paris, http://doi.org/10.1787/5jz2qck2b2vd-en.

Clarke L. et al. (2014), "Assessing Transformation Pathways", in *Climate Change 2014: Mitigation of Climate Change. Contribution of Working Group III to the Fifth Assessment Report of the Intergovernmental Panel on Climate Change*, Cambridge University Press, Cambridge and New York.

Cummins, J. and O. Mahul (2009), Catastrophe *Risk Financing in Developing Countries: Principles for Public Intervention*, World Bank, Washington DC.

Dechezleprêtre, A. and M. Sato (2015), "Asymmetric industrial energy prices and international trade", *Energy Economics*, Vol. 52, Supplement 1, pp. S130–S141.

Dessens O., G. Anandarajah and A. Gambhir (2014), "Review of Existing Emissions Pathways and Evaluation of Decarbonisation Rates", AVOID 2, London.

Dlugosch, D. and T. Kozluk (2017), "Energy prices, environmental policies and investment: Evidence from listed firms", *OECD Economics Department Working Papers*, No. 1378, OECD Publishing, Paris, http://dx.doi.org/10.1787/ef6c01c6-en.

Egert (forthcoming): "Regulation, Institution and economic growth in advanced, emerging and developing countries", *OECD Economics Department Working Paper.*

Egert, B. And P.Gal (2016), "The Quantification of Structural Reforms in OECD Countries: A New Framework", *OECD Economics Department Working Papers*, No. 1354, OECD Publishing, http://dx.doi.org/10.1787/2d887027-en.

Finus, M., E. van Ierland and R. Dellink (2006), "Stability of climate coalitions in a cartel formation game",Economics of Governance, Vol. 7/3, Springer Nature, London, pp. 271–291, http://doi.org/10.1007/s10101-005-0009-1.

Fournier J.-M. (2016), "The Positive Effect of Public Investment on Potential Growth", *OECD Economics Department Working Papers*, No. 1347, OECD Publishing, Paris, http://dx.doi.org/10.1787/15e400d4-en.

Fournier, J.-M. and Å. Johansson (2016), "The Effect of the Size and the Mix of Public Spending on Growth and Inequality", *OECD Economics Department Working Papers*, No. 1344, OECD Publishing, Paris, http://dx.doi.org/10.1787/f99f6b36-en.

Gal, P. and A. Hijzen (2016), "The Short-Term Impact of Product Market Reforms: A Cross-Country Firm Level Analysis", *OECD Economics Department Working Papers*, No. 1311, OECD Publishing, Paris, http://dx.doi.org/10.1787/5jlv2jm07djl-en.

Gerlagh, R. and Kuik, O. (2014), "Spill or leak? Carbon leakage with international technology spillovers: ACGE analysis", *Energy Economics*, Vol. 45, Elsevier, Amsterdam, pp. 381-388, https://doi.org/10.1016/j.eneco.2014.07.017.

Greenstone M., E. Kopits and A. Wolverton (2013), "Developing a Social Cost of Carbon for US Regulatory Analysis: A Methodology and Interpretation", *Review of Environmental Economics and Policy*, Vol. 7/1, MIT, Cambridge, Massachusetts, pp. 23–46, http://hdl.handle.net/1721.1/85655.

Hallegatte, S. et al. (2017), *Unbreakable: Building the Resilience of the Poor in the Face of Natural Disasters*, World Bank, Washington DC, http://hdl.handle.net/10986/25335.

Handmer, J. et al. (2012), "Changes in impacts of climate extremes: Human systems and ecosystems", in *Managing the Risks of Extreme Events and Disasters to Advance Climate Change Adaptation: IPCC Special Report*, Cambridge University Press, Cambridge and New York, www.ipcc.ch/pdf/special-reports/srex/SREX-Chap4_FINAL.pdf.

IEA (2017), Chapter 2 in *Perspectives for the Energy Transition: Investment Needs for a Low-Carbon Energy System*, International Energy Agency, Paris, www.energiewende2017.com/wp-content/uploads/2017/03/Perspectives-for-the-Energy-Transition_WEB.pdf.

IEA (2016), World Energy Outlook 2016, IEA/OECD Publishing, Paris, http://dx.doi.org/10.1787/weo-2016-en.

IRENA (2017), "Global Energy Transition and the Role of Renewables", Chapter 3 of *Perspectives for the Energy Transition: Investment Needs for a Low-Carbon Energy System*, International Renewable Energy Agency, Abu Dhabi, www.energiewende2017.com/wp-content/uploads/2017/03/Perspectives-for-the-Energy-Transition_WEB.pdf.

IPCC (2014), *Climate Change 2014: Mitigation of Climate Change. Contribution of* Working Group III to the Fifth Assessment Report of the Intergovernmental Panel on Climate Change, Cambridge University Press, Cambridge and New York.

IPCC (2013), *Climate Change 2013: The Physical Science Basis. Contribution of Working Group I to the Fifth Assessment Report of the Intergovernmental Panel on Climate Change*, Cambridge University Press, Cambridge and New York.

Jakob, M. et al. (2012), "Time to Act Now? Assessing the Cost of Delaying Climate Measures and Benefits of Early Action", *Climatic Change*, Vol. 114/1, Springer Nature, London, pp. 79-99, http://doi.org/10.1007/s10584-011-0128-3.

Kalkuhl, M. and O. Edenhofer (2016), "Knowing the Damages is not Enough: The General Equilibrium Impacts of Climate Change", CESifo, Munich.

Kozluk, T., and C. Timiliotis (2016), "Do environmental policies affect global value chains? A new perspective on the pollution haven hypothesis", *OECD Economics Department Working Papers*, No. 1282, OECD Publishing, Paris, http://dx.doi.org/10.1787/5jm2hh7nf3wd-en.

Kriegler, E. et al. (2015), "Making or breaking climate targets: The AMPERE study on staged accession scenarios for climate policy", *Technological Forecasting and Social Change*, Vol. 90/A, Elsevier, Amsterdam, http://doi.org/10.1016/j.techfore.2013.09.021.

Lis, E. M. and C. Nickel (2010), "The impact of extreme weather events on budget balances", *International Tax and Public Finance*, Vol. 17/4, Springer Nature, London, pp. 378-399, http://doi.org/10.1007/s10797-010-9144-x.

Marangoni, G. and M. Tavoni (2014), "The clean energy R&D strategy for 2°C", *Climate Change Economics*, Vol. 5/1, World Scientific Publishing, Singapore, http://doi.org/10.1142/S201000781440003X.

Meinshausen, M., S.C.B. Raper, and T.M.L. Wigley (2011), "Emulating coupled atmosphere-ocean and carbon cycle models with a simpler model, MAGICC6 — Part 1: Model description and calibration", *Atmospheric Chemistry and Physics*, Vol. 11/4, European Geosciences Union, Munich, pp. 1417-1456, http://doi.org/10.5194/acp-11-1417-2011.

Mourougane, A., et al. (2016), "Can an Increase in Public Investment Sustainably Lift Economic Growth?", *OECD Economics Department Working Papers*, No. 1351, OECD Publishing, Paris, http://dx.doi.org/10.1787/a25a7723-en.

Nordhaus, W.D. (2017), *Econonomic Policy Reforms 2017*: "Revisiting the Social Cost of Carbon", PNAS, Vol. 14/117, National Academy of Sciences of the United States of America, Washington DC, pp. 1518-1523, http://doi.org/10.1073/pnas.1609244114.

Nordhaus, W. D. (2016), "Projections and Uncertainties About Climate Change in an Era of Minimal Climate Policies", National Bureau of Economic Research, Cambridge, Massachusetts, http://doi.org/10.3386/w22933.

Nordhaus (2015), "Climate Clubs: Overcoming Free-riding in International Climate Policy", *American Economic Review*, Vol. 105/4, American Economic Association, Pittsburgh, pp. 1339-1370, http://doi.org/10.1257/aer.15000001.

OECD (2017), *Economic Policy Reforms 2017: Going for Growth*, OECD Publishing, Paris, http://dx.doi.org/10.1787/growth-2017-en.

OECD (2016a), *OECD Economic Outlook, Volume 2016*, Issue 1, OECD Publishing, Paris, http://dx.doi.org/10.1787/eco_outlook-v2016-1-en.

OECD (2016b), *Economic Survey of Indonesia*, OECD Publishing, Paris, www.oecd.org/eco/surveys/economic-survey-indonesia.htm.

OECD (2016c), *The Economic Consequences of Outdoor Air Pollution*, OECD Publishing, Paris, http://doi.org/10.1787/9789264257474-en.

OECD (2015a), *The Future of Productivity*, OECD Publishing, Paris, http://dx.doi.org/10.1787/9789264248533-en.

OECD (2015b), *OECD Environmental Performance Reviews: Brazil 2015*, OECD Publishing, Paris, http://doi.org/10.1787/9789264240094-en.

OECD (2015c), *OECD Economic Outlook*, Volume 2015, Issue 1, OECD Publishing, Paris, http://dx.doi.org/10.1787/eco_outlook-v2015-1-en.

OECD (2015d), *The Economic Consequences of Climate Change*, OECD Publishing, Paris, http://doi.org/10.1787/9789264235410-en.

OECD (2014a), *Addressing Social Implications of Green Growth: Inclusive Labour Markets For Green Growth*, Issue Note, Green Growth And Sustainable Development Forum, 13-14 November 2014, OECD, Paris, http://www.oecd.org/greengrowth/Issue%20Note%20Session%20Two_final.pdf.

OECD (2012), *Employment Outlook*. OECD Publishing, Paris, http://dx.doi.org/10.1787/empl_outlook-2012-en.

OECD (2009), *The Economics of Climate Change Mitigation: Policies and Options for Global Action Beyond 2012*, OECD Publishing, Paris, http://dx.doi.org/10.1787/9789264073616-en.

Porter, M. (1991), "America's green strategy", *Scientific American*, Vol. 264/4, Springer Nature, London.

Porter, M. E. and C. van der Linde (1995), "Toward a New Conception of the Environment Competitiveness Relationship", *Journal of Economic Perspectives*, American Economic Association, Pittsburgh, Vol. 9/4, pp. 97-118, http://doi.org/10.1257/jep.9.4.97.

Weitzman, M.L. (2012), "GHG targets as insurance against catastrophic climate damages", *Journal of Public Economic Theory*, Vol. 14/2, Association for Public Economic Theory/Wiley, Hoboken, pp. 221-244.

ANNEX 4.A1. Collective mitigation policies, world fossil fuel prices, substitution effects and innovation

Introduction

Collective global action by definition excludes free-riding and carbon leakage, and provides the scale necessary to tackle the climate change challenge.[1] However, incentives to free-ride can be substantial. Countries refraining from climate mitigation could benefit from lower global fossil fuel prices, gaining a competitive edge in carbon-intensive production. They could also be tempted to wait till low-carbon technologies become cheaper before avoiding high upfront costs. Some countries may have incentives to take the lead in terms of climate mitigation, however (Tirole, 2012). Countries will want to minimise local collateral damage such as air pollution. Moreover, the largest actors, such as China, expect to bear a non-negligible cost of climate change and officials tend to prefer policy options consistent with public opinion.

The magnitude of the necessary climate mitigation action and the feasibility of achieving climate goals are likely to depend on the size and composition of the set of countries taking up climate mitigation. What would be the consequences of partial and heterogeneous action, as opposed to a collective action? This annex addresses this question by looking at three key aspects of the global economy: the role of global fossil fuel prices, the effects via the trade channel and the role of innovation.

In practice, policy decisions are not exogenous, and they are interdependent. There is a vast literature investigating the rationale for building coalitions. Barrett (1994) emphasises credible or "self-enforcing" treaties that combine individual and collective rationality. Barrett's (2003) book on international environmental agreements discusses the "small coalition paradox" (as called by Nordhaus, 2015) according to which coalitions are either small or shallow. Dellink (2001) discusses incentives that stabilise large coalitions, and Finus, Ierland and Dellink (2006) show that a transfer scheme can help. Nordhaus (2015) strengthens the concept of coalition stability in adding rationality for each subset of the players. Lessmann, Marschinski and Edenhofer (2009) show that trade sanctions can significantly raise participation in coalitions. Lessmann and Edenhofer (2011) show that research co-operation can foster coalition stability when it focuses on research co-operation in mitigation technology rather than co-operation on augmenting productivity in the private good sector. This discussion on determinants of policy decisions goes beyond this annex that discusses their consequences.

This illustrative modelling exercise presents very simple, stylised relationships – in a world of two principal regions which are *the coalition* (where fossil fuels are subject to a carbon tax) and the *opt-out countries*, where no equivalent policies are undertaken. In practice, similar reasoning can be applied to sectors (rather than countries) and other climate policies that increases the cost of carbon emissions (e.g. emissions trading schemes). The qualitative conclusions present similarities with those derived from more detailed general equilibrium models, such as in Burniaux and Oliveira Martins (2012), reflecting a supply-side leakage related to the negative effect on fossil fuel price and a specialisation leakage effect due to changes in relative production costs (see also a discussion of leakage channels in Marschinski, Jakob and Edenhofer, 2009).

The following points are discussed:

- Carbon-pricing in a subset of countries (or sectors) can be less effective in reducing global emissions than full carbon pricing, not only because the scale of the climate change challenge requires broad action, but also because exempted entities benefit from lower fossil fuel prices and from a leakage effect.
- As climate action in individual countries spurs technological progress in low-carbon technologies, the effect on emissions may be broader due to knowledge diffusion across countries, including those not taking up climate mitigation. In the context of partial action, it may thus be even more important to assure that climate mitigation policies support both innovation and its diffusion.
- Efficiency gains in sectors with low demand elasticity (e.g. heating efficiency in advanced economies where demand is already saturated), rather than sectors with high demand elasticity are likely to result in less of a rebound effect.
- The rebound effect can be mitigated with carbon taxes.
- Should carbon-pricing cover a subset of sectors, the largest emissions reductions are expected to be reaped in sectors in which price-elasticity of demand is the highest (e.g. cases where an alternative zero-emissions option already exists). Leakage will be lowest where the elasticity of substitution with exempted items or imported items that may bear a lower carbon tax is the lowest (e.g. non-tradable goods).
- The appropriate amount of the carbon tax depends on numerous factors, including long-run fossil fuel supply-side and demand-side elasticities, which cannot be observed with precision, and the availability of technologies, suggesting that the appropriate pace of a carbon tax may need to be adjusted on a regular basis by governments informed by market developments.

Three main mechanisms are investigated in this exercise: the role of endogenous world fossil fuel prices in the leakage of emissions abroad; the role of substitution towards exempted fossil fuels; and some aspects of the role of innovation. In the first section, the computation takes into account the fossil fuel price mechanism, while ignoring the substitution effect. In the second step, the computation is augmented with a substitution effect to get an overall effect of the carbon tax on fossil fuel quantities emitted. In a third step, the role of innovation is discussed. In a final step, the theoretical channels are combined with key findings in the literature to draw policy implications.

A leadership coalition and a world fossil fuel price with a partial carbon tax and without explicit substitution between goods

In a stylised world in which agents consume or produce a composite fossil fuel good, the consumption of fossil fuels is assumed to be proportionally linked to carbon emissions. In the coalition region A, agents consuming fossil fuels are subject to a carbon tax T (or an equivalent environmental policy) that increases the end-user fossil fuel price to p^W+T, where p^W is the fossil fuel world price. In the opt-out region B, agents are also consuming fossil fuels, but are not subject to a carbon tax. All fossil fuel producers are in a third entity C. Let's denote α_T and $(1- \alpha_T)$ are the relative size of entities A and B. Fossil fuel demand is a decreasing function $d(p^W)$ of price and fossil fuel supply is an increasing function $s(p^W)$ of price. Formulated as such, this modelling is stylised, but also quite general: it can refer to any shape of demand and supply curves. Then, one can write the demand and supply functions determining quantities of fossil fuels consumed by region A (q_A), consumed by region B (q_B) and produced in the world (q_C) as:[2]

(1) $$q_A = \alpha_T d_A(p^W + T)$$

(2) $$q_B = (1 - \alpha_T) d_B(p^W)$$

(3) $$q_C = s(p^W)$$

The static equilibrium world fossil fuel consumption $q=q_A+q_B$ depends on the carbon tax T.[3] In a first step, one can see that the overall effect of the carbon tax on reducing emissions in the coalition is partly offset by the increase of consumption in response to the world (pre-tax) fossil fuel price decrease:

(4) $$\frac{\partial q}{\partial T} = \underbrace{\alpha_T d_A'(p^W + T)}_{\substack{Direct\ effect\ in\ coalition\ A, \\ holding\ world\ price\ constant}} + \underbrace{d_G'(p^W)\frac{\partial p^W}{\partial T}}_{World\ price\ effect}$$

where $d_G(p^W) = \alpha_T d_A(p^W + T) + (1 - \alpha_T) d_B(p^W)$

is the global demand function for fossil fuels.

The clearance of the fossil fuel market provides the world fossil fuel price, as the solution of the following system:

(5) $$\begin{aligned} q_A + q_B &= q_C \\ \alpha_T d_A(p^W + T) + (1 - \alpha_T) d_B(p^W) &= s(p^W) \end{aligned}$$

Hence, one can derive the effect of a carbon price on the world fossil fuel price:[4]

(6) $$\frac{\partial p^W}{\partial T} = \frac{\alpha_T dA'(p^W + T)}{s'(p^W) - d_G'(p^W)}$$

Inserting equation (6) in equation (4) provides a global effect of a coalition carbon tax on fossil fuel consumptions (and hence emissions):

(7) $$\frac{\partial q}{\partial T} = \frac{\alpha_T dA'(p^W + T) s'(p^W)}{s'(p^W) - d_G'(p^W)} < 0$$

The direct effect in coalition region A dominates the world price effect, so that fossil fuel consumption decreases at the world level. In the opt-out region B however, the world price effects pushes fossil fuel consumption up.[5] A carbon tax in the coalition will deliver larger gains the larger the size of the coalition and the higher the sensitivity of fossil fuel supply and output to fossil fuel prices. For instance, Figure A1.1 illustrates the effect of the same carbon tax in a low demand elasticity case (Panel A) and in a high demand elasticity case (Panel B).[6]

Figure A1.1. **The effect of a carbon tax on the fossil fuel world equilibrium**

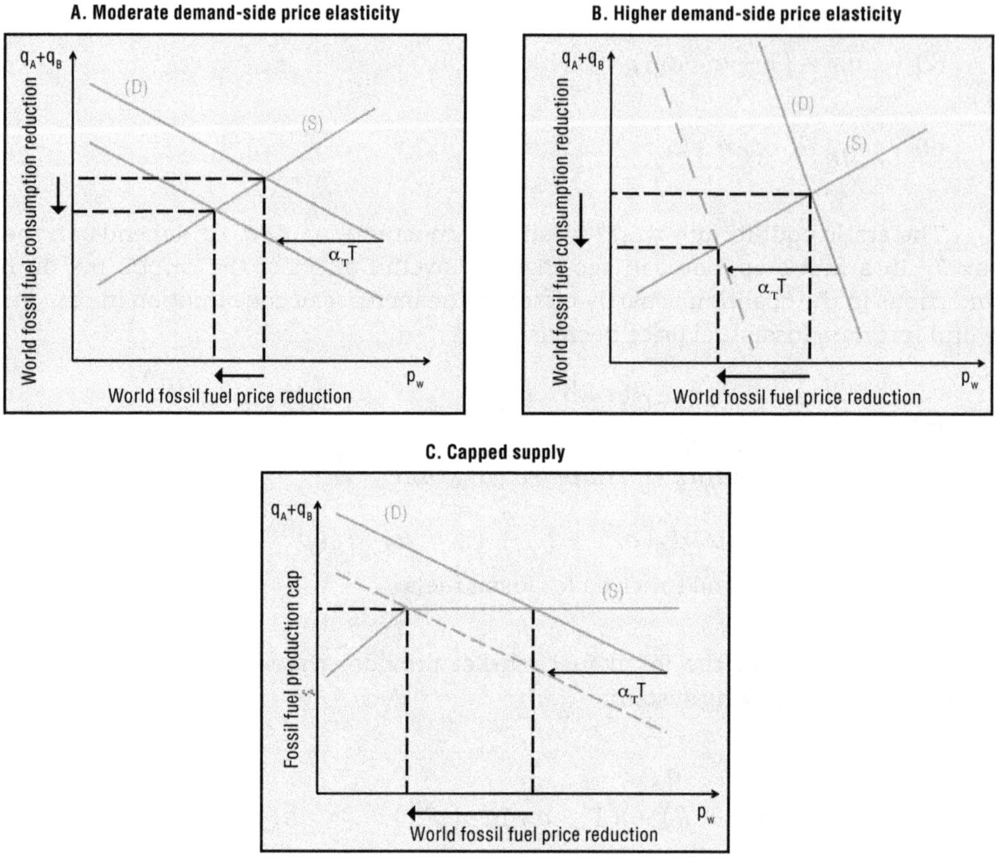

Note: (D) and (S) denote demand and supply functions, P_w the world fossil fuel price, $q_A + q_B$ the quantity of fossil fuel consumed and T a carbon tax scaled by the share α_T.

The importance of supply elasticity in emissions outcomes can be better understood with a stylised extreme case. In the extreme case where supply is inelastic: s'(p^w), quantities (and emissions) do not adjust (Figure 4.16, Panel C). This would correspond for instance to a fossil fuel supplier setting a world-level quantity cap with no reaction to the introduction of a carbon tax. In this particular case, prices are changing, so that fossil fuel consumers can capture the price premium that would otherwise be captured by the fossil fuel supplier[7]. As a result, world fossil fuel prices are more sensitive to a carbon tax, and fossil fuel quantities remain constant – with perfect leakage of emissions to the opt-out region. In practice, a fossil fuel supplier cartel can reduce the price-elasticity of supply.

A world fossil fuel price with a partial carbon tax and substitution between taxable and non-taxable fossil fuels

In practice, the demand for a good depends on the imported substitutes and their prices. This was implicit in the previous section; making the role of substitution explicit helps to think about the trade channel. The demand equations (1) and (2) are thus replaced here by equations in which demand depends both on the price in the region and on the price of the imported substitute from the alternative region, while the supply side function is left unchanged:[8]

(1') $q_A = d_A(p_A = p^W + T, p_B = p^W)$

(2') $q_B = d_B(p_A = p^W + T, p_B = p^W)$

(3') $q_C = s(p^W)$

One should then rewrite equation (4), including an additional substitution (or trade) effect:

$$(4')\quad \frac{\partial q}{\partial T} = \underbrace{\frac{\partial d_A}{\partial p_A}(p^W + T, p^W)}_{Direct\ effect\ in\ coalition\ A} + \underbrace{d'_G(p^W)\frac{\partial p^W}{\partial T}}_{World\ price\ effect} + \underbrace{\frac{\partial d_B}{\partial p_A}(p^W + T, p^W)}_{Substitution\ effect\ in\ region\ B}$$

The effect of a change in the carbon tax on the world oil price is dampened by this substitution effect, as the substitution effect mitigates the demand reduction:

$$(6')\quad \frac{\partial p^W}{\partial T} = \frac{\frac{\partial d_A}{\partial p_A}(p^W+T,p^W)+\frac{\partial d_B}{\partial p_A}(p^W+T,p^W)}{s'(p^W)-d'_G(p^W)}$$

where $d'_G(p^W) = \frac{\partial d_A}{\partial p_A}(p^W + T, p^W) + \frac{\partial d_A}{\partial p_B}(p^W + T, p^W) + \frac{\partial d_B}{\partial p_A}(p^W + T, p^W) + \frac{\partial d_B}{\partial p_B}(p^W + T, p^W)$

In sum, the overall reduction of emissions is lower, as the demand for fossil fuels is increasing in region B when prices increase in the coalition A:

$$(7')\quad \frac{\partial q}{\partial T} = \frac{\frac{\partial d_A}{\partial p_A}(p^W+T,p^W)s'(p^W)}{s'(p^W)-d'_G(p^W)} + \underbrace{\frac{\frac{\partial d_B}{\partial p_A}(p^W+T,p^W)s'(p^W)}{s'(p^W)-d'_G(p^W)}}_{positive\ substitution\ effect}$$

The relationships presented here can also be used to think about the consequence of a carbon tax (or any equivalent policy) that covers a subset of sectors only. For this purpose, regions A and B could be replaced by two sectors, one that is subject to the carbon tax, and one that is not. Of course, the calibration would then need to reflect price elasticities in each sector.

The role of innovation is uncertain a priori

Another important factor that plays a role in the move to a low-carbon environment is innovation, which leads to improvements in energy and fossil fuel efficiency, i.e. the ability to produce with lower fossil fuel inputs. The sign of the effect on improving fossil fuel and energy efficiency on carbon emissions is ambiguous. On the one hand, for a given unit of consumption, less fossil fuels and emissions are used. On the other hand, the efficiency improvements make energy-intensive goods cheaper, increasing the demand for them and hence emissions: this is the so-called "rebound effect". The rebound effect was observed in the nineteenth century by Jevons (1865): efficiency improvements to the steam engine technology had increased coal consumption.[9] The magnitude of the rebound effect is subject to much debate. In particular, Grubb (1990) argues that the rebound effect depends to the extent to which energy prices and availability constrain demand or not. He also argues this effect can be lower when price changes are policy driven and when fossil fuel prices are higher. This would be the case if policies boosting efficiency focus on price-inelastic activities where market incentives to increase efficiency are poorer.

If one assumes that fossil fuels q are used to produce a final good C with a technology A, so that $C = A\ q$, then one can write the following elasticity of fossil fuel demand to technology:

$$(8) \qquad \frac{\partial \ln(q)}{\partial \ln(A)} = \frac{\partial \ln(C)}{\partial \ln(\pi)} \frac{\partial \ln(\pi)}{\partial \ln(A)} - 1$$

where $\frac{\partial \ln(C)}{\partial \ln(\pi)}$ is the elasticity on the final good C to its price and $\frac{\partial \ln(\pi)}{\partial \ln(A)}$ the sensitivity of the price of the final good to the efficiency improvement. In other words, the rebound effect is larger when demand for the final good is price-elastic, and offsets the emissions reductions when the good is fossil fuel-intensive, such as transport. At the macroeconomic level, the long-term declining share of energy in GDP suggests implies a decreasing sensitivity of consumer price to fossil fuel prices: this second component driving the rebound effect has declined.

Similarly, the effect of carbon prices on energy efficiency innovation is ambiguous from a theoretical perspective. A stylised way to discuss the ambiguities is to look to what extent consumers' utility improves in case of innovation. Let's denote *U(C=Aq)* the utility with decreasing marginal returns. Then, the sensitivity of utility with respect to energy efficiency gain can be decomposed into two components. On one side, the price increase of fossil fuels should decrease their quantity consumed, decreasing the relevance of innovating in this field: there is a negative market size effect. On the other side, at lower consumption level, the welfare effect of a slight increase in consumption is larger. The utility is more sensitive to marginal gains, this is a positive scarcity effect:

$$(9) \qquad \frac{\partial U}{\partial A} = \underbrace{q_A}_{market\ size\ effect} \cdot \underbrace{U'(Aq_A)}_{scarcity\ effect}$$

where *U(.)* is a utility function with decreasing marginal gains.

The learning-by-doing channel presents no theoretical ambiguity about the effect of environment policy stringency (see Castelnuovo et al. 2005, for simulations with a learning-by-doing channel). In this channel, the pace of technology improvement increases with the size of the market. As a result, a carbon-pricing policy will increase innovation in zero-carbon energy, and as this innovation spills over, this should lead to worldwide gains. Di Maria and van der Werf (2008) build on the spillover argument to argue that the induced-technology effect of climate policy reduces carbon leakages. In this channel, first movers can benefit from this learning-by-doing effect, but they may also support a short-term cost of starting to use new methods before the technology has matured: early investment can be depreciated faster.

Insights from the empirical literature and policy implications

The empirical literature provides evidence on the pollution haven hypothesis, namely that highly polluting industries can concentrate in countries with low environmental policy stringency – after controlling for other factors, such as capital intensity, endowments or geography (see for instance Aichele and Felbermayr, 2015; and Koźluk and Timiliotis, 2016). This illustrates the importance of the substitution effect across countries. At the same time, Koźluk and Timiliotis (2016) show a significant but small shift in specialisation: in countries with high environmental policy stringency, firms tend to specialise in low-pollution goods so that overall net exports are not affected by disparities in environmental policies. The pollution haven hypothesis can raise the issue of the net gains of openness. Kuik and Gerlagh (2003) find that welfare gains of freer trade outweigh the costs of abating the trade-induced leakage. In addition, Gerlagh and Kuik (2014) argue that the positive technological spillover effect can dominate the carbon leakage effect.

OECD (2016) shows that effective carbon prices are highly uneven within countries. Effective carbon prices are typically high on final goods that consumers cannot substitute with carbon-tax exempted goods, such as gasoline, and low on other tradable goods, such as manufacturing goods. This is consistent with a rational non-co-operative behaviour in which governments prefer to avoid a carbon tax in cases where the elasticity of substitution is the highest.

Fossil fuel consumption need not always be proportional to carbon emissions. In some cases, carbon emissions can be reduced without reducing fossil fuel consumption (e.g. carbon capture and storage, reforestation). In these cases, the world fossil fuel price channel does not operate.

Many academic studies have investigated the price elasticity of energy demand. End-user price elasticity estimates are summarised in a meta-analysis by Labandeira, Labeaga and Lopéz-Otero (2017), and Fournier et al. (2013) provide an overview of selected empirical studies of the oil price elasticity. There is large uncertainty surrounding the estimates. In addition, long-term elasticity is much larger than short-term elasticity. This is expected because some of the adjustments take time to materialise (e.g. switching to low-consumption cars). Coal supply elasticity is typically assumed to be larger, and plays a critical role as carbon intensity of coal is high (Burniaux and Oliveira Martins, 2012). As regards magnitudes, there is a difference between end-user price elasticity and world prices elasticity: a 1% change in a world price does not translate into a 1% change in end-user price in the presence of an excise tax. Long-term demand elasticity to end-user energy prices is about -0.5 (ranging from about -1.2 to about 0 in a selected sub-sample of estimates) according to Labandeira, Labeaga and Lopéz-Otero (2017), and long-term demand elasticity to oil prices is about -0.2 (from about -0.6 to about 0) according to Fournier et al. (2013).

The few studies estimating supply-side elasticity provide mixed results (e.g. Lin, 2008). Such studies are scare, in part because it is even more difficult to isolate demand shocks than supply shocks. The global financial crisis provides a particular experiment, with the sharp demand-driven collapse of the oil price at the end of 2008 suggesting that short-term supply elasticity is very low. However, this should be read with great care as this is a single event, during which financial market disruptions could also have affected the world price; the shale gas revolution could also have played a role. To overcome this problem, some papers (e.g. Dées et al., 2007) model the supply of oil using a curve-fitting technique along the lines of Hubbert (1962). However, such techniques are inherently difficult and tend to be consistent with a wide range of possible supply paths, particularly if peak production is unknown.

The fact that short-term supply and demand-side elasticities are much lower than long-term elasticities has a policy implication on the appropriate tax level to reach tighter emissions targets. Rapid increases in the stringency of environmental policies are likely to require a particularly large carbon price, supporting the case for a gradual increase in environmental policies over a long period. Should short-term supply-side elasticity be much lower than short-term demand-side elasticity, then the short-term effect of a sharp increase in a carbon tax could be a sharp collapse of fossil fuel prices, dampening substantially the emissions effect. In particular, this raises concerns about any delayed action scenario that necessitates a quick and large rise of carbon price to compensate for the delay. The pace of increase of carbon prices needs to remain reasonable; it is all the more necessary to start tightening emissions targets early. This also means that mitigation policies need to be announced in advance with credible commitments, so that agents can anticipate changes that need time to be implemented.

Recent literature suggests that there is evidence of a rather limited rebound effect – that is, that energy efficiency gains are only partially offset by demand increases. In particular, Dimitropoulos, Oueslati and Sintek (2016) find in a recent meta-analysis that the direct rebound effect in transport is around 12% in the short run and 32% in the long run. This analysis does not include indirect macroeconomic effects, so the overall rebound effect may be larger but is still likely to be limited.

The technological spillover effect is a strong argument in theory to claim that a leadership coalition can be successful in mitigating climate change, but this spillover should not be taken for granted. In practice, stylised facts suggest that spillovers take time. The dispersion of energy efficiency across the world is large and stable over a long period. More broadly, Barro (2015) investigates the pace of GDP per capita convergence. He provides evidence in support of the "iron law of convergence" according to which countries eliminate gaps in levels of real per capita GDP at a rate of around 2% per year. Convergence at a 2% rate implies that it takes 35 years for half of an initial gap to vanish. This suggests that for the technological spillover effect to help in mitigating climate change, additional policy action would be useful. For example, policies in favour of foreign direct investment (FDI) may help to transfer technologies; Meyer and Sinani (2009) provide a meta-analysis to understand when and where FDI generates positive spillovers.

Notes

1. For instance, Kriegler et al. (2015) explore a scenario in which a front runner coalition embarks on immediate ambitious climate action while the rest of the world makes a transition after 2030. They find that the resulting climate outcome is unlikely to be consistent with the goal of limiting global warming to 2 degrees.
2. This set-up ignores the effect of emissions on temperatures, which is turn is expected to reduce heating demand. It is reasonable to ignore this link in the short to medium run as global temperature is a function of the stock of past emissions and hence a change in emissions takes a long time to translate into a change of the pace of temperature. For a model in which oil price demand depends on temperature, see Cho et al. (2016).
3. The intertemporal dimension is ignored here for the simplicity of the exposure. Sinn (2008) finds a similar world fossil fuel channel in an intertemporal supply-side perspective, introducing the "green paradox": suppliers can even boost current supply in the presence of climate mitigation policies that depress future fossil fuel demand. Edenhofer and Kalkuhl (2011) show this green paradox occurs for carbon taxes that increase at a rate higher than the effective discount rate of the resource owners.
4. This derivation is computed with the implicit function theorem.
5. Strictly speaking, the overall effect includes an implicit substitution effect as demand depends on available imported substitutes. This is ignored for the sake of the simplicity of the presentation at this stage and is introduced in the next section.
6. Equation (7) shows the role of derivatives with respect to the level of prices. One can convert such derivatives into elasticities by multiplying by the price to quantity ratio.
7. If one assumes that this supplied quantity is optimised after the environmental policy change, one can show that the supplier side optimal quantity would be shifted downward, so that overall fossil fuel consumption is decreased.
8. For the sake of simplicity, entities A and B are assumed to have the same share here.
9. The Jevons paradox has been revisited by Khazzoom (1980) and Brookes (1990) and renamed Khazzoom–Brookes postulate by Saunders (1992).

References

Aichele R. and G. J. Felbermayr (2015), "Kyoto and Carbon Leakage: An Empirical Analysis of the Carbon Content of Bilateral Trade", The Review of Economics and Statistics, Vol. 97/1, MIT Press, Cambridge, Massachusetts, pp. 104-115, http://doi.org/10.1162/REST_a_00438.

Barrett, S. (1994), "Self-Enforcing International Environmental Agreements", Oxford Economic Papers, Vol. 46, Oxford University Press, Oxford, pp. 878-894.

Barrett, S. (2003), Environment and Statecraft: The Strategy of Environmental Treaty-Making, Oxford University Press, Oxford and New York.

Barro, R. (2015), "Convergence and Modernisation", The Economic Journal, Vol. 125, Wiley, Hoboken, pp. 911-942, http://doi.org/10.1111/ecoj.12247.

Brookes, L. (1990), "The greenhouse effect: the fallacies in the energy efficiency solution", Energy Policy, Vol. 18/2, Elsevier, Amsterdam, pp. 199-201, https://doi.org/10.1016/0301-4215(90)90145-T.

Burniaux, J.-M. and J. Oliveira Martins (2012), "Carbon leakages: a general equilibrium view", Economic Theory, Vol. 49/2, Springer Nature, London, pp. 473-495, http://doi.org/10.1007/s00199-010-0598-y.

Castelnuovo et al. (2005), "Learning-by-Doing vs. Learning by Researching in a model of climate change policy analysis", Ecological Economics, Vol. 54/2-3, Elsevier, Amsterdam, pp. 261-276, https://doi.org/10.1016/j.ecolecon.2004.12.036.

Cho, C.-H. et al. (2016) "The effect of global activities and warming on oil price", Energy Sources, Part B: Economics, Planning, and Policy, Vol. 11/3, Taylor and Francis, Abingdon, United Kingdom, pp. 274-281, http://doi.org/10.1080/15567249.2011.596896.

Dées et al. (2007), "Modelling the world oil market: Assessment of a quarterly econometric model", Energy Policy, Vol. 35, Elsevier, Amsterdam, pp. 178-191, https://doi.org/10.1016/j.enpol.2005.10.017.

Kriegler, E. et al. (2015), "Making or breaking climate targets: The AMPERE study on staged accession scenarios for climate policy", *Technological Forecasting and Social Change*, Vol. 90/A, Elsevier, Amsterdam, http://doi.org/10.1016/j.techfore.2013.09.021.

Dellink, R. (2001) “Drivers of Stability of Climate Coalitions in the STACO Model”, Climate Change Economics, Vol. 2/2, World Scientific Publishing, Singapore, pp. 105–128, http://doi.org/10.1142/S2010007811000231.

Di Maria, C. and E. van der Werf (2008), “Carbon leakage revisited: unilateral climate policy with directed technical change”, Environmental and Resource Economics, Vol. 39, Springer Nature, London, pp. 55–74, http://doi.org/10.1007/s10640-007-9091-x.

Dimitropoulos, A., W. Oueslati and C. Sintek (2016), “The Rebound Effect in Road Transport: A Meta-analysis of Empirical Studies”, *OECD Environment Working Papers*, No. 113, OECD Publishing, Paris, http://dx.doi.org/10.1787/8516ab3a-en.

Edenhofer, O. and M. Kalkuhl (2011), “When do increasing carbon taxes accelerate global warming? A note on the green paradox”, Energy Policy, Vol. 39/4, Elsevier, Amsterdam, pp. 2208–2212, https://doi.org/10.1016/j.enpol.2011.01.020.

Finus, M., E. van Ierland and R. Dellink (2006), “Stability of climate coalitions in a cartel formation game”, Economics of Governance, Vol. 7/3, Springer Nature, London, pp. 271–291, http://doi.org/10.1007/s10101-005-0009-1.

Fournier, J.-M. et al. (2013), “The Price of Oil – Will it Start Rising Again?”, *OECD Economics Department Working Papers*, No. 1031, OECD Publishing, Paris, http://dx.doi.org/10.1787/5k49q186vxnp-en.

Gerlagh, R. and Kuik, O. (2014), “Spill or leak? Carbon leakage with international technology spillovers: A CGE analysis”, Energy Economics, Vol. 45, Elsevier, Amsterdam, pp. 381-388, https://doi.org/10.1016/j.eneco.2014.07.017.

Grubb, M. J. (1990), “Communication – Energy Efficiency and Economic Fallacies”, Energy Policy, Vol. 18/8, Elsevier, Amsterdam, pp. 783-785, https://doi.org/10.1016/0301-4215(90)90031-X.

Khazzoom, J. D. (1980), “Economic Implications of Mandated Efficiency in Standards for Household Appliances”, The Energy Journal, Vol. 1/4, International Association for Energy Economics, Cleveland, pp. 21-40.

Kuik, O. and R. Gerlagh (2003), “Trade Liberalization and Carbon Leakage”, The Energy Journal, Vol. 24/3, International Association for Energy Economics, Cleveland, pp. 97-120.

Hubbert, M. K. (1962), “Energy Resources: A Report to the Committee on Natural Resources of the National Academy of Sciences – National Research Council”, National Academy of Sciences – National Research Council, Washington DC.

Jevons, W. S. (1865), “The Coal Question: Can Britain Survive?”, first published in 1865, reprinted by Macmillan in 1906.

Koźluk, T. and C. Timiliotis (2016), “Do environmental policies affect global value chains? A new perspective on the pollution haven hypothesis”, *OECD Economics Department Working Papers*, No. 1282, OECD Publishing, Paris, http://dx.doi.org/10.1787/5jm2hh7nf3wd-en.

Kriegler, E. et al. (2015), “Making or breaking climate targets: The AMPERE study on staged accession scenarios for climate policy”, Technological Forecasting and Social Change, Vol. 90/A, Elsevier, Amsterdam, http://doi.org/10.1016/j.techfore.2013.09.021.

Labandeira, X., J.M. Labeaga and X. Lopéz-Otero (2017), “A meta-analysis on the price elasticity of energy demand”, Energy Policy, Vol. 102, Elsevier, Amsterdam, pp. 549-568, https://doi.org/10.1016/j.enpol.2017.01.002.

Lessmann, K. and O. Edenhofer (2011), “Research Cooperation and International Standards in a Model of Coalition Stability”, Resource and Energy Economics, Vol. 33/1, Elsevier, Amsterdam, pp. 36-54, https://doi.org/10.1016/j.reseneeco.2010.01.001.

Lessmann, K., R. Marschinski and O. Edenhofer (2009), “The effects of tariffs on coalition formation in a dynamic global warming game”, Economic Modelling, Vol. 26/3, Elsevier, Amsterdam, pp. 641-649, https://doi.org/10.1016/j.econmod.2009.01.005.

Lin, C.Y.C. (2008), “Estimating Supply and Demand in the World Oil Market”, The Journal of Energy and Development, Vol. 34/1-2, International Research Center for Energy and Economic Development, Boulder, Colorado.

Marschinski, R., M. Jakob and O. Edenhofer (2009), “Analysis of carbon leakage in an extended Ricardo-Viner model”, presented at the Conference on the International Dimensions of Climate Policies, Bern, 21-23 January 2009.

Meyer, K. E. and E. Sinani (2009), "When and Where Does Foreign Direct Investment Generate Positive Spillovers? A meta-analysis", Journal of International Business Studies, Vol. 40/7, Springer Nature, London, pp. 1075-1094, http://doi.org/10.1057/jibs.2008.111.

Nordhaus (2015), "Climate Clubs: Overcoming Free-riding in International Climate Policy", American Economic Review, Vol. 105/4, American Economic Association, Pittsburgh, pp. 1339-1370, http://doi.org/10.1257/aer.15000001.

OECD (2016), Effective Carbon Rates: Pricing CO_2 through Taxes and Emissions Trading Systems, OECD Publishing, Paris, http://dx.doi.org/10.1787/9789264260115-en.

Saunders, H. D. (1992), "The Khazzoom-Brookes Postulate and Neoclassical Growth", The Energy Journal, Vol. 13/4, pp. 131-148, International Association for Energy Economics, Cleveland.

Sinn, H.-W. (2008), "Public policies against global warming: a supply side approach", International Tax and Public Finance, Vol. 15/4, Springer Nature, pp. 360-394, http://doi.org/10.1007/s10797-008-9082-z.

Tirole, J. (2012), "Some Political Economy of Global Warming", Economics of Energy and Environmental Policy, Vol. 1/1, International Association for Energy Economics, Cleveland, http://dx.doi.org/10.5547/2160-5890.1.1.10.

ANNEX 4.A2. Brief description of the models

The YODA model

The Yoda model encompasses structural features (such as hysteresis, impact of credit risks premium faced by governments on public debt) and some international dimensions. By incorporating some non-linearities, this model specifically depends on the current state of economies, and in particular their position in the business cycle. For instance, the extent to which hysteresis affects economies will depend not only on the degree of labour-market rigidity but also on the extent to which the economy is experiencing a shortage of demand. This is an important difference from standard macroeconomic models. This model also allows for monetary policy to be constrained at the zero-lower bound.

In the current version the major advanced economies, major emerging-market economies and the rest of the world are modelled. Each country model comprises about 20 key reduced-form equations. In total, the model comprises about 700 equations.

Countries and regions are connected through trade volume linkages. One main advantage of this instrument is that it is flexible and can be easily amended to address specific issues such as the treatment of structural reforms.

With this model, simulations can be run up to 2050.

Specification of the main equations

This section reports the key equations that are driving the simulations.

Economic growth, which is modelled in reduced form, depends on potential growth, real interest rates and discretionary fiscal policy.

$$(1) \quad \Delta y_t = \Delta y_t^* + a_{y,gap} gap_{t-1} + a_{y,r} \Delta r_t + \lambda_1 \Delta ig_t + \lambda_2 \Delta cg_t - \lambda_3 \Delta tax_t - \theta_p \Delta ep + \varepsilon_{y,t}$$

With y_t the log of actual output, $y_t{}^*$ the log of potential output, r_t the real long-term interest rate, ig, cg and tax are respectively public investment, public consumption and tax in percentage of potential GDP, and the output gap. λs fiscal multipliers. Ep is the log of (after tax) energy prices. The gap term captures the effect of other market mechanisms and stabilisation policies that are not explicitly modelled (*e.g.* unconventional monetary policy).

International trade spillovers are introduced in the growth equation when the model is simulated jointly for several countries (linked mode).

Potential output is affected by past developments in demand. Hysteresis has a permanent impact on the level of potential:

$$(2) \quad \Delta y_t^* = \Delta y_{t-1}^* + \mu * \mathrm{Min}(gap_{t-1}, 0) + \frac{\epsilon}{(1-deprec)} \Delta ig_t + \delta * (\Delta y_{t-1}^* - \Delta y_{ss}^*) + \varepsilon_{y^*,t}$$

with μ>0 the degree of labour market hysteresis, ε is the elasticity of public capital in the production function, deprec the depreciation rate, δ the speed of convergence of potential output to the steady state, y^*_{SS} and error term $\varepsilon_{y^*,t}$ is a supply shock.

Inflation is driven by an expectation-augmented Phillips curve where expectations are anchored to an inflation target.

$$(3) \quad \pi_t = a_{\pi,\pi}\pi_{t-1} + (1 - a_{\pi,\pi})\pi_t^T + a_{\pi,gap} * gap_t + \varepsilon_{\pi,t}$$

With π_t inflation, π_t^T inflation target and $\varepsilon_{\pi,t}$ an inflation shock. The specification assumes dynamic homogeneity (*i.e.* that the coefficients on past and expected inflation sum to unity).

Monetary policy settings follow a Taylor rule:

$$(4) \quad i_t = Max(\theta_1 i_{t-1} + (1 - \theta_1) * (i^* + \sigma_1(\pi_t - \pi_t^T) + \sigma_2 * gap_t), \bar{\imath})$$

With i_t nominal short-term interest rate, $\bar{\imath}$ a lower threshold under which i_t cannot go and i^* the neutral rate which varies over time. The neutral rate is always consistent with targeted inflation and potential output developments. In euro area countries, monetary policy responds to euro area-wide inflation and output gap, so that country-specific inflation and output gap affect monetary policy to the extent of the weight of the respective country in euro area nominal GDP.

The long-term nominal interest rate on public debt is assumed to follow the short-term rate with a term premium and a fiscal risk. Fiscal risk increases by φ basis points for each additional percentage point of gross debt. The implicit assumption here is that financial markets impose a risk premium on the interest rate applied to debt, that is function of the level of debt.

$$(5) \quad irl_t = i_t + term_t + risk_f + \varepsilon_{i,t}$$

with

$$(6) \quad term_t = \vartheta term_{t-1} + term$$

and

$$(7) \quad risk_f = \varphi d_{t-1}$$

where irl_t is the long-term nominal interest rate bearing on public debt, *term* the term premium, $risk_f$ fiscal risk, d_t public debt-to-GDP ratio, and $\varepsilon_{i,t}$ a shock. The term premium is time-varying, with an auto-regressive component, and in the medium term it converges to its historical average (term).

The real interest rate is computed as the difference between the nominal interest rate and inflation.

$$(8) \quad r_t = irl_t - \pi_t$$

Public balance is broken down into a structural component and a cyclical one, which moves with the output gap.

$$(9) \quad pb_t = \overline{ig}_t + \overline{cg}_t + \overline{tax}_t + (\alpha_{cg} + \alpha_{tax})gap_t + \varepsilon_{pb,t}$$

where pb_t is the public balance, in percentage of GDP and α semi-elasticity of the respective fiscal variable to the output gap. $\overline{pb}_t$ is the cyclically-adjusted primary balance and comprises cyclically-adjusted public investment, public consumption and tax. One option in the model is to activate a fiscal reaction function whereby the primary balance is derived to stabilise the debt-to-GDP ratio over the long term.

Finally, the debt-to-GDP ratio is calculated using a standard debt accumulation formula.

$$(10) \qquad \Delta d_t = \frac{(r_t - \Delta y_t)}{(1 + \Delta y_t)} d_{t-1} - pb_t$$

Parameters and calibration

The model has been constructed for selected OECD economies (Australia, Canada, France, Germany, Italy, Japan, United States) and emerging economies (Brazil, China, India, Indonesia, the Russian Federation and South Africa).

Parameters have been estimated, wherever possible. This is in particular the case for the growth and Phillips curve equations (Table A2.1). Those coefficients have been estimated using annual Economic Outlook data and IMF WEO data.

Some parameters were calibrated using existing literature. Fiscal multipliers have been calibrated using Coenen et al. (2012).

The elasticity of public capital is calibrated using a meta-analysis on the impact of public investment on growth (Bom and Ligthart, 2014). Following Fournier (2016), it is also assumed that the lower the initial stock of public capital in the country, the higher this elasticity. In practice, this means it is lower in Japan and higher in emerging economies.

The hysteresis parameter measures the effect of persistent weak demand on potential output. It is calibrated following Kapadia (2005) and Delong and Summers (2012) to 0.1 in English-speaking economies and 0.2 in continental European countries and in Japan. These values are consistent, though on the low side, with those estimated by Mourougane (2016) using a panel of OECD member countries. Hysteresis is set to zero in emerging economies, which are characterised by widespread informality.

Although there is now a broad recognition that it is important to incorporate the feedback effect of financial markets, no consensus has emerged on the best way to model fiscal risks. The approach adopted in this model is to opt for simplicity and assume the premium depends on the level of the debt-to-GDP ratio.

The parameters entering the Taylor rule are standard. The inflation target is set at the official inflation target in OECD and emerging economies. Central banks are assumed to avoid abrupt jumps in the policy rate by smoothing its adjustment. It is assumed the ECB reaction function is consistent with its de jure mandate and that the central bank targets only inflation.

The cyclical part of the budget is calculated using the elasticities of budget items to the output gap derived in Price, Dang and Botev (2015). The resulting budget semi-elasticity ranges from 0.41 in Japan to 0.61 in France. It has been estimated with error-correction models using disaggregated spending and revenue data.

The steady-state term premium is computed using the average of the observed difference between short and long-term rates over the period 1999 to 2014 in the euro area countries and 1995 to 2014 in the other economies.

Table A2.1. **Calibration**

	Parameter or variable	Value	Source
		Equation g*	
μ	Degree of labour-market hysteresis	0.1 in the United States, the United Kingdom and Canada 0.2 in European countries and in Japan	Calibrated using Kapadia (2005); Delong and Summers (2012)
ε	Elasticity of public capital in the production function	Between 0.05 to 0.25 depending on the initial stock of capital	Bom and Ligthart (2014) and Fournier (2016)
Deprec	Depreciation rate	5%	Average general government capital in the United States using data from the Statistical office (4.6%)
δ	Potential output speed of convergence	-0.3	Calibrated
y^*_{ss}	Steady state rate of potential output growth	2% for the United States, 0.5% for Japan and 1% for the euro area countries	Calibrated
λ_1	Fiscal multiplier public investment	1.1 (0.7 for Japan)	Calibrated using Gechert et al. (2015) and Auerbach and Gorodnichenko (2014)
λ_2	Fiscal multiplier other public spending	1	
λ_3	Fiscal multiplier tax	0.3	
θ_p	Elasticity of energy prices on g	Country specific	Calibrated using energy consumption
		Taylor rule and interest rates	
θ_1	Inertia in interest premium in the Taylor rule	0.5	Calibrated
σ_1	Weight on inflation in the Taylor rule	1	Calibrated
σ_2	Weight on the gap in the Taylor rule	0.5	Calibrated
π_t^T	Inflation target	2% for OECD countries. Set to official objectives in other	Calibrated
$\bar{I}$	Lower limit on the interest rate	-5%	Calibrated
$\varphi_{\square}$	Influence of debt on interest premium	0.5 basis point; 0.1 basis point for Japan	Calibrated
Term	Steady-state term premium	Average difference between long and short-term rates	
		Public deficit	
$\alpha_{cg}, \alpha_{tax}$	Elasticity of fiscal variables to the output gap	Country-specific value, their sum is around 0.4-0.6 and takes into account changing share of each component in GDP and in total government spending/ revenue	Calibrated using Price, Dang and Botev (2015)
Equation g	Panel with same coefficient for all the countries , λs calibrated		SURE Estimation 1990-2014 – OECD countries and emerging economies are estimated in two different panels
Equation Phillips	Panel with same coefficient for all the countries		SURE Estimation 1990-2014 – OECD countries and emerging economies are estimated in two different panels

The Oxford Economics' Global Economic Model (GEM)

The Oxford Economics' Global Economic Model (GEM) is a quarterly system of macro-econometric interlinked equations for a broad set of economic variables. The GEM includes around 100 different countries, both advanced and emerging economies.

The overall economic structure of the GEM consists of a Keynesian paradigm in the short term and a monetarist point of view in the long run. Output is driven by demand-side factors in the short run and supply-side factors in the long run. The long-run output is determined by a Cobb-Douglas production function that uses the capital stock, labour supply and technological progress as inputs. In the simulations presented in this chapter, the potential output equation has been modified to introduce an explicit effect of public capital in output in the long term.

Large G20 economies models pool more than 600 variables and those of smaller economies 400 to 500 variables. The rest of the world is modelled as six regional aggregates of smaller size (100 to 250 variables).

The detailed breakdown of economies includes labour, financial and energy markets. This breakdown and the rich set of countries enable examination of a multitude of various shocks and systemic interactions between the most important economies. In addition, GEM enables sector-level analysis since aggregate output and employment are split into various manufacturing and services industries. In particular, one of the main advantages of this model is that the energy sector (oil, coal and gas) is extensively detailed for the major economies, so consistency between energy prices and supply/demand balances is ensured. Carbon taxes are also incorporated in the model.

The macro-econometric equations are single-variable error-correction model (ECM) estimated using historical data. The ECM makes it possible to disentangle short- and long-run dynamics in a concise manner. In the majority of cases, the functional form for equations is similar across countries and differences in simulation outcomes will reflect mostly differences in estimated parameters. There are some exceptions though, where countries are heavily dependent on a particular sector, such as oil, or where FDI plays a major role, for instance in emerging economies.

The model encompasses a number of options in terms of monetary policy. It usually assumes adaptive expectations, though it can introduce rational expectations for some variables.

Country/region models are linked through trade, prices, exchange rates and interest rates, capital flows and commodities prices.

With the 25-year horizon version of this model, it is possible to run simulations up until 2045 .

The GEM is a widely used macroeconomic model with clients including the IMF, World Bank and a couple of private sector entities, and has been used for modelling economic consequences of climate change before (e.g. University of Cambridge Institute for Sustainability Leadership, 2015).

References

Auerbach, A.J. and Y. Gorodnichenko (2014), "Fiscal Multiplier in Japan", *NBER Working Paper*, No. 19911.

Bom, P. and J. Ligthart (2014), "What Have we Learned from Three Decades of Research on the Productivity of Public Capital?", *Journal of Economic Surveys*, Vol. 28/5, Wiley, Hoboken, pp. 889-916, http://doi.org/10.1111/joes.12037.

Coenen, G., et al. (2012), "Effects of Fiscal Stimulus in Structural Models", *American Economic Journal: Macroeconomics*, Vol. 4/1, American Economic Association, Pittsburgh, http://doi.org/10.1257/mac.4.1.22.

Delong, J. and L. Summers (2012), "Fiscal Policy in a Depressed Economy", *Brookings Papers on Economic Activity*, Brookings Institution, Washington DC, www.brookings.edu/bpea-articles/fiscal-policy-in-a-depressed-economy/.

Fournier, J.-M. (2016), "The positive effect of public investment on potential growth", *OECD Economics Department Working Papers*, No. 1347, OECD Publishing, Paris, http://dx.doi.org/10.1787/15e400d4-en.

Gechert S., A. Hughes Hallet and A. Rannenberg (2015), "Fiscal Multipliers in Downturns and the Effects of Eurozone Consolidation", *CEPR Policy Insight*.

Kapadia, S. (2005), "Optimal Monetary Policy under Hysteresis", Department of Economics, University of Oxford, www.economics.ox.ac.uk/Department-of-Economics-Discussion-Paper-Series/optimal-monetary-policy-under-hysteresis.

Mourougane, A. (2016), "Crisis, Potential Output and Hysteresis", *International Economics*, Vol. 149, Elsevier, Amsterdam, http://doi.org/10.1016/j.inteco.2016.07.001.

University of Cambridge Institute for Sustainability Leadership (2015), "Unhedgeable risk: How climate change sentiment impacts investment", Cambridge.

Price, R.W., T. Dang and J. Botev (2015), "Adjusting Fiscal Balances for the Business Cycle: New Tax and Expenditure Elasticity Estimates for OECD countries", *OECD Economics Department Working Papers*, No. 1275, OECD Publishing, Paris, http://dx.doi.org/10.1787/5jrp1g3282d7-en.

ANNEX 4.A3. Robustness checks on the model simulations

To obtain insights into the sensitivity of the results to model specification, the simulations of a decisive transition have also been run using the large-scale macroeconomic Oxford model (Figure A3.1). Outcome differences between the two tools appear to be small, well within the margin of error. Differences are more noticeable for the short to medium term for emerging economy fossil fuel importers, reflecting the different pace of adjustment embodied in the two models. Differences are larger over the long term, notably for some fossil fuel net exporters, underlining the large uncertainties of the results at this time horizon. They reflect differences in model specifications, notably regarding the reaction of business investment in both models to the fiscal initiative.

This apparent similarity does, however, mask large uncertainties. In particular, GDP outcomes remain sensitive to the assumptions on fiscal multipliers in the short term and the rate of return of public capital in the long run. For instance, in the Yoda model, the output impact could range from 0.1% to 1.25% in the short to medium term in an advanced net fossil fuel importer, depending on how large the short-term fiscal multiplier is in the range between 0 and 2 (Figure A3.2). At this horizon, the effect could be negative in some net fossil fuel exporters, assuming a typical pro-growth package is implemented. A package that is more adapted to the country's needs would enhance the positive counterbalancing effect. Uncertainties are increasing over the long term, especially for emerging economies.

Figure A3.1. **Output impact of a decisive transition to decarbonisation using different models**

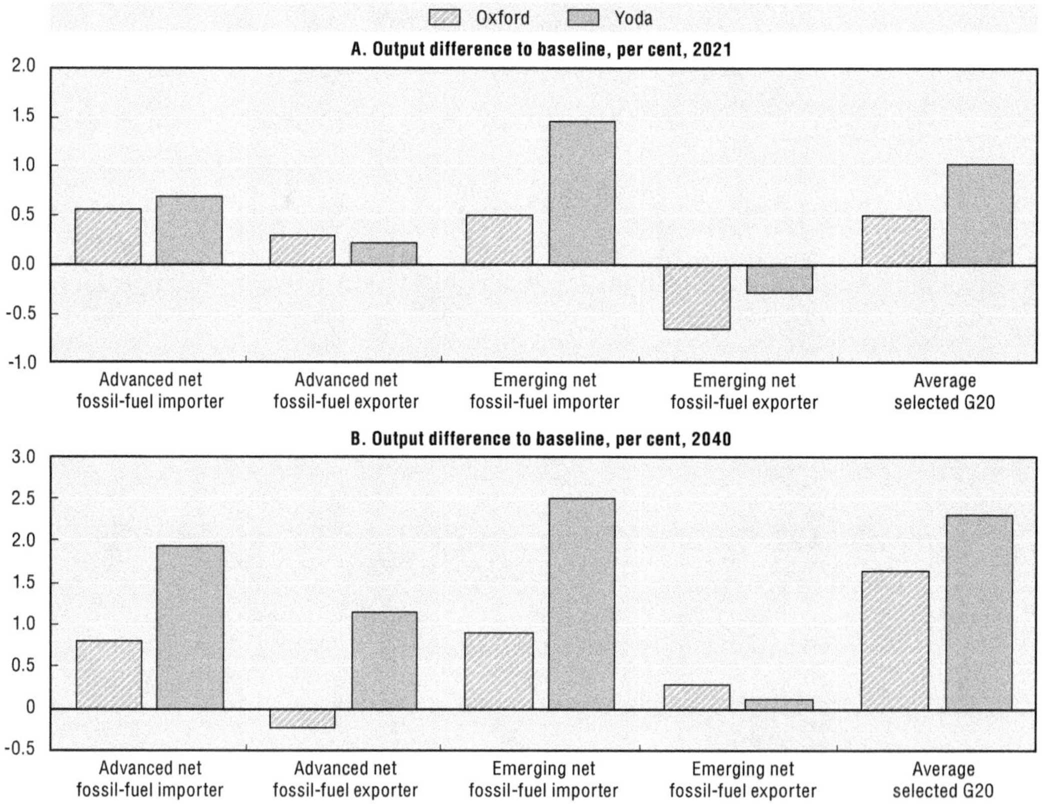

Note: See Figure 4.5.

Figure A3.2. **Uncertainties around output impact of a decisive transition to decarbonisation**

Output difference to baseline, per cent

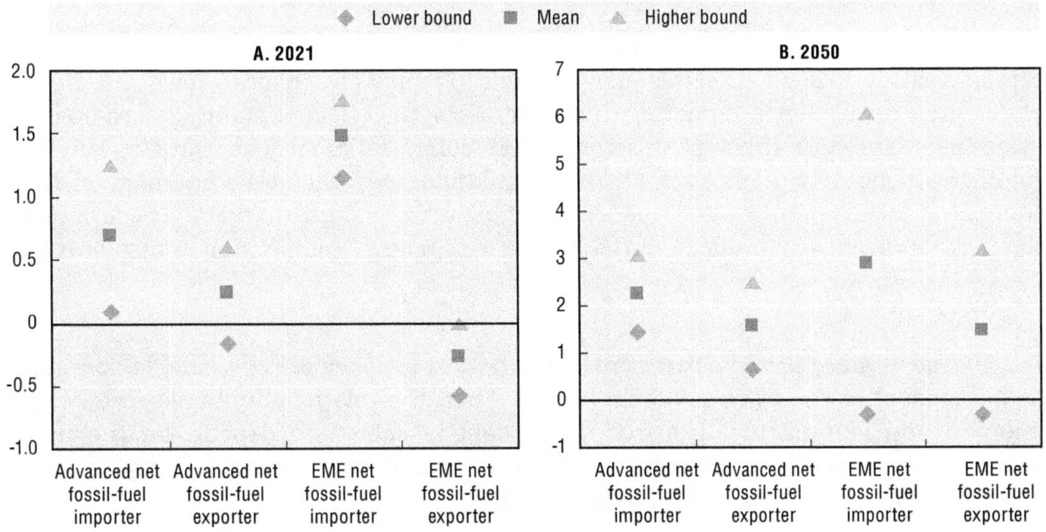

Note: Bounds were computed by varying the short-term fiscal multiplier from 0 to 2, and multiplying the long-term rate of return of public capital by +/-10%. Both assumptions are consistent with estimates from the economic literature.

ANNEX 4.A4. Flexible economies adapt better to climate-change mitigation action

With policies stimulating private investment and innovation in low-carbon technologies, firms are likely to need a flexible environment to be able to adapt to new incentives. This annex presents evidence validating this hypothesis: the effect of energy prices on business investment and the effect of environmental policy stringency on productivity can be significantly positive if product market regulations are "flexible", or, more specifically, conducive to competition and the entry of new firms into markets. These effects are significantly negative if product market regulations restrict competition or firm entry.

Methodology and data

Two separate approaches are considered here to show evidence that a flexible economy helps to adapt to the climate change action. First, the relationship between energy price inflation and manufacturing firms' investment is shown to depend on the flexibility of domestic product market regulations. In a second one, a similar story is found for manufacturing industries' productivity growth.

Manufacturing investment, energy prices and regulations

The evidence on manufacturing investment effects is built on the specification and the data used by Dlugosch and Koźluk (2017) to estimate the effect of energy prices on business investment at the firm level. Their baseline specification (DK hereafter) is estimated with a linear fixed-effect least square estimator. In particular, the time fixed effect rules out worldwide demand shocks that can affect both energy prices and investment. It is augmented with an interaction term between the product regulation indicator and energy price inflation:

$$(1) \quad \frac{I_{isct}}{K_{isct}} = \beta_1 * \Delta EPI_{sct-1} + \beta_2 * \mathrm{PMR}_{ct-1} + \boldsymbol{\beta_3} * \mathbf{PMR_{ct-1}} \boldsymbol{\Delta EPI_{sct-1}} + \beta_4 \mathrm{X}_{isct} + \alpha_i + \theta_t + \varepsilon_{isct}$$

where:

- $\frac{I_{isct}}{K_{isct}}$ is the ratio of investment to the capital stock using the book values of capital expenditure (I) and capital stock (K) of firm i in sectors s and c countries at time t.
- ΔEPI is the three-year moving average of energy prices inflation.[1]
- PMR denotes the OECD's product market regulation indicator
- X_{isct} is a vector of additional control variables
- α_i is a firm-specific intercept, and θ_t a year fixed effect.

Productivity, environmental stringency and regulations

The effect of environmental stringency on productivity is analysed at the sector level, following the specification and data used in Albrizio, Koźluk and Zipperer (2017).[2] For this analysis, the specification is augmented with the OECD's Product Market Regulation (PMR) indicator as follows:

$$\Delta \ln MFP_{cit} = \alpha_1 + \alpha_2 \, (ED_{i\,1987} \Delta EPS_{ct}^{3MA}) + \alpha_3 \, gap_{cit-1} \, (ED_{i\,1987} \Delta EPS_{ct}^{3MA}) + \alpha_4 \, gap_{cit-1}$$
$$+ \alpha_5 \, \Delta \ln \widehat{MFP}_{it} + \boldsymbol{\alpha_6} \, \boldsymbol{PMR_{ct-1}} \, (\boldsymbol{ED_{i\,1987} \Delta EPS_{ct}^{3MA}}) + \alpha_7 \, PMR_{ct-1} + x_{cit}\gamma + \eta_t + \delta_{ci} + \epsilon_{cit}$$

where:

- $\Delta \ln MFP_{cit}$ is the multifactor productivity growth for each combination of country c and industry i.
- ΔEPS^{3MA} is a three-year moving average of the change of the country EPS and captures the tightening of country's environmental policy stringency. A change in a country's EPS is interacted with pre-sample industry environmental dependence (*ED*).[3]
- The third term allows for nonlinear effects of the policy as a function of the technological gap, defined as the distance to the country-industry frontier $gap = \ln\left(\frac{\overline{\text{MFP}}_i}{\text{MFP}_{ci}}\right)$.
- The fourth term is the distance to the productivity frontier, which allows for technological catch-up effects.
- The fifth term is the growth in the leader MFP and represents the technological pass-through.
- x_{cit} is a vector of additional country and industry controls, which varies across the econometric specifications considered (see section "Results").
- PMR denotes the product market regulation indicator. Higher values indicate a higher degree of restrictiveness.
- Finally, aside from the above, the specification controls for a common time trend , the financial crisis, output gap and country-industry fixed effects (δci) or, alternatively, country and industry-fixed effect are included separately: δc and δi.

Data

Data used for the investment estimates are described in detail in Dlugosch and Koźluk (2017). Firm level data are from Thomsons Reuters Worldscope, which reports mandatory data for listed companies. A main advantage of data reported mandatorily lies in its reliability. However, there are two caveats. First, listed companies may differ from non-listed firms, which cannot be observed. However, Dlugosch and Koźluk (2017) find a striking similarity in practice between manufacturing industry numbers from the OECD STAN database and aggregates from Worldscope, supporting the notion that a large part of outcomes can be explained by a small number of large firms (Gabaix, 2011). Second, investment figures include investment in foreign subsidiaries that is likely to mitigate the role of domestic regulation, so one can assume that the true importance of regulation is larger than the one identified here. The energy price index is constructed by Sato et al. (2015) for 12 sectors in each country over 1995-2011, by weighting country-level IEA fuel prices for four different types of fuel – oil, gas, coal and electricity – by the consumption of these fuels in each country-sector.

Data used for the productivity estimates are described in detail in Albrizio, Koźluk and Zipperer (2017). The index of multifactor productivity is the residual from a log Cobb-Douglas production function, with the labour factor intensity equal to two-thirds, constructed using the OECD Structural Analysis database (STAN) and the Productivity Database By industry (PDBi). The technological frontier at the industry level as well as the distance to frontier (DTF), are constructed following Bas et al. (2016). The environmental policy stringency (EPS) is measured with a new composite index developed by the OECD. This index covers 24 OECD countries over the period 1990-2012 and summarises environmental policy stringency across selected instruments, primarily for energy and transport (for a detailed description, see Botta and Koźluk, 2014).

The stringency of product market regulation is measured here with the OECD product market regulation (PMR) indicator. This indicator comprises three high-level components: state control, barriers to entrepreneurship, and barriers to foreign trade and investment, and several subcomponents (Figure A4.1). It was developed in 1998 (Nicoletti et al., 2000), is available every five years, and was last updated for 2013 (Koske et al., 2015). In OECD countries, the economy-wide indicator shows a decline in regulation between 1998 and 2008, and broad stabilisation since then.

Figure A4.1. **The tree structure of the PMR indicator set**

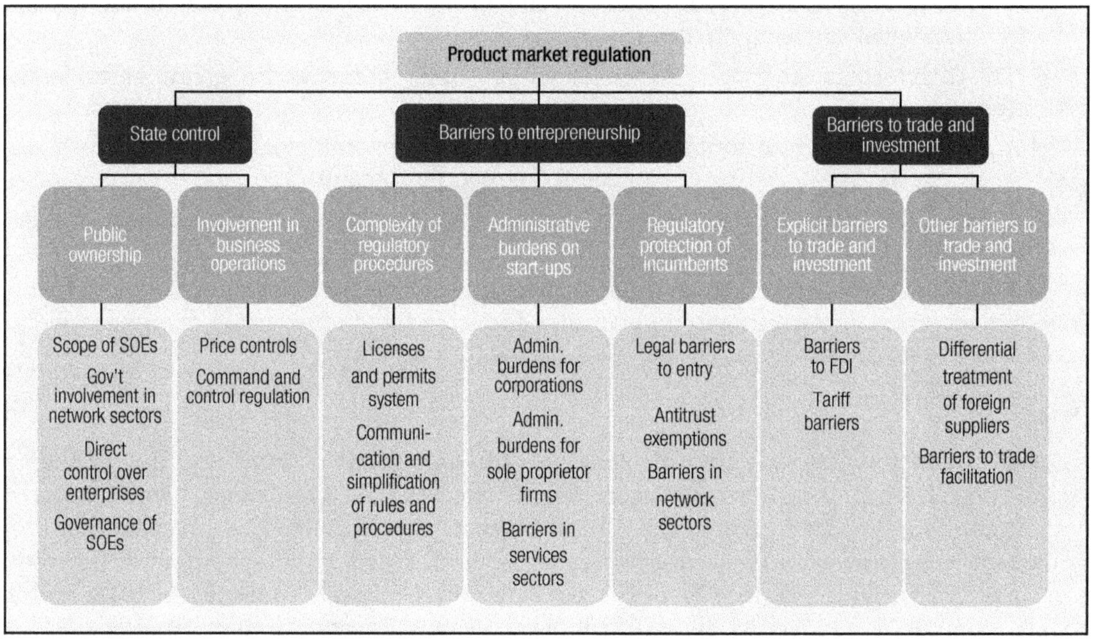

Source: Koske et al. (2015).

Empirical results

The interaction between product market regulation and energy price inflation significantly affects manufacturing investment (Table A4.1). The first column is the baseline specification without product market regulation (Dlugosch and Koźluk, 2017, Table A2.1). The second column shows the effect of shrinking the sample to years for which the product market regulation indicator is available. The third to sixth columns provide the evidence that the interaction between product market regulations and energy price inflation matters. First, results are reported with interpolated product market regulation data to preserve the year coverage. This setting is used to illustrate that the effect could be slightly larger if one only considers periods of energy inflation that rise above a typical core inflation target (2%): the adaptation issue should be more prominent when energy prices increase. If a more rigid economy implies postponing or spreading the necessary adjustment over time rather than avoiding it, one should expect a bounce-back effect in a highly regulated economy. Such an effect cannot be identified in the estimates (column 5).[4] Columns 3 to 5 need to be interpreted with care as interpolation can generate serial autocorrelation, which would lead to underestimated standard errors. Column 6 thus provides a more reliable baseline, in which the magnitude of the interaction effect is similar to the one observed in the other columns.

Table A4.1. **Interaction between energy price inflation and overall product market regulation stringency**

Dependent variable: Investment / Total Assets	(1) Baseline (DK)	(2) Baseline with PMR sample	(3) Interact. PMR (interpolated)	(4) Interact. PMR (interpolated)	(5) Interact. PMR (interpolated)	(6) Interact. PMR (non-interp.)
EPI Inflation sample	all	all	all	EPI>2%	all	all
PMR (t-1) * EPI (MA) (t-1)			**-0.0860*****	**-0.0965*****	**-0.1213*****	**-0.1116*****
			(0.0189)	**(0.0292)**	**(0.0216)**	**(0.0368)**
PMR (t-1) * EPI (MA) (t-4)					-0.0030	
					(0.0206)	
PMR (t-1)			-0.0013	0.0146***	0.0132***	0.0033
			(0.0022)	(0.0033)	(0.0030)	(0.0048)
EPI (MA) (t-1)	-0.0107*	0.0177	0.1215***	0.1777***	0.1947***	0.1915***
	(0.0057)	(0.0154)	(0.0291)	(0.0433)	(0.0330)	(0.0597)
EPI (MA) (t-4)					-0.0054	
					(0.0311)	
Layoff rates *	-0.0036***	-0.0114***	-0.0035***	-0.0012*	-0.0021***	-0.0101***
Employment Protection (t-1)	(0.0006)	(0.0012)	(0.0006)	(0.0007)	(0.0006)	(0.0014)
Dependency External Fin. *	0.0120*	0.0611***	0.0098	-0.0069	0.0141*	0.0560***
Fin. Dev. (t-1)	(0.0065)	(0.0185)	(0.0065)	(0.0069)	(0.0072)	(0.0185)
Observations	68,334	12,359	68,332	46,679	54,638	12,359
Adj. R2	0.412	0.354	0.412	0.429	0.420	0.355

Note: All models include firm- and time-fixed effects, sales over total capital, lagged out gap and lagged real interest rates as further controls. Estimated coefficients are not shown due to brevity of presentation. EPI inflation (MA) denotes the three-year moving average of changes in the energy price indicator. Energy intensity is the share of electricity, water and gas inputs in total inputs to the production of each industry. Low and high levels are defined as being above or below the pooled median. The energy intensity has been demeaned before application. Low energy-intensive sectors thus have a negative sign. Time sample: 1999-2011. Firm clustered standard errors in parentheses. *, **, *** denote significance at the 10, 5, and 1% level respectively.

The setup allows some insights into what types of regulations may matter most. Estimates with disaggregated PMR indicators suggest that barriers to entrepreneurship and international barriers to trade and investments matter the most, while there is a lack of evidence on the role of the state involvement in business operation (Table A4.2, column 1). Among these two main fields of regulations, there are five sub-items. Their relative roles are explored in two different ways: either by including all the interaction terms together (Table A4.2, column 2), or including them one by one and reporting cases where the interaction term is significant at least at the 10% level (Table A4.2, columns 3 to 5). There is evidence that administrative burdens on start-ups, complexity of regulatory procedure and explicit barriers to international trade and investment are more likely to matter. Notably, the results using disaggregated PMR indicators should be interpreted with caution as it is hard to disentangle the role of each regulation; in practice, there is a substantial correlation across subdomains.[5]

Table A4.2. Interaction between energy price inflation and subdomain product market regulation stringency

Dependent variable: Investment / Total Assets	(1)	(2)	(3)	(4)	(5)
Interpolated PMR	no	no	no	no	no
EPI Inflation sample	all	all	all	all	all
State Control (t-1) * EPI (MA) (t-1)	**0.0237**				
	(0.0398)				
Barriers to Entrepreneurship (t-1) *	**-0.0743****				
EPI (MA) (t-1)	(0.0365)				
Administrative burdens on start-ups (t-1)		-0.0593*	-0.0476**		
* EPI (MA) (t-1)		(0.0314)	(0.0194)		
Complexity of regulatory procedures (t-1) *		-0.0177		-0.0335**	
EPI (MA) (t-1)		(0.0144)		(0.0130)	
Regulatory protection of incumbent (t-1) *		0.0087			
EPI (MA) (t-1)		(0.0541)			
Barriers to trade and investment (t-1) *	**-0.0752****				
EPI (MA) (t-1)	**(0.0356)**				
Explicit barriers to trade and investment (t-1) *		-0.1187***			-0.0941***
EPI (MA) (t-1)		(0.0361)			(0.0254)
Other barriers to trade and investment (t-1)		0.0466			
* EPI (MA) (t-1)		(0.0363)			
State control (t-1)	0.0071**				
	(0.0028)				
Barriers to entrepreneurship (t-1)	-0.0038				
	(0.0042)				
Administrative burdens on start-ups (t-1)		-0.0000	-0.0005		
		(0.0044)	(0.0025)		
Complexity of regulatory procedures (t-1)		-0.0022		-0.0020	
		(0.0019)		(0.0017)	
Regulatory protection of incumbent (t-1)		-0.0064			
		(0.0062)			
Barriers to trade and investment (t-1)	0.0029				
	(0.0051)				
Explicit barriers to trade and investment (t-1)		-0.0091*			-0.0081**
		(0.0047)			(0.0041)
Other barriers to trade and investment (t-1)		0.0035			
		(0.0044)			
EPI (MA) (t-1)	0.1850**	0.1796***	0.1081***	0.1055***	0.0760***
	(0.0760)	(0.0688)	(0.0412)	(0.0296)	(0.0218)
Observations	12.359	12.359	12.359	12.359	12.359
Adj. R2	0.356	0.357	0.354	0.355	0.355

Note: All models include firm- and time-fixed effects, Layoff rates * Employment Protection, Dependency External Fin. * Financial development, sales over total capital, lagged out gap and lagged real interest rates as further controls. Estimated coefficients are not shown due to brevity of presentation. EPI inflation (MA) denotes the three-year moving average of changes in the energy price indicator. Energy intensity is the share of electricity, water and gas inputs in total inputs to the production of each industry. Low and high levels are defined as being above or below the pooled median. The energy intensity has been demeaned before application. Low energy intensive-sectors thus have a negative sign. Time sample: 1999-2011. Firm clustered standard errors in parentheses. *, **, *** denote significance at the 10, 5, and 1% level respectively.
Source: OECD calculations.

There is similar evidence that the interaction between product market regulation and environment policy stringency significantly affects industry productivity (Table A4.3). The so-called Porter Hypothesis – according to which strict environment regulations can induce incentives for innovation, efficiency improvements and within-firm reallocation that may lead to higher productivity (Porter, 1991; Porter and van der Linde, 1995) – is controversial (Ambec et al. 2011; Lanoie et al. 2011; Kozluk and Zipperer, 2015), but in more flexible economies, firms should be more inclined to reap such gains. This would imply that the less regulation impedes competition, the higher the multifactor productivity gains reaped from environmental policy incentives. Empirical results appear in line with the Porter Hypothesis, using an interpolated PMR indicator (columns 1 and 2), the initial PMR indicator (columns 3 and 4) and a dummy that takes value one if the product market regulation indicator is above the sample median of the given year (columns 5 and 6). Last, a disaggregate product market regulation indicator provides tentative evidence that barriers to international trade and investment could have an important role in hindering productivity gains (columns 7 and 8).

Table A4.3. **Interaction between environment protection stringency and overall product market regulation stringency**

Dependent variable: Mfp growth Interpolated PMR PMR dummy	(1) yes no	(2) yes no	(3) no no	(4) no no	(5) no yes	(6) no yes	(7) no no	(8) no no
Leader MFP growth	0.17***	0.13***	0.31***	0.37***	0.31***	0.37***	0.31***	0.37***
	(0.033)	(0.033)	(0.10)	(0.099)	(0.10)	(0.098)	(0.10)	(0.100)
Distance to frontier	0.22***	0.10***	0.16***	0.11***	0.15***	0.11***	0.15***	0.11***
(lagged)	(0.029)	(0.016)	(0.049)	(0.030)	(0.048)	(0.029)	(0.049)	(0.029)
EPS tightening (MA)	0.63***	0.48***	0.96**	0.80*	0.42***	0.27**	0.86**	0.67*
	(0.19)	(0.15)	(0.44)	(0.41)	(0.13)	(0.11)	(0.42)	(0.38)
Effect of gap on EPS tightening (MA)	-0.16	-0.18**	-0.71**	-0.43	-0.67**	-0.40	-0.79**	-0.49*
	(0.12)	(0.087)	(0.29)	(0.28)	(0.30)	(0.28)	(0.32)	(0.27)
PMR (t-1)* EPS	**-0.31*****	**-0.22****	**-0.41***	**-0.40***	**-0.29*****	**-0.26*****		
tightening (MA)	**(0.10)**	**(0.089)**	**(0.22)**	**(0.22)**	**(0.087)**	**(0.082)**		
State control (t-1) *							-0.061	-0.037
EPI (MA) (t-1)							(0.12)	(0.11)
Barriers to entr. (t-1)							-0.073	-0.084
* EPI (MA) (t-1)							(0.23)	(0.099)
Barriers to trade and							-0.39	-0.37**
inv. (t-1) * EPI (MA) (t-1)							(0.45)	(0.15)
PMR (t-1)	0.058***	0.030	0.069	0.070	0.028*	0.031*		
	(0.022)	(0.021)	(0.055)	(0.054)	(0.017)	(0.017)		
State control (t-1)							0.0099	0.010
							(0.027)	(0.025)
Barriers to entrepre-							0.0090	0.0094
neurship (t-1)							(0.034)	(0.030)
Barriers to trade and							0.10	0.10*
investment (t-1)							(0.074)	(0.059)
Fixed effects								
Country*Industry	Yes	No	Yes	No	Yes	No	Yes	No
Country	No	Yes	No	Yes	No	Yes	No	Yes
Industry	No	Yes	No	Yes	No	Yes	No	Yes
N	1.541	1.541	366	366	366	366	366	366
adj. R2	0.17	0.18	0.25	0.23	0.25	0.23	0.25	0.23

Note: Robust standard errors in parentheses and they are clustered at country-industry level; *** denotes statistical significance at the 1% level, ** significance at 5% level, * significance at 10% level. (MA): denotes the moving average of the EPS change over three-years-lags. Changes in a country's EPS are interacted with pre-sample industry environmental dependence. All specifications include the following controls: output gap, dummy for crisis and year trend, employment protection legislation (OECD EPL), and country's degree of capital account openness (Chinn-Ito Index).

Simulation inputs for the modelling

Under a number of simplifying assumptions,[6] the effect of energy prices on PMR on manufacturing investment as a share of GDP can be simulated as follows:

$$effect_t^{v0} = (\beta_1 * \Delta EPI_{ct-1} + \beta_3 * \text{PMR}_c \Delta EPI_{ct-1}) * (\frac{I}{GDP})/(\frac{I}{K})$$

where β_1 and β_3 are the coefficients estimated in the sixth column of Table A4.1, ΔEPI_{ct-1} is a country level change in end-user energy price for the manufacturing sector used in the scenarios, PMR_c is the PMR level in 2013 (2008 in Indonesia), the investment to capital ratio is the average over 2009-11 observed in the firm-level database used for the regressions (from Worldscope), and the manufacturing investment to GDP ratio is reported in national accounts. For countries for which the manufacturing investment data is not available, the manufacturing investment to GDP ratio is taken from the GTAP8 database. This computation can cover 17 out of the G20 countries.

The resulting simulated effect is a change in manufacturing investment relative to a baseline scenario, expressed as a share of GDP. This is included in the modelling as an additional investment shock, reflecting the path of energy price of the scenario.

Notes

1. A moving average specification has been used by Albrizio, Koźluk and Zipperer (2017) and builds on the argument that investment is usually planned ahead; a reaction to energy prices may thus take some time.
2. This analysis is not performed at the firm level because of limited country coverage: Albrizio, Koźluk and Zipperer (2017) firm level estimations cover 12 countries. Lack of cross-country information is a particular concern here because the product market regulation is observed at the country level.
3. Industry environmental dependence is an index of industry pollution intensity ranging from 0 to 1 and is used to proxy for industries' exposure to environmental regulation. Pre-sample observations are used to ensure exogeneity (see Albrizio, Koźluk and Zipperer, 2017).
4. Adding an additional lag of the energy price inflation does not lead to any identification of a bounce back effect either.
5. Insignificance can also be due to the inherent uncertainty surrounding such detailed estimates, and selection results can vary if the selection method is modified. For instance, regulatory protection of incumbents would be identified as strongly significant is a similar exercise with interpolated PMR data, and this can reflect the interest of incumbents in deterring transitions.
6. In particular, the control variables used in the regression are assumed to be held constant in the future, and the coefficient estimates derived from the Worldscope sample are assumed to hold for the whole manufacturing sector.

References

Albrizio, S., T. Koźluk and V. Zipperer (2017), "Environmental Policies and Productivity Growth: Evidence across Industries and Firms", *Journal of Environmental Economics and Management*, Vol. 81, Elsevier, Amsterdam, pp. 209-226, https://doi.org/10.1016/j.jeem.2016.06.002.

Bas, M., et al. (2016), "The effects of input tariffs on productivity: panel data evidence for OECD countries", *Review of World Economics*, Springer Nature, Vol. 152/2, pp. 401-424, http://doi.org/10.1007/s10290-016-0247-z.

Botta, E. and T. Koźluk (2014), "Measuring Environmental Policy Stringency in OECD Countries: A Composite Index Approach", *OECD Economics Department Working Papers*, No. 1177, OECD Publishing, Paris, http://dx.doi.org/10.1787/5jxrjnc45gvg-en.

Dlugosch, D. and T. Kozluk (2017), "Energy prices, environmental policies and investment: Evidence from listed firms", *OECD Economics Department Working Papers*, No. 1378, OECD Publishing, Paris, http://dx.doi.org/10.1787/ef6c01c6-en.

Gabaix, X. (2011), "The Granular Origins of Aggregate Fluctuations", *Econometrica*, Vol. 79/3, Wiley, Hoboken, pp. 733-722, http://doi.org/10.3982/ECTA8769.

Johansson, Å., et al. (2008), "Taxation and Economic Growth", *OECD Economics Department Working Papers*, No. 620, OECD Publishing, Paris, http://dx.doi.org/10.1787/241216205486.

Koske, I. et al. (2015), "The 2013 update of the OECD's database on product market regulation: Policy insights for OECD and non-OECD countries", *OECD Economics Department Working Papers*, No. 1200, OECD Publishing, Paris, http://dx.doi.org/10.1787/5js3f5d3n2vl-en.

Nicoletti, G., S. Scarpetta and O. Boylaud (2000), "Summary Indicators of Product Market Regulation with an Extension to Employment Protection Legislation", *OECD Economics Department Working Papers*, No. 226, OECD Publishing, Paris.DOI: http://dx.doi.org/10.1787/215182844604.

Porter, M. (1991), "America's green strategy", *Scientific American*, Vol. 264/4, Springer Nature, London.

Porter, M. E. and C. van der Linde (1995), "Toward a New Conception of the Environment Competitiveness Relationship", *Journal of Economic Perspectives*, American Economic Association, Pittsburgh, Vol. 9/4, pp. 97-118, http://doi.org/10.1257/jep.9.4.97.

Sato, M. et al. (2015), "International and sectoral variation in energy prices 1995-2011: how does it relate to emissions policy stringency", Grantham Research Institute on Climate Change and the Environment, London School of Economics, London.

Chapter 5

Policies for scaling up low-emission and resilient investment

Triggering the investment needed for low-emission and resilient economic growth requires a co-ordinated constellation of policies, spanning structural pro-growth reforms, climate policies and the broader investment environment. This chapter first considers how structural reforms can kick-start growth while also supporting the low-carbon transition. It then focuses on climate-change policies, including carbon pricing. Next, the chapter examines how policies making up the broader investment environment may be misaligned with climate objectives. If investment conditions are not conducive to low-carbon investments, even the best-designed climate policy is unlikely to be effective. Finally, the chapter considers how policy can be used to better orient public infrastructure planning and implementation towards low-carbon options, both at national and sub-national levels.

Quality investment, whether from the public or private sector or a combination of the two, is a key driver of low-emission and resilient growth. A successful transition requires policies that influence investment decisions across the entire fixed capital stock – long-lived infrastructure and buildings, and shorter-term production capital and machinery – as well as individual and business consumption decisions. This chapter shows how triggering the right investment while stimulating economic growth requires a co-ordinated constellation of policies, spanning both core climate policy and structural pro-growth reforms.[1]

The chapter starts by considering how structural reforms can kick-start growth while also supporting the low-carbon transition. It then focuses on policies targeted specifically at climate change objectives, including carbon pricing. Next, the chapter examines how policies making up the broader investment environment may be misaligned with the objectives of climate policies, building on *Aligning Policies for a Low-carbon Economy* (OECD/IEA/NEA/ITF, 2015) and the OECD *Policy Framework for Investment* (OECD, 2015a). If investment conditions are not conducive to low-carbon investments – or if policy conditions favour carbon-intensive investments – even the best-designed climate policy is unlikely to be effective. Figure 5.1 shows the interaction of structural, climate and investment policies. Finally, the last section of the chapter considers how policy can better orient public infrastructure planning and implementation towards low-carbon options. Public sector decision-making has a key role in infrastructure investment at both the national and sub-national levels. In some countries, the government remains the key actor and investor in infrastructure, whether through traditional public procurement or through state-owned enterprises.

Several policy themes run throughout the chapter: the need for country specificities to be taken into account, the need for policies to be inclusive and progressive (see also Chapter 6) and the role of innovation. Innovation is a crucial component of low-carbon growth and is influenced by policy in many ways, beyond core public funding of research and development. In addition, in this chapter policies are considered in particular for how they influence investment decisions for infrastructure. Different or additional incentives may be required to influence infrastructure investment decisions, as opposed to consumption or operational choices, for example due to the longer-term nature of infrastructure projects or their contribution to the public good.

Figure 5.1. Structural reform policies underpinning growth and resilience

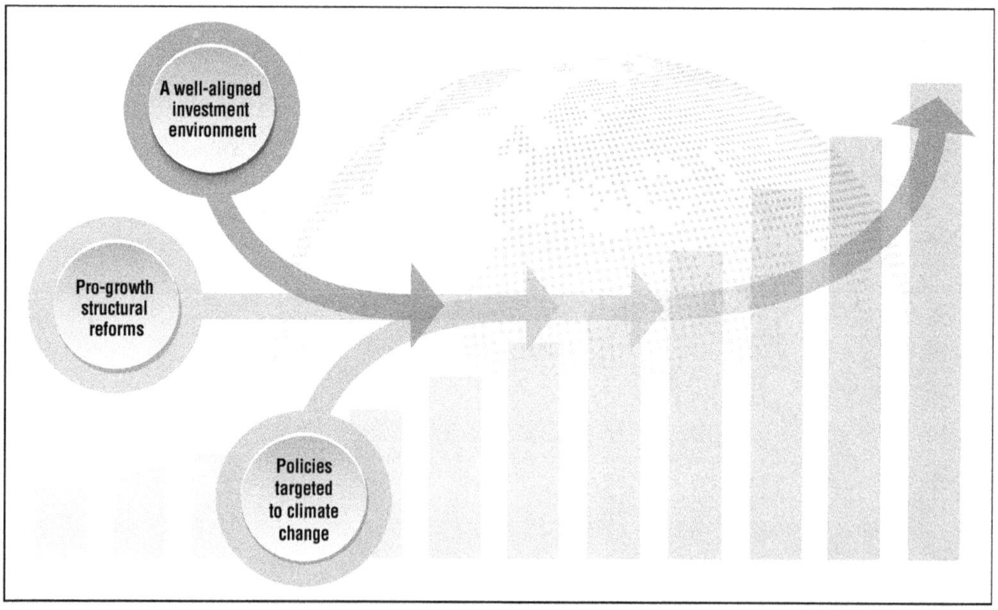

Each country's economic structure, starting point and broader policy conditions will influence the policies required for attracting investment for the transition, their interaction with one another and with existing policies, and the best form of implementation. Chapter 2 indicated how greatly country contexts differ, within the G20 and more widely. Factors with an important bearing on policy options include:

- The nature of the existing infrastructure and other capital stock, and the sectoral investment needs required by low-emission pathways. Infrastructure will need to be expanded in emerging economies, and replaced and maintained in advanced economies (see Chapter 3).
- The structural factors that affect the agility and responsiveness of an economy to a growth stimulus, such as labour market flexibility, access to education and social protection (see Chapter 4).
- The openness of infrastructure sectors to private investment. The public and private sectors have different roles depending on a country's level of development (mobilisation of private finance is discussed in Chapter 7).
- Countries' social, economic and political landscapes, and governments' political will to act on climate, which are heavily influenced by factors such as poverty, access to energy and dependence on fossil fuels (see Chapters 2 and 6).
- The availability of resources, both financial and human, the strength of institutions, and capacity of the government and regulatory and business institutions (including state-owned enterprises, SOEs) to implement the needed reforms and uphold the rule of law.

Structural reform policies with benefits for both growth and the low-carbon transition

To create the conditions for low-emission growth, governments need first to ensure that the economy as a whole is conducive to growth and open to competitive investment. Structural reform policies that promote growth include product and labour market reforms, reforms that improve access to education and training, and those that increase knowledge-based capital. Such pro-growth reforms can also support the low-carbon transition by making economies more flexible, adaptable and resilient, including to climate impacts.

Reforms leading to well-functioning product and labour markets, as well as policies that do not trap resources in inefficient firms, can facilitate the reallocation of resources to their most productive use. Such structural policies play a key role in boosting productivity, the main determinant of long-term economic growth (OECD, 2015b). They can also help ensure that resources are allocated in a way that is consistent with moving to low-carbon economic activities, and minimise adjustment costs of getting there. Policies that help diffuse technology can also boost productivity, allowing more firms to approach the technological frontier.

Knowledge-based capital has become a key determinant of long-term productivity growth. It is particularly important in the context of the low-emission transition, as it accelerates the adoption of new technologies (OECD, 2015c; Andrews and Criscuolo, 2013). Among the different forms of intangible knowledge capital, R&D is vital. As a key driver of innovation, it reduces the costs of the transition to low-emission pathways. Taking advantage of knowledge-based capital to accelerate low-emission development depends on market reforms that allow reallocation of labour and capital to their most productive uses, as modelled in Chapter 4. This is because the initial cost incurred in deploying knowledge-based capital and R&D typically does not increase when it is combined with increasing amounts of other inputs (labour, capital) in the production of goods or services.

Provisions in employment protection legislation need to ensure that labour markets are flexible enough while ensuring a "just transition" for workers. As discussed in Chapter 6, the modest aggregate effect on jobs of the transition hides substantial job losses and geographical dislocation in some sectors, in addition to significant creation of new jobs, some of which require new skills. While employment protection legislation is instrumental in guaranteeing a just transition, it should not impose heavy or unpredictable costs on hiring and firing (Andrews and Cingano, 2014). In some emerging economies, employment legislation discourages hiring on formal contracts and keeps workers in low-productivity activities in the informal sector, thus excluding them from social safety nets and hampering worker mobility. Simpler and more flexible labour laws and broader safety nets can help workers move from the informal to the formal sector. In Turkey, for example, employment protection rules nurture a large informal sector (OECD, 2017a), and the social safety net for displaced workers could be improved by making public support for retraining and job search more reliable.

Identifying and addressing skills bottlenecks can help create a pipeline of low-carbon, growth-enhancing investment projects. When low-emission technologies create demand for skills, new vocational training programmes can be developed and top-up training offered to the existing workforce. The recent increase in environmental patenting underlies the importance of preparing the workforce for a period of rapid eco-innovation, including by raising science, technology and engineering skills (OECD, 2012a). There may be a significant role for local labour market institutions in identifying and satisfying specific training needs, for example, with respect to transport, urban development and waste management (OECD, 2014a).

Policies facilitating mobility of workers can boost growth and smooth impacts of the transition. Improving access to low-cost housing and assisting workers to relocate can help them to benefit from economic opportunities where they arise. Speeding up administrative procedures for building permits can boost responsiveness of housing supply, which can benefit the transition if it is combined with efforts to accelerate energy efficiency investment in housing. Transaction costs affecting the buying and selling of dwellings – such as stamp duties, acquisition taxes or other fees – lower mobility. On the other hand, policies that improve access to low-cost quality housing – for example, by introducing housing benefits or boosting the supply of low-rent social housing – improve well-being of low-income households, especially if the housing is made available in areas well-connected to jobs. Regulation that balances landlords' and tenants' interests can encourage the take-up of rented housing, encouraging mobility. Flexibility in the regulation of rents can also improve incentives for energy-efficiency investment, helping to overcome landlords' reluctance to make such investment (Andrews et al., 2011; Ameli and Brandt, 2015).

Policies that improve access to education make the most of the potential of youth from disadvantaged socio-economic backgrounds and improve intergenerational mobility across countries, boosting the inclusiveness of economic growth. They are also important to boost low-emission growth and to meet demand for new skills. Higher enrolment rates in early childhood care and education, and higher spending on childcare and early education, for example, tend to lower the influence of socio-economic background on students' secondary education achievement (Causa and Johansson, 2009). In emerging economies, access to education is also important to strengthen growth and ensure more equitable spread of benefits, including adult education to make it easier for workers to move to new jobs (OECD, 2017a). Cash transfers to low-income families, conditional on school attendance, have proven effective to improve education access and outcomes in some countries, such as Brazil (OECD, 2013a).

Some aspects of financial policy and regulation can also be considered as structural reforms important for the transition. Bankruptcy regimes that do not sanction small-business failure too severely can foster experimentation with new, riskier technologies, including low-carbon technologies. This increases the ability of economies to learn from new innovations, making it more likely that entrepreneurial ventures are brought to the market. Reducing the cost of winding down a business also makes it less likely that (inefficient) firms with low growth potential and outdated technologies, including high-carbon business models, are prevented from "exiting" the market and instead can release resources and valuable skills to more innovative business ventures.

Legal systems need to support efficient resource allocation and raise the returns to innovation, in part by limiting the duration of legal procedures such as bankruptcy (Andrews et al.; OECD, 2015c). In Mexico, improvements in the legal system could enhance the efficacy of contracts and the security of property rights, allowing the entry of firms with high potential to grow.

Reforms to improve the efficiency of lending and capital markets are also important for the efficient reallocation of resources, notably policies to avoid bailing out banks, including too-big-to-fail banks (Chapter 7). Reluctance to pursue such policies may retain capital in high-income countries, where bail-out expectations are likely to be more credible, while investment needs to achieve decarbonisation and sustainable development are biggest in emerging economies.

The rapid economic, technological and structural changes needed to meet the Paris Agreement objectives require responsive and inclusive economic and social systems. Pro-growth structural reforms may appear secondary to policy interventions focused on GHG mitigation or climate resilience. Yet they are critical for a more effective growth-oriented transition.

Strong and coherent climate policy as the basis for the transition

Core climate policies are the basic building block to shift investment and decision-making towards low-emission and climate-resilient options. They include carbon pricing, the removal of fossil fuel subsidies, and policies that complement carbon pricing. Such complementary policies include targeted investment incentives (e.g. feed-in tariffs or tenders for renewables), standards and direct regulation (e.g. to overcome barriers to energy efficiency, such as landlord-tenant situations) and information provision (e.g. energy efficiency labelling). Core climate policies also include measures to enhance adaptation to climate change impacts.

Carbon pricing: an essential emissions mitigation policy

Carbon pricing currently delivers weak investment signals.

Putting a price on GHG emissions is an essential climate mitigation policy. Pricing emissions, through taxes or tradable permits, encourages emitters to seek cost-effective abatement options (Box 5.1). The introduction and strengthening of carbon prices can also signal strong policy commitment, which may have knock-on effects on behaviours and activities not directly subject to the price.

Box 5.1. **Carbon pricing instruments in practice**

Carbon pricing includes taxes on energy use, carbon taxes and carbon emission permit prices, all of which increase the price of carbon emissions relative to other products. Carbon taxes and carbon emission permit prices (from emissions trading systems) are sometimes called "explicit" carbon prices as they are directly aimed at influencing decisions based on carbon content. Taxes on energy use are sometimes referred to as "implicit" carbon prices as they implicitly price the carbon contents of fuels, even if their primary (and political) purpose is not related to climate change.

Carbon pricing encourages energy users to reduce energy consumption from carbon-intensive sources and switch to cleaner alternatives. These prices can be set to reflect marginal external costs of carbon emissions or to attain abatement targets, or they can be the result of other policy objectives while still having the same behavioural effects as prices implemented for climate reasons.

Taxes and emissions trading are cost-effective policy tools to reduce emissions for three reasons. First, emitters have an incentive to cut emissions as long as this is cheaper than paying the price, and this equalises marginal abatement costs across emitters, ensuring overall cost-effectiveness. Second, carbon prices decentralise abatement decisions, thus overcoming the asymmetry of information between the government and polluters: the regulator does not need to stipulate which emissions should be reduced using which technologies. Third, they provide an ongoing incentive to cut emissions, thus stimulating innovation (OECD, 2016a).

Emissions trading and taxes offer different levels of certainty about total emissions and price levels. Emissions trading systems generally cap total emissions but the price of permits fluctuates following market dynamics. Such volatility can be reduced by setting a minimum and a maximum price. Goulder and Schein (2013) argue that the policy objective for emissions trading is cost-effectiveness, i.e. achieving a predefined cap and minimum costs. Taxes can send a stable price signal, but give no guarantee on the level of emissions. Net benefits are maximised under a tax when marginal abatement costs increase faster than environmental damages (Weitzman, 1974). The evidence suggests that abatement costs generally increase faster than environmental damages as long as economies are sufficiently far from irreversible tipping points.

From a public administration perspective, taxes are easy to implement if they can be grafted onto existing excise tax systems. Emissions trading systems generally require a whole new set of instruments. Monitoring and enforcing emission caps is crucial.

The Paris Agreement allows parties to co-operate to meet their targets (Article 6) and 95 (I) NDCs state intentions to use international carbon pricing in some form. International co-operation on carbon pricing can take different forms, from agreeing on a minimum carbon price to linking emissions trading systems, and may provide numerous benefits. These include a more consistent price signal, a lower abatement cost for a given level of emissions (cost-effectiveness), lower price volatility and higher market liquidity. In addition, international co-operation can reduce concerns about carbon leakage – which occurs when companies move their production to countries with less ambitious climate measures, leading to a rise in emissions – due to different carbon prices between countries. New research emphasises the importance of establishing trust and reciprocity for international co-operation on climate policy (Cramton et al., 2017), and argues that minimum carbon prices are a natural focal point (Cramton, Ockenfels and Stoft, 2017). Monitoring countries' use of carbon pricing can encourage participation through transparency.

Box 5.1. **Carbon pricing instruments in practice** *(cont.)*

Making international links through carbon pricing does present some challenges. Under emissions trading, countries may be concerned about surrendering control of the domestic carbon price to an international market – and the possible impact on inflation – or the potential scale of international transfers generated. It is easier to link emissions trading systems when markets are characterised by a similar size, share comparable levels of ambition, and show other common features, such as the types of price controls and allocation methods leveraged. The linked emission trading systems in California and Québec show that these technical challenges can be overcome. There is no experience to date of jurisdictions agreeing to a common carbon price level using taxes, but the European Union has adopted a set of minimum taxes for mineral oil products.

The potential of carbon pricing is far from being realised. At the same time, lower oil prices in recent years have reduced the impact of carbon taxation policies. Within the G20, most CO_2 emissions are not priced at all, and less than 10% are priced at EUR 30 or more per tonne of CO_2 (a conservative estimate of the lowest social costs that would result from a tonne of CO_2 emissions), measured on a basis of "Effective Carbon Rates" (ECRs)[2].

Calculating ECRs is a means to combine the carbon prices resulting from emissions trading, carbon taxes and the carbon price equivalent of specific taxes on energy use (OECD, 2016a).[3] These three components all increase the price of CO_2 emissions compared with other spending items, so they capture the economically relevant contribution of tax and emissions trading policies to the cost of emitting CO_2.

Currently, high effective carbon rates occur mostly in road transport, due to excise taxes on motor fuels. Such taxes are generally implemented as revenue raising instruments rather than to price carbon or change behaviour, but they do provide an incentive to reduce CO_2 emissions, as well as helping to curb air pollution, congestion and other external costs related to car use. Therefore, high rates in transport may well be justified. They have not been enough to decarbonise road transport, however, indicating that even where prices work, they cannot by themselves drive a structural shift away from low-occupancy, car-oriented mobility patterns.

In G20 countries, 84% of energy-related CO_2 emissions occur outside road transport, where they face very low or zero effective carbon rates, with only 2% facing a price of EUR 30 or more. ECRs consist mostly of excise taxes, although carbon taxes and emissions trading systems play a significant role in some sectors and countries. There is large variation in the levels and coverage of ECRs across the main economic sectors in different G20 members (Figure 5.2). The ten G20 members with the highest average ECRs (Argentina, Australia, France, Germany, Italy, Japan, Korea, Turkey, United Kingdom and the European Union) account for 18.1% of G20 emissions, whereas the nine countries with the lowest rates (Brazil, Canada, China, India, Indonesia, Mexico, Russia, South Africa and the United States) account for 81.8% of these emissions. The Netherlands is the only country surveyed that prices more than 50% of emissions from non-road sectors at EUR 30 per tonne of CO_2 or more (OECD, 2016a).[4]

Table 5.1. **Proportion of emissions subject to a positive effective carbon rate**
by price instrument across all G20 countries (biomass emissions included)

		Taxes			Effective carbon rate
	ETS	Carbon tax	Specific taxes	Combined tax	
Industry	10%	6%	21%	21%	26%
Residential and commercial	3%	6%	21%	21%	19%
Electricity	15%	4%	27%	27%	36%

Figure 5.2. **Proportion of CO_2 emissions from energy use priced at different levels**
in the ten G20 members with the highest average effective carbon rates (upper row)
and the nine G20 members with the lowest (lower row)

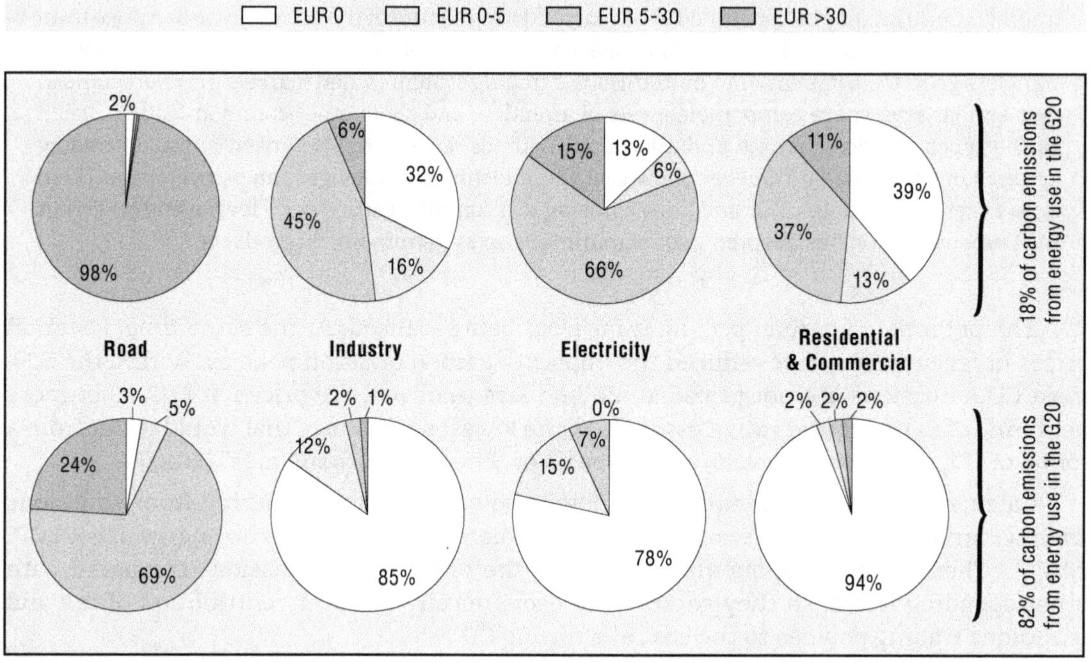

Note: Biomass emissions included.
Source: Adapted from OECD (2016a).
StatLink http://dx.doi.org/10.1787/888933484338

Carbon pricing needs careful design to send strong low-carbon investment signals

Certain design features of carbon pricing are necessary to reduce emissions and ease implementation. First, prices need to be sufficiently high and apply to the broadest possible range of emissions.

The larger the share of emissions covered by pricing, the stronger the incentive for cost-effective abatement. Coverage of energy and carbon taxes can be broadened by taxing fuels that have previously been untaxed (e.g., coal is untaxed in many countries and natural gas for heating often benefits from preferential or zero rates). Sector coverage can be broadened too. Some trading systems, such as New Zealand's, cover fuels directly at the stage of production and import, achieving broad coverage.

Second, any transitional support given to firms or households should avoid weakening abatement incentives in general, and low-emission investment signals in particular. For carbon pricing to influence investment decisions effectively, it is important to avoid preferential rates and free allocation of permits. Preferential tax treatment weakens all mitigation incentives, including for low-carbon investment. Granting permits to firms for free can create windfall profits and can also affect firms' technology choices, especially in sectors with limited competition.[5]

Full auctioning of permits sends a stronger signal to invest in low-emission technologies and to develop new ones, as it avoids the creation of rents that can favour carbon-intensive investment choices. Full auctioning can be phased-in to accommodate the transition of the existing emission-intensive capital stock. A commitment to full auctioning should also reassure firms that competing activities have to decarbonise, strengthening low-carbon innovation incentives. It also sends a strong positive signal to existing low-carbon firms.

To influence investment strongly, commitment to carbon prices needs to be long-term, and prices need to be predictable and subject to limited volatility. A stable price guarantees firms a minimum return on investment and innovation in clean technologies, and predictability facilitates risk management. Price fluctuations in trading systems create uncertainty; without a price floor, a minimum return on clean investment is not guaranteed (Goulder and Schein, 2013). The minimum auction price in California, which rises by 5% plus inflation each year, signals to firms that they will need to innovate in order to produce with low-carbon technologies. Taxes tend to generate more predictable carbon prices than emissions trading systems, although they too can change.

Productive use of revenue from carbon pricing is essential

How public revenues are used is also a crucial aspect of carbon pricing policy design. Revenue raised through carbon pricing can be large and can add to the direct benefits of pricing, but poor revenue-use can weaken these benefits. Pricing all emissions at EUR 30 per tonne would generate revenue of around 1% of GDP, on average across G20 countries, at current emissions levels.

Decisions on how to best use revenue from carbon pricing should strike a balance between social benefits and ensuring widespread support for carbon pricing. Depending on circumstances, the revenue can be deployed to stimulate growth (e.g., via higher spending or the reduction of more growth-inhibiting taxes), to increase inclusiveness (e.g., by increasing transfers), to compensate for potential loss of competitiveness, or to support low-carbon R&D. These options affect economic efficiency, welfare and social support for carbon pricing differently. A thorough examination of revenue use from carbon pricing should be part of an assessment of its payoffs. Distributing permits for free, awarding concessionary rates or setting tax-free thresholds should be considered as implicit forms of spending potential revenue (Bowen, 2015; ICMM, 2013).

Carbon tax revenues have so far more often returned to taxpayers through tax cuts whereas auction revenues from ETS permits are more often spent on green investment (OECD, 2017b forthcoming). Among G20 countries, carbon-pricing revenues are recycled in a variety of ways, including support for the transition, other green spending and broader tax cuts (Box 5.2) but clear assessment underlying these choices is not usually available.

Box 5.2. **How G20 countries use revenue from carbon-pricing policies**

In the **European Union,** at least half of EU ETS auction revenues for each participating country should be used for "climate and energy-related purposes" (European Commission, 2016).

Various provinces in Canada have implemented carbon-pricing policies. In **British Columbia,** the carbon tax is offset by tax cuts and credits for households and businesses, to obtain revenue neutrality. In **Alberta,** revenues from a carbon levy provide rebates to low- and middle-income households; since 1 January 2017, they have also funded a reduction of the small business income tax to 2%. Revenues are also reinvested in renewable energy projects, green technologies and infrastructure, and support energy efficiency through a new provincial agency (Alberta, 2016). In **Quebec**, all revenues from the GHG emissions cap-and-trade system are paid into a Green Fund to boost green spending and investment. In **Ontario,** auction proceeds will fund projects to promote pollution abatement, such as housing retrofits and incentives to promote electric-vehicle usage, as well finance the Green Investment Fund (Ontario, 2017).

Box 5.2. **How G20 countries use revenue from carbon-pricing policies** *(cont.)*

In **China,** auction revenue from ETS in Beijing, Shanghai, Shenzhen, Tianjin, Chongqing and Hubei may be used to compensate firms faced with competitiveness and leakage concerns. The government also has signalled an intention to set aside revenues to support "corporate carbon reduction, carbon market regulation and the carbon trading market" (Carl and Fedor, 2016).

In **France**, revenues from carbon taxation (implemented in addition to the EU ETS) finance a tax credit to boost business competitiveness (MEEDM, 2017). As of 2017, over half of revenues are allocated to an Energy Transition account to promote renewable energies.

In **India**, a tax on coal production (not strictly a carbon price) is used to fund the National Clean Energy Fund to promote low-carbon innovation and green activities and technologies, as well other ministerial budgets (Bowen, 2015).

In **Japan**, revenues from **Tokyo's** cap-and-trade emissions system are largely targeted at offsetting additional costs and price surges affecting small businesses and households, through tax reductions, credits and low-interest loans to encourage green spending (Bureau of the Environment, Tokyo Metropolitan Government, 2010.)

In **Mexico**, revenues from the carbon tax flow into the general government budget (OECD, forthcoming).

In the **United Kingdom**, revenues from a carbon price floor (implemented in addition to the EU ETS) contribute to the general budget.

In the **United States**, several states have implemented carbon-pricing policies. Revenues from the north-eastern **Regional Greenhouse Gas Initiative (RGGI)** promote energy efficiency, clean and renewable energy, green technologies and research. Households also receive rebates on their energy bills (RGGI, 2016). In **California**, ETS auction proceeds fund projects and programmes including low-carbon transportation and energy- and resource-efficient housing (CARB, 2017). Additionally, California requires electrical utilities to fully auction their emission allowances and to return all proceeds to households and businesses, to offset higher electricity bills (CARB, 2013).

South Africa is scheduled to implement a carbon tax in 2017. It has been indicated that revenues should return to households through tax shifting, tax incentives and financial assistance. Additionally, revenues would be earmarked to programmes that promote a low-carbon economy as well as specific mitigation measures for low-income households (The Davis Tax Committee, 2015).

Carbon taxation in **Australia** was repealed in 2014 after two years of implementation. Half of revenues were directed towards households through tax cuts, pension and allowance increases and other financial assistance, with a particular focus on low-income households. Revenues were also used to invest in green spending and technologies, and assist and provide funding to small businesses (Commonwealth of Australia, 2012).

With careful policy design, carbon-pricing revenue use can sometimes generate a double dividend: an emission dividend and a growth dividend through the reduction of other distortionary taxes, such as on labour or corporate income. Opportunities for a double dividend can also arise because of the low administration costs of upstream carbon taxes and their potential to cover emitting activities in the informal part of an economy. Revenues are frequently earmarked (Box 5.2). Strong earmarking – assigning all of the revenue to a stated purpose – is often deemed inefficient as it does not allow policymakers the flexibility to redirect spending when needed. Soft earmarking, or statements of policy intent, can help garner social support for implementation without the disadvantages of strong earmarking.

Higher energy prices strengthen some firms' ability to compete, creating carbon-neutral growth potential

Higher carbon prices may adversely affect the ability of some firms to compete, although the available empirical evidence reveals little to no effects to date. There is growing econometric evidence that the immediate competitiveness impacts of existing carbon-pricing mechanisms have been negligible (Arlinghaus, 2015; Partnership for Market Readiness, 2015). While this can partly be explained by the low prices and free allocation prevailing in most mechanisms, these same prices have reduced emissions, and windfall profits have occurred, so it is not the case that prices have always been ineffective environmentally or trivial economically. Rather than impinging upon strong firms' capacity to realise productivity growth, enhanced environmental policy stringency tends to be followed by an increase in short-run productivity growth of the most productive industries and firms (Albrizio et al., 2014). In the longer run, substitution possibilities are larger and stronger firms will be able to exploit them, potentially resulting in improved competitiveness.

Where measures to alleviate competitiveness concerns are used, support to industry is often not well targeted to address the risk that companies will move production to countries with less stringent carbon policies (i.e. carbon leakage). For example, over the current trading period of the EU ETS, 43% of allowances are freely allocated to firms deemed at risk of carbon leakage, but the criteria used to distribute allowances to firms are not always closely related to real risks of relocation or downsizing (Martin et al., 2014). In particular, the trade intensity metric, which implies a free allocation to 75% of the subsectors in the EU ETS, ignores that most of the firms exposed to trade have a low carbon footprint and are therefore immune to a carbon price.

International co-ordination on carbon pricing – such as agreeing on a minimum carbon price or linking emissions trading systems – would alleviate the risk of carbon leakage (Box 5.1). The benefits of co-ordination are recognised in the private sector, leading to several recent calls from industry for the G20 to co-ordinate on carbon prices (*e.g.* BDI, 2016; Investors on Climate Change, 2016, or the Carbon Pricing Leadership Coalition).

Carbon prices can help countries to reduce the size of the informal sector

Higher carbon prices can reduce the incentive for economic activity in the informal sector, complementing the structural reforms described above. Carbon and energy taxes can be collected upstream from the few firms that are importing or extracting fuels, ensuring low-cost compliance. Likewise, an emissions trading system can require importers and extractors of fossil fuels to acquire permits for the entire carbon content of the fuels they sell. In a competitive market,[6] importers and extractors will pass the carbon price through to fuel users. Thereby all sectors of the economy will pay the tax, both the formal as well as the informal. In addition, if revenues from energy taxes are recycled to reduce personal income or corporate profit taxes, it becomes less profitable to work in the informal sector.

Recent simulation studies for the United States and Spain suggest that the economic gains from reducing the informal sector through higher energy taxation are substantial (Bento et al., 2014, and Markandya et al. 2013). A countervailing effect could be increased attempts to evade energy taxation. While simulations for India and China show this may happen, the gains from reducing the informal sector through higher energy taxation dominate (Bento et al., 2014).

Strategies for reforming subsidies for fossil fuels

Subsidies to fossil fuels are incompatible with policies that attempt to internalise the costs of carbon emissions through pricing. Although no complete global accounting of subsidies has been compiled, subsidies that artificially maintain low prices for electricity and fuels — particularly petroleum products used in transport, kerosene, natural gas, fuel-inputs to electric power plants — are the most common (Table 5.2 and Figure 5.3). While subsidies have recently declined, this is partly due to falling oil and fossil fuel prices. Many of the policies that maintain these low prices have been around for decades. Investments in energy-consuming capital — vehicles, home appliances, factories — have in turn been made on the basis of these artificially low prices, locking in inefficient patterns of consumption.

Table 5.2. Government support to fossil fuels in G20 countries

USD billions

	2012	2013	2014
Production	28	21	18
Consumption	398	380	354
Of which, market transfers to consumers	*247*	*236*	*213*
Of which, other transfers (direct payments and tax preferences)	151	144	141
General Services	4	6	4

Note: 1. Including subsidies related to the under-pricing of electricity.
Sources: IEA (2015) and OECD (2015d).

Some governments also favour domestic production of fossil fuels, and their processing into products or electricity, through policies including concessional credit, loan guarantees, infrastructure and special tax features, such as accelerated depreciation. Some multilateral development banks have also financed fossil-fuel-related projects in developing and emerging countries, justifying these projects by the additional export earnings they will bring for the country, or investment in new industries they will attract. The effect for the climate, however, may be to encourage a development model that diverges from a low-emission track or that risks stranding a lot of fossil-related assets, like oil pipelines and coal-fired power plants (see also Chapter 7).

The fiscal impact of fossil-fuel subsidies, particularly those that keep fuels at low prices, can be large. Before Mexico embarked on its reform of gasoline and diesel price subsidies in 2014, the fiscal burden of those subsidies reached as high as 1.4% of GDP. Morocco's energy subsidies reached 5.5% of its GDP in 2011, before reforms were implemented (World Bank, 2012). And in 2012, before Indonesia embarked on its recent reforms, its fossil-fuel consumption subsidies were 4.1% of GDP – four times total government expenditure on health (AsDB, 2015).

Developing more complete and accurate information on subsidies can not only help make the public case for reforms, but also provide the basis for improving the targeting of subsidies (see also Chapter 6). In Germany, an official biannual Subsidy Report is released, including a "sustainability check" of all subsidies, and the Federal Environment Agency also carries out ongoing analysis of environmentally harmful subsidies, including in the energy sector.[7] Italy releases an official annual "Catalogue of environmentally friendly and harmful subsidies" aimed at helping the Parliament and the Council of Ministers define environmental policies that take into consideration international and European recommendations (Ministero dell'ambiente, 2016).

When subsidies are provided through interventions that force state-owned companies to sell energy below cost, investment often suffers, and the quality of the service declines over time (Fattouh and El-Katiri, 2013). Rationing may be another result. In Argentina, for example, electricity tariffs were set below production costs for over a decade, leading to significant underinvestment in the sector and frequent blackouts (Di Bella et al., 2015). To address this issue, the current government has embarked on an ambitious utility subsidy reform (Box 5.3).

Box 5.3. Electricity subsidy reform in Argentina

For over a decade after its 2001 economic crisis, Argentina froze prices for public services, including electricity, to support poorer households (Lakner et al., 2016). The years of accumulated price distortion led to residential consumers only paying 10% of average electricity generation costs (Bidegaray, 2015). The total cost of energy subsidies in Argentina amounted to an average of 3.9% of GDP (2.1% on fuel, in addition to 1.8% on electricity) between 2011 and 2013, among the highest in the region (Di Bella et al., 2015). Besides the fiscal burden, the subsidy was not evenly distributed, with the majority being assumed by wealthier households (Lakner et al., 2016).

In 2015, an extensive energy- subsidy reform was launched, covering oil and gas as well as electricity, seeking to tackle the fiscal deficit and rebalance the economy. The reform has led to rapidly increasing electricity tariffs, for example between 61% and 148% in Buenos Aires, depending on consumption level. Despite the sharp increases, the tariff still only recovers 47% of generation costs (Buenos Aires Herald, 2017). A "social tariff" on electricity is available to protect the 2 million poorest consumers. The wider energy reforms face significant public resistance due to fears about job losses and increasing poverty. In August 2016, the Supreme Court reversed increases in gas prices for residential consumers until public hearings on tariff increases are held (Financial Times, 2016). The government faces the challenge of implementing the reform while minimising the social costs and smoothing the transition.

Figure 5.3. Support to fossil fuels in OECD and key partner* countries

Total support for fossil fuels in OECD countries (left) and selected partner economies (right) by year and type of fuel (billions of current USD)

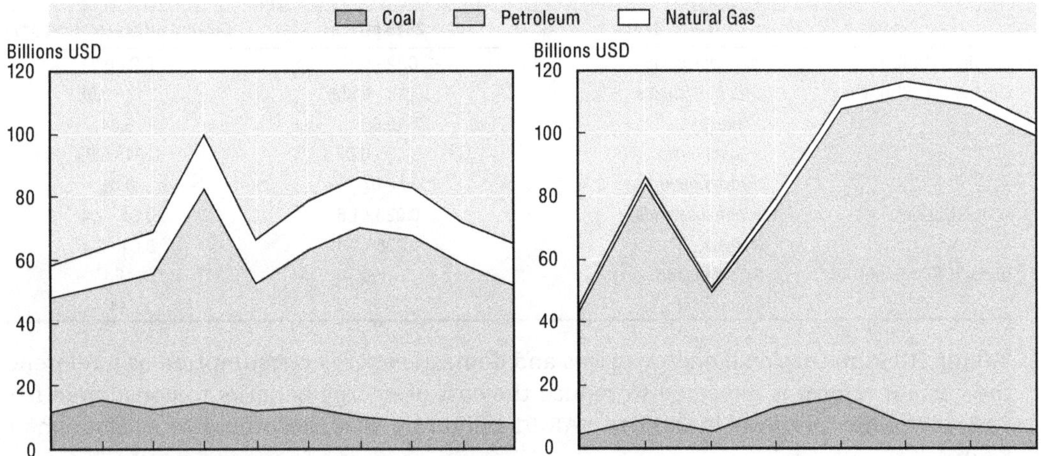

Notes: *Brazil, the People's Republic of China, India, Indonesia, the Russian Federation and South Africa.The charts are based on an arithmetic sum of the individual support measures identified in the Inventory. Along with direct budgetary support, it includes the value of tax relief measured under each jurisdiction's benchmark tax treatment. The estimates do not take into account interactions that may occur if multiple measures were to be removed at the same time. Because they focus on budgetary costs and revenue foregone, the estimates for partner economies do not reflect the totality of support provided by means of artificially lower domestic prices. Particular caution should therefore be exercised when comparing these estimates with those reported by the IEA (2014a) for these countries.

Source: OECD (2015d).

StatLink http://dx.doi.org/10.1787/888933484346

Several countries could benefit from the unilateral reform of their fossil-fuel subsidies. Substantial energy subsidies to consumers are in place in emerging and developing countries, while OECD countries use a combination of tax expenditures and direct budgetary support to fossil-fuel producers and consumers (Coady et al., 2015; OECD, 2015d; IEA, OPEC, OECD and World Bank, 2010 and 2011). Reforms of fossil-fuel subsidies can bring many benefits. They can alleviate economic inefficiencies and trade distortions, tackle distributional inequity, and boost energy security, resource conservation and environmental protection (Burniaux and Château, 2011; Burniaux et al. 2011; OECD, 2014b). They can also support economic diversification, such as reforms now underway in Saudi Arabia (Box 5.4). At the global level, revenue gains from fossil-fuel subisdy reforms could amount to 4% of global GDP and CO_2 reductions could be more than 20% (Coady et al., 2015).

Box 5.4. **Energy and water price reform in the Kingdom of Saudi Arabia**

The government of Saudi Arabia, recognising the risks posed by an economy over-reliant on oil exports and with rapidly growing domestic energy consumption, has recently launched an ambitious whole-of-government reform programme called Vision 2030. One pillar of Vision 2030 is the Fiscal Balance programme, a significant energy and water price reform with the joint objective of improving fiscal stability and initiating a low-emission transition, by progressively increasing domestic fuel prices.

During the first phase of price reform, implemented in 2016, tariffs have increased on average by 35% for households and 46% for industry (Table 5.3). In the proposed second phase, scheduled to start in 2017, domestic energy prices are expected to be linked to a reference export price. While the methodology for calculating the reference price and the mechanism for varying domestic prices have not yet been announced, the stated objective is to adopt a domestic pricing mechanism based on international market prices, rather than on domestic production costs.

Table 5.3. **Energy and water prices in Saudi Arabia, 2016 and March 2017**

Product	Sector	2016 prices	Current prices (March 2017)
Gasoline (USD/l)	*- for households*	0.12 - 0.16	0.2 - 0.24
Electricity (USD/kWh)	*- for households*	0.013 - 0.069	0.13 - 0.08
	- industry	0.03	0.04
	- Commercial	0.03 - 0.07	0.04 - 0.08
	- government	0.07	0.09
Water (USD/m³)	*- for households*	0.026 - 1.6	0.04 - 2.4
	- others	0.026 - 1.6	0.04 - 2.4
Diesel (USD/barrel)	*- for transport*	10.6	19.1
	- for industry	9.12	14

Taking 2015 international energy prices and domestic energy consumption as a reference, the current reform is expected to reduce the cost of energy benefits to consumers from SAR 300 billion per year in 2015 to SAR 91 billion by 2020, according to estimations by KAPSARC.

Source: KAPSARC (2017).

Given that subsidies encourage the over-consumption of fossil fuels and increase GHG emissions, each country has an interest in other countries reforming their subsidies as well. For this reason, several groups of countries have recently made joint commitments to phase out inefficient fossil-fuel subsidies. The G20 did so first in September 2009, followed the Asia-Pacific Economic Co-operation (APEC) forum. Another grouping, the nine-nation Friends of Fossil Fuel Subsidy Reform, has made similar commitments. Belonging to such a group provides additional legitimacy for domestic reform efforts, and can facilitate information and experience sharing.

The G20 and APEC have followed similar paths in their common objectives to "rationalise and phase out over the medium term inefficient fossil fuel subsidies that encourage wasteful consumption", while recognising "the importance of providing those in need with essential energy services, including through the use of targeted cash transfers and other appropriate mechanisms". Both started with voluntary reporting of fossil fuel subsidies (all subsidies for some, for others just those that in their view would qualify as inefficient). From 2013, both groups agreed to conduct voluntary peer reviews of countries' reform efforts. Only a minority of G20 and APEC countries have completed peer reviews, however, or are in the process of completing them, and the process is time-consuming.

The resulting published reviews have expanded the public information available on fossil fuel subsidies, as well as improving understanding within the government. China, for example, prepared for its reciprocal G20 peer review with the United States by investing more than a year in identifying and understanding the aspects of its energy policy that favoured fossil-fuel production or consumption – consulting with academics, internal and foreign experts, and its energy industry – before submitting its report. By participating in the review teams, country experts have also come to better understand the diversity of support mechanisms, and the challenges other countries faced in designing and implementing reforms that will be politically acceptable and endure. One of the defining characteristics of peer reviews is that they are conducted on a non-adversarial basis, and are dependent on the existence of mutual trust among the parties involved in the review (Pagani, 2002).

Experiences from past attempts to reform agricultural subsidies suggest that, in order to maintain momentum for reform, there is a need for regular, up-to-date, publicly available and comparable information on progress in all the participant countries. This approach is institutionalised through the regular notification of domestic support to agriculture by the members of the World Trade Organization (WTO) and, among the OECD countries and their key partners, by the maintenance of a database on government support and the publication annually of an *Agricultural Policy Monitoring and Evaluation report* (OECD, 2016b). The WTO's Agreement on Agriculture committed all its members to the cessation of export subsidies and phased reductions of certain other forms of support. In contrast, the OECD policy monitoring and evaluation process is, formally, voluntary.

A similar regular and concurrent process could be considered as a mechanism to underpin international efforts to reform inefficient fossil-fuel subsidies. Comprehensive data on support have already been gathered for most of the G20 and APEC countries (OECD, 2015d), and several years' worth of estimates of the level of consumption subsidies, by fuel, have been compiled on an individual country basis for most of the rest of the world (IEA, 2015; Coady et al., 2015; IDB and World Bank, 2017 forthcoming). What remains to be done is to develop a common understanding of the effects of different types of fossil-fuel subsidies, which types (and in what combinations) are the most inefficient, and what approaches to reform work best in each country.

Critical in most fossil-fuel subsidy reform is the design of accompanying measures to compensate the rise in prices of basic commodities. Chapter 6 describes how countries have managed this transition, including minimising the possible regressive impacts of reform.

Beyond carbon pricing: interactions with other incentives and regulations

Well-designed and targeted economic instruments to promote reduction in GHG emissions can usefully complement carbon-pricing measures by addressing additional market failures and barriers, such as split incentives, information asymmetries, path dependency in innovation, policy uncertainty and country risks. Measures include specific investment incentives (in both infrastructure and land-use sectors), standards and mandates (e.g. efficiency standards) and information instruments (e.g. energy labelling) (Table 5.3). Where multiple measures are used, managing policy interactions is important.

Table 5.4. **A wide range of policy instruments can complement carbon pricing**

Policy type	Examples	Most commonly employed for:		
		Renewable energy	Energy efficiency	Low-carbon transport
Targeted investment incentives (or technology support policies)	• Feed-in tariffs (premium or "fixed") and other renewables incentives (e.g. net-metering for households) • Tradable permits (e.g. UK renewable obligation certificates) • Capital grants • Tax incentives (e.g. tax rebates, depreciation rules)	Yes	Yes	Yes
Standards and mandates	• Technology standards (e.g. biofuel blend mandate) • Performance standards (e.g. fleet average CO_2 emissions, energy performance standards) • Renewable portfolio standards	Yes	Yes	Yes
Information approaches	• Rating and labelling programmes • Public information campaigns		Yes	Yes

Sources: OECD/IEA/NEA/ITF (2015); Hood (2011); de Serres, Murtin and Nicoletti (2010).

Investment incentives need to be stable and predictable, with clear, non-discretionary criteria for policy exit.[8] For example, many governments have introduced mechanisms providing fixed tariff contracts (power purchase agreements). While these provide revenue certainty for investors, setting appropriate tariff levels is challenging. If financial incentives do not adapt to the decreasing costs of maturing technologies, scarce government resources might end up funding unnecessary rents. For this reason, several governments are adopting new policy design features, such as declining feed-in tariffs, or turning to auctions to control the cost of support incentives (Box 5.5). Other countries have opted for market-based mechanisms, such as renewable portfolio obligations (RPOs) coupled with tradeable renewable energy certificates; monitoring and enforcement of obligations is critical for such systems to work (Haas et al. 2011; Shrimali and Tirumalachetty, 2013).

Environmentally motivated tax incentives – tax reductions or subsidies aiming to encourage environmental behaviour – are another form of investment incentive but should be used sparingly. They are poor substitutes for policies that directly discourage pollution in a cost-effective manner, such as carbon pricing. This is partly because it is difficult to target incentives correctly. There is a risk of subsidising non-intended choices that may be more carbon-intensive; for example, subsidising public transport while not increasing the costs of car use may encourage some households to locate further away from work and increase commuting distances. Tax incentives can also support investment that would have been made even in the absence of the tax incentive (poor additionality).[9] In addition, tax incentives are often costly and regressive (Appelt et al, 2016; Greene and Braathen, 2014).[10]

Box 5.5. **Evolving technology incentives for renewable electricity**

Fixed tariffs for renewable electricity off-take have stimulated deployment effectively but it has proven difficult for governments to control the total cost of these subsidies. Feed-in tariffs (FIT) support the deployment of renewable energy by offering long-term purchase agreements. They have played a major role in promoting investment in renewable energy projects. However, governments have found it difficult to control FIT programme costs as market uptake and technology cost reductions have been underestimated (Couture, T. et al., 2010). Some countries have introduced automatic mechanisms to control FiT programme costs, with the incentives declining based on the level of deployment.

Countries are moving away from pre-defined feed-in tariffs and adopting tendering processes to set tariffs for renewables. In theory, tenders provide numerous benefits, including competitive determination of tariff levels, control over budget allocated to renewables financial support and installed capacity. At least 67 countries had adopted

Box 5.5. Evolving technology incentives for renewable electricity *(cont.)*

renewable energy auctions by November 2016, up from 6 in 2005 (IRENA and CEM, 2015) (Figure 5.4). For instance, South Africa abandoned its renewable energy feed-in tariff in 2011, without a single contract being signed or negotiated in the two years of its existence, in favour of a competitive tender process, which is considered largely successful in channelling private sector finance into the renewable energy sector (Eberhard et al., 2014). The four rounds completed to date have cumulatively generated ZAR 192.6 billion in private sector investment, with foreign investment accounting for 28% (DOE South Africa, 2015). Mexico held its first auction open to private bidders in 2016 (Bloomberg, 2016a). The United Kingdom's renewable obligation certificate trading scheme is being replaced by a contract for difference mechanism,[11] where the variable top-up on the market price is determined through tendering. Some G20 countries are also experimenting with competitive tendering frameworks to support energy efficiency, as in the case of the German "STEP up!" scheme (STEP up, 2017).

Figure 5.4. **Number of countries/states/provinces with renewable energy policy, by type**

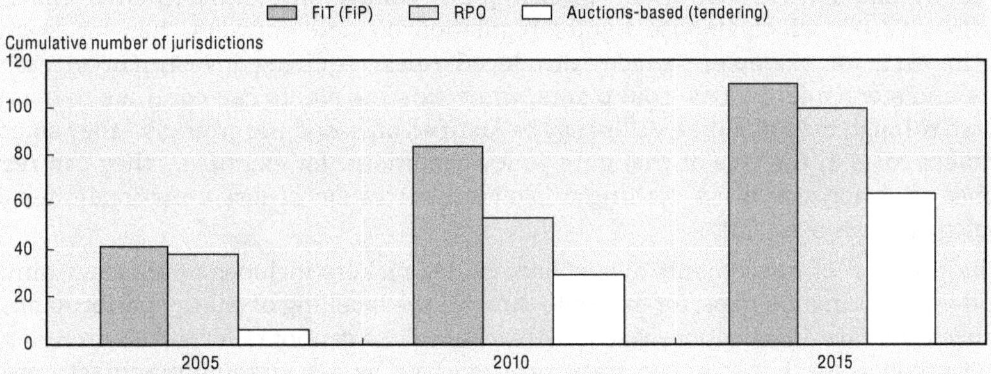

Source: REN21 (2016).

The cost-effectiveness of renewable-energy auctions depends on design and the level of competition. In technology-neutral auctions, different technologies compete to be the least-cost option (in Brazil, for example, renewables were competing with natural gas). In contrast, technology-specific auctions support the development of targeted technologies, as in India. The segmentation can also target generation profiles (baseload electricity, peaking electricity and non-peaking intermittent) as in California. In the United Kingdom renewables are clustered into technology "pots" reflecting their level of maturity. The volumes and frequency of auctions can also affect competition: if demand outstrips supply, then competition is severely reduced, as was reportedly the case in the 2011 South African tender (IRENA and CEM, 2015).

The introduction of technology- or experience-specific qualification requirements (financial guarantees), while important to determine the adequacy of bidders, may restrict market entry and hamper competition, for example by posing significant barriers to the participation of SMEs. The division of public procurement contracts in smaller lots can help but has to be balanced with the economies of scale expected from auctioning large lots. The overall tendering process should be designed to avoid price manipulation and collusive behaviour.

The bankability of the off-taker has a key influence on the attractiveness of auctions (IRENA and CEM, 2015). For instance, the creditworthiness of Indian government-owned utilities (Discom) varies widely from state to state; this has been cited as an important factor behind the significant difference in the auctions' ability to attract investors. Useful mechanisms in these circumstances are the creation of specific funds to ensure compliance with PPAs (e.g. Argentina) or policy risk insurance (e.g. USA OPIC) (IRENA, 2016, IASS, 2016).

Sources: OECD (2015e) and as cited.

Standards and mandates, such as technology and performance standards, have also been widely used as part of climate policy, in particular to promote energy efficiency. By failing to put a price or opportunity cost on the negative externality, however, standards and mandates generally do not ensure that environmental objectives are achieved at the least economic cost. Their use may be appropriate when barriers or information asymmetries stifle price elasticity. Performance standards can be successful in overcoming split incentives: energy efficiency standards for buildings can ensure that landlords invest in energy efficiency, which they otherwise would not do as tenants pay the energy bill.

Performance standards allow firms to search for the cheapest options to meet requirements. Technology standards are more prescriptive but may be the best option in specific circumstances, notably when the administrative costs of performance standards are too high or when abatement costs are relatively homogeneous across agents (de Serres, Murtin and Nicoletti, 2010). The effectiveness of technology standards can also rival that of prices when regulators and emitters have access to similar information. Performance standards for power plants may be one example.

Some standards might have the perverse impact of imposing higher costs on new (and more environmental friendly) firms. An example is vintage differentiated environmental regulations (VDR), which impose tighter regulation on new entrants than on existing firms; in 2015, for example, Canada introduced rules requiring investment in carbon capture and storage for all new coal plants, while existing plants can continue to operate unabated through to 2030. While VDRs may be justified on economic grounds – they smooth adjustment costs in the face of changing policy conditions, for example – they can result in higher costs for new firms, leading to slower penetration of less emissions-intensive technology (Coysh et al., 2017).

The third set of instruments available to policy makers includes approaches aiming at removing information gaps, for example through the labelling of energy performance of appliances and cars. Information-based instruments alone cannot reflect all environmental costs of product use, but they can transform markets by encouraging manufacturers to compete on this newly visible attribute (de Serres et al., 2010). Information approaches have also proven effective where actors may be unaware of either risks or available incentives related to climate change, such as in agriculture.

Climate policy instruments are usually not used in isolation, and managing interactions is important. Overlapping policies can be appropriate when several failures and market barriers need to be addressed. For example, several energy efficiency policies (e.g. labelling, standards, tax credits) may be introduced because there are several barriers (e.g. lack of information, split incentives) that prevent energy savings investment from being made. Targeted investment incentives, which in some cases overlap with carbon pricing, can be justified if they stimulate and lower the cost of low-carbon technologies, contributing to lowering the future cost of climate mitigation (de Serres et al., 2010; Acemoglu et al. 2012). Even where there is a rationale for overlapping policies, however, their interaction with quantity-based carbon pricing instruments such as emissions trading needs to be carefully evaluated (Hood, 2013).

Balancing climate and economic policy for agriculture and land use

Policy makers also need to take into account the synergies and trade-offs between productivity, mitigation and adaptation in agriculture and other land-use sectors, given the importance of land sectors for GHG emissions. Agriculture GHG emissions are increasingly decoupled from agricultural production in OECD countries (OECD, 2014c). This can be largely attributed to improved technologies and farm management practices, combined with incentives to lower emissions. Some policy incentives have reduced on-farm energy consumption (OECD, 2016c, 2016d, 2015b, 2015f, 2015g). Several other measures could reduce farm GHG emissions cost-effectively, such as increasing efficiency of fertiliser use and improving cattle breeding. These could be encouraged by information or incentive-based measures. Some market instruments like manure rights can be used and expanded to encourage a shift towards more profitable, emission-efficient sectors. Incentives that encourage emissions abatement and natural asset sustainability could facilitate the transition if governments help communities address their capacity gaps.

Policies to reduce agricultural GHG emissions can have significant impacts on water management and on water quality (though, absent mitigation, the impact of agriculture on water is likely to be more important with climate change; OECD, 2014d). Mitigation practices can affect water quality through their impacts on soil erosion rates, fertiliser and pesticide input uses, and the amount and nutrient content of animal manure. In addition, the development of bioenergy feedstocks may cause additional water use and, in some places, expand the use of water for irrigation. The synergies and trade-offs between mitigation and agricultural management practices are site-specific, however, and in many cases not yet known. It is nevertheless important to recognise these linkages in the design of mitigation policies to reduce the risk of conflict between mitigation and water policy objectives and to maximise potential synergies.

Combining sustainable agriculture and land-use policies with free trade principles is also important, as removing barriers to trade in agriculture products can optimise land use and reduce the overall demand for land in countries that have a comparative advantage in producing agricultural goods. In turn, it can alleviate pressures on forests. To avoid additional deforestation pressure from agricultural free trade agreements in resource-abundant countries, parallel policy instruments are needed to promote conservation efforts in accessible forest areas, as well as stronger institutions (Robalino and Herrera, 2010). Providing incentives to reduce emissions in other land-use sectors is equally important. In an attempt to halt deforestation and reduce emissions from land-use change, Brazil has introduced a number of policy instruments, including tradable forest quotas and low-interest loans for sustainable agriculture practices (Box 5.6).

Box 5.6. **Land-use policies to combat emissions from deforestation in Brazil**

Over the past decade, emissions from land-use change in Brazil have fallen more than five-fold from a peak in 2005, while emissions from other sources (including energy, agriculture, waste and industrial processes) have been increasing. As a result, emissions from land-use change as a share of total emissions have fallen from 68% in 2005 to 31% in 2015 (Government of Brazil, 2016; La Rovere et al., 2017). There has also been over 70% reduction in the deforestation rate in the Amazon between 2004 and 2016 (INPE, 2017).

Central to the efforts to reduce deforestation have been measures to address the underlying factors, such as the lack of clear land rights – which was resulting in rural conflicts and forest clearing as a way to define ownership – and enforceability of legal instruments to protect forests. The new Forest Code, adopted in 2012, introduces several changes to address these issues. These include the use of high-resolution satellite imaging to monitor forest cover and mandatory registration of all rural properties, with detailed geo-referenced information (OECD, 2015h). The code requires landowners to preserve a certain share of native vegetation of their land, with some flexibility allowed through trading.

Box 5.6. Land-use policies to combat emissions from deforestation in Brazil *(cont.)*

To address emissions from the agriculture sector, in 2010 Brazil adopted the National Plan for Low Carbon Emissions in Agriculture (ABC Plan), which provides low-interest loans to encourage farmers to adopt sustainable agriculture practices. The ABC Plan has set ambitious 2020 targets, which include reducing annual GHG emissions from agriculture by 160 million tonnes of CO_2 equivalent, and rehabilitating 15 million hectares of degraded pasture land (Angelo, 2012). While the uptake of the programme was initially slow, and slowed again in 2015-16, 28 500 loans have been issued since its inception, totalling more than BRL 13.2 billion (Newton et al., 2016).

Policies to improve climate change adaptation in agriculture and resilience of infrastructure

Climate change policy discussions often focus on reducing GHG emissions, but policies to adapt to climate change and improve resilience are also vital, as Chapter 2 highlighted. The majority of G20 countries have now developed national adaptation plans to co-ordinate their policy response to climate change. These can strengthen countries' capacity to manage climate risks (OECD, 2015i; OECD, 2015j). Key areas for achieving this are raising awareness and filling knowledge gaps on adaptation; mainstreaming climate risk management into decision-making; identifying necessary adaptation actions for key sectors and monitoring progress in building resilience.

Adaptation in agriculture: a key sector, vulnerable to climate impacts

Governments have an important role to play in encouraging adaptation in the agricultural sector, all the more so as sustainable, productive and climate-friendly agricultural systems will be needed to reach food security objectives for a growing world population (Chapter 2). As the benefits of most adaptation measures are local and directly captured by farmers, policy interventions may be needed to align privately profitable actions with socially desirable outcomes (Ignaciuk, 2015; OECD, 2015k). Smallholder farmers are less likely than better educated and better capitalised medium- and large-scale farmers to have access to the knowledge of sustainable agriculture pathways, meaning that adaptation efforts may fall short of what is best for society. Especially in less-developed countries, smallholder farmers face limited access to markets, credit, extension advice, weather information, risk management tools and social protection (FAO, 2016).

Agricultural policies are becoming more climate-aware.[12] Many OECD countries are promoting adaptation in agriculture and making national plans for climate change adaptation, especially in water policy, climate risk management, and extension services (OECD, 2016c, 2016d, 2015b, 2015f, 2015g). However, there is room for further improvement, in particular to better align strategic policy objectives and remove disincentives for farmers' adaptive actions.

Policy misalignment may send contradictory signals and dilute the impact of adaptation initiatives. For instance, although well-devised insurance programmes are a relevant policy instrument to help agriculture become more resilient to weather-related disasters, subsidised crop insurance premiums may induce more risky farming practices. Countries that support their agriculture through farm payments based on current production, land acreage or animal numbers may be creating incentives to continue producing subsidised commodities that are not well adapted to gradually changing weather conditions or extreme weather variability. Overall, a general lack of capacity to access and make use of relevant climate data can prevent farmers from integrating climate risk into their daily practices. Budgets for adaptation to climate change remain small at this stage.

Further, adaptation efforts in the agriculture sector need to be aligned with water policies, and consider both climatic uncertainty and regional specificity. Effects of climate change on agriculture occur through crop water requirements, availability and quality of water, varying across a range of scales from local to regional to continental (OECD, 2014d). Furthermore, the frequency and severity of extreme events such as floods and droughts may increase as a result of climate change and cause substantial damage to agricultural production (OECD, 2016e).

As a result, risk management instruments such as prevention and insurance can play a major role in managing the increased risk of floods and droughts. At the watershed level, well-designed, flexible and robust water-sharing rules and economic instruments, such as water pricing and water trading, can foster adaptation of water systems. Short-run incentives like public buyback of water rights on spot markets allow farm systems to cope with intra-seasonal volatility of water supply and reallocate water to its most efficient uses within the growing season. Long-run incentives like individual or group water rights and administered water pricing will allow agriculture to adapt to continuously changing conditions of water supply and other factors like growing population, increasing demand from urban areas, and the need to reserve some water for the needs of natural ecosystems. Policies aimed at improving water management can also orient agriculture towards increased productivity gains, which will make agriculture more input-efficient, and thus help mitigate the impact of the sector on climate (OECD, 2014d, 2016e).

To strengthen the agricultural sector's resilience to climate change, there are a number of entry points available for government intervention (Ignaciuk, 2015; OECD, 2015k, 2016h, 2016i). First, there is a need to scale up investment in public and semi-public agricultural research and innovation programmes. Improved access to research results through training and education then helps farmers and other stakeholders make rational decisions and scale up these innovations. Second, providing accurate and detailed information along with technical assistance on the risks and consequences of climate change, as well as opportunities for adaptation, allows stakeholders to make timely and informed adaptation decisions. Third, by encouraging adaptation planning and offering temporary financial assistance in clearly defined circumstances, public policy can help smooth the costs of switching to climate-adaptive practices (e.g., building the carbon storage capacity of agricultural soils through no-till cropping coupled with the permanent use of a land cover plant species and crop rotation). Finally, creating a well-functioning free trade system for agricultural and food products may support adaptation by compensating regional changes in productivity induced by climate change and enhance the four pillars of food security: accessibility, availability, utilisation and stability. Reducing tariffs and subsidies on agricultural products would help decrease agricultural prices.

Promoting resilient infrastructure

To achieve the overall goals of national climate change adaptation plans, it is crucial to ensure the resilience of all new and existing infrastructure, rather than simply to construct new infrastructure specifically intended to address climate risks. The vulnerability of new infrastructure investments to climate impacts will depend on their design, location and operation over time.

Climate resilience measures need to be tailored to specific contexts, and range from rerouting transport links out of flood plains to operating regimes for hydropower facilities. In general, owners and operators of infrastructure are best placed to decide on the appropriate measures. There is an incentive for them to do so, as they will reap the benefits of increased reliability, increased lifetime and reduced maintenance costs. However, in practice, there are several constraints on adaptation that can lead to a sub-optimal level of resilience.

Infrastructure owners and operators may be unaware of the potential impacts of climate change, lack the data or tools to make decisions or face distorted incentives, meaning they may not be in a position to measure and value relevant risks. Sources of distortion include misaligned policies, including regulatory frameworks, and unpriced externalities. Governments can use four tools to help overcome these barriers and facilitate the climate resilience of infrastructure networks (Vallejo and Mullan, 2017):

- **Improving risk information and assessment to support decision making:** Governments can support this process by ensuring that suitable data are available on climate-related hazards, including historical weather and flood losses, and encourage the provision of high-quality climate projections.
- **Screening and factoring climate risks into public investments:** Governments can influence resilience measures by considering climate resilience in their procurement and lending criteria. Multilateral development banks are playing a pioneering role in implementing risk screening of their lending decisions. For infrastructure implemented through public-private partnerships (PPPs), the allocation of risks should be examined to ensure that it is conducive to adaptation. Well-designed insurance requirements for PPPs can help to manage governments' potential liabilities while providing an incentive for risk reduction.
- **Enabling infrastructure resilience through policy and regulation:** Decisions relating to resilience will be shaped by the broader regulatory environment (including spatial planning, economic regulation and technical standards). Governments can address misalignments and support sub-national entities' capacity to implement relevant policies.
- **Encouraging climate risk disclosure:** There are a growing number of voluntary initiatives in support of climate risk disclosure. Governments can encourage the disclosure of climate risks where there are gaps in existing initiatives. If such mechanisms are appropriate, they should avoid duplication, provide incentives for reporting and be enforced.

Climate policies can boost innovation for the low-carbon transition

Many of the core climate policies presented above will play a role in driving innovation that generates low-emission and climate-resilient solutions. Policies that stimulate innovation and orient it in this way are vital. Markets do not provide sufficient incentives, even with carbon pricing. Low-carbon innovation is hindered by market failures that are related to the lack of carbon pricing and that affect innovation more broadly. The different types of climate policies considered above unlock the market for low-emission products and business practices –by internalising CO_2 costs, for example, or by providing incentives for environmental friendly behaviour – while defining the returns from follow-up innovations (GGGI, 2014). Recent evidence suggests that carbon pricing stimulates innovation rapidly and significantly (Dechezleprêtre, Martin and Bassi, 2016). However, not all environmental policies provide the same dynamic incentives, so governments should evaluate the innovation potential of climate policies according to several criteria, including (Johnstone et al., 2010):

- Dynamic efficiency: Does the policy create incentives to continuously search for cheaper abatement options?
- Predictability: What effect does the policy have on investor uncertainty?
- Flexibility: Are potential innovators free to identify the best way to meet the objective?
- Incidence: Does the policy target the environmental objective as closely as possible?

Tax incentives are often used to promote private R&D and can be targeted towards low-carbon technologies. Current market conditions and path-dependence concentrate innovation in high-carbon technologies. Preferential treatment of low-carbon innovation is justified to overcome this drawback and push economies to a low-carbon growth path. Further, the knowledge spillovers from patents for low-carbon technology are among the highest, as these patents are frequently cited (Dechezleprêtre, Martin and Mohnen, 2014). Large spillovers suggest that focusing on low-carbon innovation enhances economic growth.

However, the mere existence of large spillovers does not imply that targeted tax incentives are justified in every circumstance. Tax incentives may work best for bringing near-to-market solutions closer to broader market dispersion, whereas direct support for R&D or technology demonstration or deployment is likely to give a greater boost to innovation upstream in the supply chain. Upstream support has larger pay-offs, so it may be better to use instruments other than tax incentives (Dechezleprêtre et al., 2016). However, direct support requires expertise to ensure research projects are well selected and executed. If targeted tax incentives for green R&D, or direct support, are implemented, they need to be carefully designed so as not to create new path dependency. Additionality also needs to be considered: the tax incentive should create additional environmental benefits that would not have occurred otherwise.

Tuning broader investment conditions for low-emission and resilient infrastructure

Climate policies are introduced in the framework of existing policies and regulations sometimes geared towards carbon-intensive activities. These existing policies might inadvertently weaken the incentives provided by core climate instruments – and as a result weaken the business cases for investment and innovation in low-emission and climate-resilient infrastructure. For example, Box 5.7 points to recent evidence suggesting that policy interactions between climate and other policies have affected investment (and innovation) in renewable power. To align broader investment conditions with climate goals, policy makers need to assess whether both the general investment environment and specific policies in areas such as competition, land-use planning, trade and tax provide an unwarranted advantage to carbon-intensive technologies.

Addressing misalignments: Applying the OECD Policy Framework for Investment to low-emission infrastructure

The OECD Policy Framework for Investment (PFI) provides a systematic checklist of key policy issues for governments to create better conditions for attracting private investment (Table 5.4).[13] This section focuses on selected policy areas most relevant to supporting investment in low-emission infrastructure. Many of these policies also feature in the OECD Product Market Regulation index, used in Chapter 4 to model the effects of policy on private investment (see Annex 4.A3 in Chapter 4), so the policy suggestions in this section are supportive of the "decisive transition" in Chapter 4.

Table 5.5. Relevant policy areas for low-emission infrastructure from the OECD Policy Framework for Investment

Policy Domains	Selected policy areas	Relevance for low-emission infrastructure investment		More pressing for
		Energy sector	*Public transport*	*Advanced or emerging*
Investment policy	Restriction on foreign direct investment.		High barriers constrain flows of low-carbon capital and access to "best in class" technologies.	Emerging economies
	Contract enforcement and policy stability.	Retroactive policy changes and contract enforcement issues with off-taker in case of tendering (i.e. bankability issues, limited creditworthiness) undermine investment attractiveness.	Long-term contracts with a public authority require specific mechanisms for contract renegotiation and enforcement.	Both
	Land rights acquisition.	As opposed to fossil generation, renewable power plants need to be located close to the exploited renewable energy source.	Construction works require dealing with several owners both in urban and extra-urban projects.	Emerging economies
	International investment agreements.	Their impact is uncertain. They might help to attract foreign funds but might constrain the relocation of capital.		Both
Investment promotion and facilitation	Complex and uncertain permitting procedures.	Lower the investment risk-return profile for both on- and off-grid generation projects.		Both
	Investment promotion agencies.	Opportunity to increase the capital flows and to target investors (i.e. long-term investment funds).		Both
	Special economic zones.	Can become the testbeds for new policies and help resource rich countries to diversify.		Emerging economies
	Rule of law.	Corruption risk undermines efforts to secure an attractive investment environment for capital-intensive projects.		Both
Competition policy and design of regulated markets	Governance and competition with SOEs.	May crowd out private investment or increase the cost of capital due to a perceived higher risk.		Emerging economies
	Independent competition authority.	Limited competition and market rigidities can favour fossil-fuel incumbency. In addition, regulators need to be skilled in new low-carbon technologies to manage cases where environmental regulations and competition may not be aligned.		Both
	Electricity market design.	The increasing penetration of renewable generation challenges the current design of wholesale power market, given its low marginal cost.		Advanced economies
Land planning	Integrating land-use with transport planning and resource assessment.	Areas rich in renewable energy sources might require adjustment to their land-use destination.	Higher densities make the deployment of large-scale public transport systems more feasible.	Both
Trade policy	Tariffs and non-tariff measures on imports of low-carbon equipment.	Increase the (domestic) cost of low-carbon technologies, thereby slowing investment in cleaner energy.		Both
	Barriers to trade in environmental related services.	Limit access to the expertise associated with installing and operating low carbon equipment, thereby increasing costs and investment risk.		Both
Tax policy	Tax incentives.	Can have unintended environmental effects (e.g. provision of company cars for private use and the deductibility of commuting costs from personal income taxes).		Both
	Corporate income tax and technology bias.	Restrictions on loss carryovers tend to disadvantage technologies with a high share of upfront investment costs, including renewables.		
	Accounting rules for energy efficiency contracts.	Accounting rules may not capture the economic nature of energy performance contracts.		Both
Policies influencing responsible business conduct	Governments schemes on climate-related information disclosure.	Government-sponsored schemes are limited in the scope of required disclosure and their methodologies are insufficiently comparable across countries.		Both

Box 5.7. Investment conditions and core climate policies combine to boost investment and innovation in renewable electricity generation

New research by the OECD analyses data on more than 70 explanatory variables across OECD and G20 countries – climate mitigation policies, investment environment variables and control variables – to investigate determinants of investment flows and innovation in renewable power from 2000 until 2014.

The results support the hypothesis that beyond the need to set stronger, coherent climate mitigation policies, policy makers also need to align the broader investment environment in order to effectively mobilise investment in renewable power. Across OECD and G20 countries, renewables investment was primarily driven by targeted investment incentives, i.e. feed-in tariffs (FiTs), renewable certificates (RECs) and public tenders. FiTs and RECs have been effective in advanced OECD and G20 countries, while tenders have been effective in emerging G20 and OECD countries, and for wind-power investments across OECD and G20 countries. Support measures for fossil fuels used in power generation appear to deter renewable investment for emerging economies. Results also suggest that climate mitigation policies are enhanced when they are combined in certain ways. In emerging economies, for example, public RD&D spending in the renewable sector enhances the impact of explicit carbon prices or RECs.

Results suggest that investment flows in renewable power also depend on the attractiveness of the broader investment environment, particularly in emerging economies, as well as in wind and solar energy. Significant factors include the overall ease of doing business; perceived levels of corruption; investment policy (e.g. policies on registering property; and regulatory quality for solar energy); investment facilitation (e.g. licenses and permit systems for wind energy); competition policy (e.g. direct control of the state over enterprises) and trade policy (ease of trading across borders in the EU and for solar power). Financial market policy is also a key factor, for example access to domestic credit for the private sector, sovereign credit ratings, and implementation of Basel III leverage ratio requirements. Separately, results also confirm the important role played by core climate policies in promoting innovation, in particular that feed-in tariffs and public RD&D spending have stimulated patenting activity in renewable power technologies across OECD and G20 countries.

Source: Ang, Röttgers and Burli (2017).

Investment policy

Core investment principles such as not discriminating against foreign investors and ensuring the transparency and predictability of regulations are important to stimulate investment in low-emission infrastructure. Beyond the impact on total available capital, any restriction on foreign investment is also likely to limit access to best-in-class technologies (OECD, 2015m), thus slowing down the uptake of less carbon-intensive technologies and practices. Restrictions on foreign direct investment (FDI) in infrastructure sectors have fallen recently, as measured by the OECD FDI index.[14] China has lifted several restrictions on FDI, including for renewable power projects (NDRC, 2015). Such restrictions are above average in the transport sector across G20 countries, however.

Land acquisition is a pressing issue for renewable generation projects as they need to be located where natural resources such as wind and geothermal are best. Conflicts may emerge if sites are already occupied or are protected for nature, biodiversity or tourism reasons. In cities, complexity and uncertainties in acquiring land rights can delay the development of mass public transportation systems. When formal property rights for land are lacking, particularly in remote rural areas in some developing countries, these issues are likely to be exacerbated.

Tools that can reduce barriers in property regularisation include increased co-ordination between the different institutions involved in the property registration process and introduction of a land lease process (OECD, 2016f). For renewable energy, mapping natural renewable resources has proved instrumental, in part by reducing due diligence costs. To mobilise support from local communities, Denmark introduced an obligation for developers to offer a certain percentage of project shares to local residents (ENS, 2015).

Investment promotion and facilitation policies

Investment promotion and facilitation policies include laws, policies and regulations that promote a country or region as an investment destination for international investors, such as targeted investment incentive schemes, or that ease the burden for investors to establish or expand their existing investments, for example by streamlining administrative processes.

Low-emission infrastructure projects typically face high transaction costs due to cumbersome approval procedures. Streamlining licensing and permitting procedures facilitates investment in low-carbon infrastructure projects, especially for renewable power generation, and could be beneficial both in OECD countries (e.g. several European countries) and in other economies (e.g. China, India and Indonesia). Intricate administrative procedures can delay utility-scale renewable projects up to several years, while smaller projects might simply not go ahead because of the administrative uncertainties and costs (OECD, 2015l). One-stop-shops (OSSs)– offices in charge of issuing all required permits for project approval – can ease the regulatory burden. In Denmark, the three licences required to establish an offshore wind farm are all granted by the Danish Energy Agency. In Indonesia, authority for licensing geothermal power generation is now held by the central government while it was previously held by either central or local government depending on location (IASS, 2016; AsDB/World Bank, 2015). However, OSS effectiveness can be hindered if their role is limited to collecting permits issued by other offices (OECD, 2015l). Standardisation of contracts can also help reduce due diligence costs and favour aggregation of projects for refinancing by larger investors (CEFF, 2016).

Permitting issues also hinder the development of off-grid electrification. For instance, environmental authorities may subject mini-grid projects to the same process as power stations, requiring prohibitively expensive environmental impact assessments that may not be commensurate with the scale of potential project risks (Reiche, Tenenbaum and Torres de Mästle, 2006). Some countries are experimenting with a permanent exemption from obtaining a licence or permit for small mini-grids, as in Tanzania (up to 100 kW) and Mali (up to 20 kW), or developing an integrated licence for all activities connected to mini-grid operations (REN21, 2014).

The overall regulatory burden for large-scale infrastructural projects can also be reduced by tendering pre-approved sites, a common practice for transport projects that is now being adopted for low-emission infrastructure projects. In the Netherlands, the power grid operator obtains the necessary permits for future off-shore wind sites and performs a general environmental impact assessment for the sites, leaving bidders to submit only a project-specific environmental impact assessment. A similar approach has recently been introduced in the United States, where the Department of the Interior is encouraging developers to build solar farms on public land by completing preliminary (non-project specific) environmental impact statements before opening the competitive leasing procedure (Bloomberg, 2016b; Solar PEIS, 2017). Other mechanisms to lower administrative burdens include establishing a legal time limit for permit approval (Kozluk, 2014).

Investment promotion agencies (IPAs) can play an important role in attracting foreign capital by emphasising low-carbon options and carefully targeting and packaging projects. This requires a focus on specific subsectors whose selection should be based on criteria such as production costs, supplier-client proximity and availability of technologies and skills (UNCTAD, 2013). Focusing on projects that are bankable – politically supported, regulatory prepared and packaged in suitable sizes for relevant target groups – can be an effective strategy. For example, the Danish IPA (invest-in Denmark) has taken steps in this direction by focusing on specific clean technologies, including wind and bioenergy.

Other concerns regarding a country's investment environment, such as political instability, sovereign credit rating and corruption, may increase financing costs and deter foreign capital. Corruption particularly undermines efforts to secure an attractive investment environment for low-carbon projects given their capital intensity and reliance on public procurement. It is vital that governments follow best practice in fighting illegal business conduct, including due diligence and presence of appropriate reporting channels (OECD, 2016g; OECD, 2011; OECD, 2009; OECD, 2006).

Competition policy and design of regulated markets

Effective competition is essential to develop a business environment that can attract private investment in infrastructure sectors traditionally dominated by governments and state-owned enterprises. As elements of natural monopoly are present in sectors crucial for the low-carbon transition, such as energy and transport, good regulation is especially important, including through an independent competition authority with well-delineated responsibilities whenever several public agencies are present (energy authority, transport authority).

When formerly integrated activities are liberalised, good market design is crucial to ensure alignment with low-carbon objectives (OECD/IEA/NEA/ITF, 2015). G20 countries differ in the degree of competition in infrastructure sectors. Many developed countries have adopted market approaches in the electricity sector, unbundling generation, transmission and distribution and creating wholesale electricity markets open to competition. Other countries, including in the G20, retain a more integrated electricity sector. In the transport sector, competition is generally even more limited. In rail transport, other operators have entered the market in only a few countries, even though competition is technically possible in others (Casullo, 2015).

In markets where state-owned (or formerly state-owned) incumbents retain a dominant position, governments should ensure a level playing field. If SOEs receive undue preferences or advantages – in addition to direct compensation for their public service objectives – this may crowd out private investment and increase the cost of capital for new entrants. In Brazil, for instance, equal access by all firms to funding by the state-owned development bank could improve competition and resource allocation. In South Africa, exemptions for state-owned enterprises from competition laws has allegedly hindered competition. In China, removing implicit guarantees to SOEs could improve the allocation of capital, and potentially reduce emissions by slimming excess energy-intensive manufacturing capacity.

The OECD guidelines on SOE governance lay out the importance of effective corporate governance for SOEs as well as ensuring that SOEs are not exempted from general laws and regulations, including stringent financial reporting and disclosure requirements (OECD, 2015n). In the case of renewables, SOEs or incumbent utility firms may receive faster or cheaper access to the electricity grid due to close relationships with system operators. They may also be less likely to face curtailment when supply exceeds demand, and could be preferred bidders in tenders for new capacity. In many G20 countries, incumbent energy utilities, which are often publicly owned and carbon-intensive for historical reasons, can have a significant influence on energy policy. In some circumstances they have blocked energy efficiency programmes(Amon and Holms, 2015). However, other evidence suggests SOEs have been prominent investors in renewable electricity (Prag et al., 2017 forthcoming).

Where markets have been liberalised, investment decisions are meant to be driven by the price signals delivered by the wholesale power market, but market arrangements need to be improved in order to sustain competition and provide effective investment incentives. The progressive decarbonisation of electricity generation is challenging the current design of energy-only markets. Many liberalised electricity markets are no longer delivering accurate price signals because they were not designed to mix conventional power generation with high proportions of zero-marginal-cost renewable electricity. In addition, the variable and non-dispatchable nature of some renewable technologies can make it more difficult to balance demand and supply. Addressing this challenge requires equipping power markets with features such as high-resolution prices, pricing system reliability, better demand response, improved allocation of transmission and distribution networks costs, and the integration of storage (IEA, 2016a). More inter-connected grid systems can also significantly reduce duplication of resources, thus freeing investment for more efficient use elsewhere.

Another challenge for liberalised electricity markets is the need to maintain electricity generation capacity margins for security purposes at times of scarcity. Countries are developing different options to guarantee security of supply, including capacity mechanisms that ensure that plants will be available to run in times of scarcity. Some have criticised capacity mechanisms recently for prolonging the life of fossil-fuel plants that would otherwise be retired, therefore slowing down the process of decarbonisation. Others argue that capacity mechanisms are necessary to ensure adequate supply and flexibility as the share of variable renewable energy increases significantly.[15] Governments need to carefully monitor the effects of these measures as they look for an economical approach to decarbonising electricity while ensuring a reliable electricity system (IEA, 2016a).

In emerging markets with fast-growing electricity demand, the majority of electricity systems remain vertically integrated or without wholesale electricity markets. In China, market prices are regulated but transmission has been separated from generation. While some provinces have recently launched "market pilots" as part of the power sector reform efforts started in 2015 (RAP, 2016b),[16] revenues of power generators are essentially still based on administratively determined power prices. Similarly, in India about two thirds of the generation capacity is government-owned (OECD/IEA/NEA/ITF, 2015) and power traded on the wholesale market constituted only about 10 per cent of the total electricity generation in 2015-16 (CERCIND, 2016; Chattopadhyay, 2014). Such integrated and highly regulated electricity systems pose different alignment challenges. For instance, the administratively determined power prices paid to Chinese power generators are generally set to cover average cost. In turn, this requires generators to produce the yearly quantity of electricity planned by the public authorities in order to equalise revenues and average costs. This has reportedly led to heavy curtailment of wind power generation in order to maintain running hours of fossil fuel plants.[17] Furthermore, regulated prices make it difficult for utilities in some countries, such as China and India, to pass on the higher prices they pay for renewable PPAs, meaning integrated utilities face a disincentive to accepting more variable renewable energy supply on the system. Tariff issues, along with factors like non-technical losses (theft), have eroded the creditworthiness of several Indian distribution companies, potentially weakening the attractiveness of tendered PPAs. Recent measures to reform the Indian energy sector are likely to improve the overall business environment.

In general, rigid provisions on generation, transmission and distribution may hinder off-grid renewable-based electrification. For many low-income households in developing countries, off-grid solar has become an economically attractive alternative to kerosene for lighting purposes. Several companies are entering this market, either selling (or increasingly leasing) solar household-level systems or building community-level mini-grids powered by renewable or diesel generators. Even if a vertically integrated structure is maintained, the legal framework should accommodate distributed generation, or off-grid solutions will remain informal, foreclosing access to financing and limiting their growth (REN21, 2014).

Land-use planning

Land-use planning policies are important levers to reduce GHG emissions over the longer term because they can prevent the locking-in of energy- and carbon-intensive behaviour, particularly in urban areas. Cities in developing and developed countries are facing similar challenges from congestion and urban growth. However, these concerns are more pressing in developing countries, where urban growth is far more rapid due to urbanisation and motorisation processes that have not yet peaked (Srinivasan, 2001). Further, evidence suggests that as cities become richer, they tend to experience urban sprawl and declining density. This has been particularly the case in Asia (Box 5.8).

Sprawl tends to be associated with increasing road traffic and cost of infrastructure development. Integrating land-use and transport planning to address urban sprawl and create more compact cities can thus make the deployment of large-scale public transport systems more feasible, increase the share of public transport and encourage non-motorised travel. By combining pricing and regulatory policies to slow down the ownership and use of personal vehicles with strict land use control, governments could decrease cars' share of transport from 50% to 25%, depending on the region (OECD/ITF, 2017).

With numerous stakeholders and authorities involved in transport and urban planning, alignment is challenging. Consolidating planning agencies' responsibilities to include all transport modes as well as urban planning would facilitate such alignment. There are several successful examples of multimodal transport agencies, such as Singapore's land transport authority, which co-ordinates metro lines, buses and road projects. The institutions responsible for urban and transport planning are usually separate, however, and sometimes operate at different administrative scales. A possible solution is to promote fully integrated agencies operating at the metropolitan level rather than the jurisdiction of urban authorities.

Transit Oriented Development (TOD) aims to create dense neighbourhoods around transit stations, and has been successful in some places, notably in the United States (Cervero et al., 2004). Relying on a partnership between private property developers and public authorities, TOD can be seen as a more local and project-specific approach to integrating land and transport planning. If such projects are not accompanied by appropriate policies to preserve affordable housing in areas close to transit stations, however, the ensuing real estate premiums can lead to a "gentrification" process. In turn, this may lead to lower than expected ridership since wealthier households are more likely to prefer private vehicles to public transport, especially in developing countries (Siemiatycki, 2006).

While the existing built environment limits the options available to existing cities, new cities can be built with a low-carbon approach at their heart. Flagship eco-city projects are being built across the globe, including Masdar in Abu Dhabi, Tianjin Eco-City in China, and

Songdo eco-city in Korea. These are developed according to such principles as sustainable mixed-use development and integration of transport and land use. For instance, the Tianjin masterplan aims at providing public transit options for 80% of the population within 800m of their homes (GIZ and ICLEI, 2014). However, such projects are not exempt from criticism, including the risk of being "enclaves of the rich" (Caprotti, 2014), and still exposed to pollution generated by neighbouring sources.

Box 5.8. **Urban planning: alarming trends in city density**

Planning tools and strategic documents for urban planning exist in most countries, but they have proven to be insufficient to control urban sprawl. Despite land control policies, most developed countries have witnessed urban expansion since the 1950s when many people moved from city centres to suburbs. This resulted in urban sprawl with lower population density. This trend continues: between 1990 and 2000, the urbanised area in Europe increased by 18.4%, while population density fell by 9% (Oueslati et al., 2014). The two main determinants of urban extent are population and GDP per capita (OECD/ITF, 2017). Urban extent increases at a slower pace than population, thus bigger cities tend to be denser. However, richer cities tend to be more sprawling, meaning the average density of cities decreases (Table 5.6). This is true for all regions of the world, although to different degrees. The density decrease is particularly sharp in Asia, where rising GDP per capita will drive urban expansion.

Moreover, the combined effects of urban extension, population and income growth will result in a surge in road traffic, and require significant road investments. The situation is particularly dramatic for Asian cities, where the drop in density is sharp, 19% between 2010 and 2050, while road traffic rises by 532%. Maintaining an efficient transport system would require an increase of 295% of the trunk road network. Strict land use rules would drastically reduce the need for road infrastructure building, thus allowing the funding of large transit systems. By avoiding excessive sprawl, the costs of new road construction could globally be decreased by at least USD 10 trillion between 2015 and 2050.

Table 5.6. Projected variation in density, traffic and trunk road needs in cities
(% of variation between 2015 and 2050 for different world regions in business as usual scenario)

	Density	Traffic	Trunk road needs
Africa	-8%	325%	158%
Asia	-19%	532%	295%
Europe	-7%	40%	39%
Latin America	-8%	152%	92%
Middle East	-2%	228%	98%
North America	-1%	68%	36%
OECD Pacific	-8%	12%	24%

Source: Chen and Kauppila, 2017. and as cited.

Land-use planning when combined with land value capture tools can also help cities fund large-scale transport systems. A municipality can increase the value of land either through legal actions (e.g. providing permits for building facilities that will increase the value of existing buildings, such as a mall) or through improved access (e.g. new metropolitan lines). Through land value capture, the local authorities and the private actors agree beforehand how to share the capital gain, thus helping cities to mobilise financing for the initial construction and long-term operation and maintenance. Examples of land value capture tools include tax increment financing, development charges, development rights and joint development. Land value capture is not new, but its application for large investments in metropolises is growing. New York, Washington, London and Paris have all recently used some form of land value capture scheme and this example could be followed in other large G20 cities (see Box 7.3 in Chapter 7). A major difficulty of these schemes is the ex-ante evaluation of the value created by public investment and to calibrate the tax accordingly. Areas with weak property rights, such as slums, are also less amenable to land value capture.

Table 5.7. **Funds raised or to be raised by land value capture for selected large projects**

City	Project	Fund raised or projected	Percent of project cost
London	Crossrail	4.1 billion GBP	32%
Washington, D.C.	New York Avenue Metro Station	25 million USD	28%
Washington, D.C.	Dulles Metro-rail Silver Line Expansion	730 million USD	14%
New York	Subway 7 Line Extension	2.1 billion USD	88%

Source: Adapted from Salon (2014).

Implications of international trade and trade policy

When countries are linked through international trade, their climate policies affect and are affected by other countries' climate policies. The resulting fear of loss in competitiveness and "carbon leakage" can be an important political consideration where the stringency of climate policies differs across countries. Trade is also a source of GHG emissions in its own right, partly due to the energy consumption of international shipping and aviation – which are difficult to regulate due to the international nature of bunker fuels – as well as road transport. However, trade is also a mechanism for diffusion of low-emission technologies and services can and help countries to adapt to changing production conditions, including due to climate impacts. This section focuses on the role of trade in supporting investment in low-emission infrastructure and technologies.

An open, rules-based trade regime is one means of achieving the investment necessary to achieve the low-carbon transition. A clear recent investment pattern has been the growing fragmentation of production across borders, underscoring the necessity for countries to maintain an open, predictable and transparent regime for both trade and investment, as the two have become increasingly interlinked.[18] More and more, companies rely on networks of goods and services suppliers to source the inputs they need at lower cost (OECD, 2017c). Trade and investment remain the primary channels for technology diffusion (Keller, 2004; Alvarez et al., 2013), including for low-emission technologies.

In the past decade, global trade in environmental goods has outpaced overall merchandise trade (Figure 5.5), driven in large part by demand from the renewable-energy sector and for solar photovoltaic (PV) panels in particular. This reflects the growing urgency of climate change and the increasingly stringent policies that many governments have put in place to mitigate their countries' emissions of GHGs and other pollutants.

Policy obstacles still hinder trade in low-carbon goods and associated services. Many countries levy tariffs on imports of low-emission equipment. These make low-emission technologies more expensive than they need to be, thereby slowing investment. Several OECD countries and emerging economies are negotiating a plurilateral agreement liberalising trade in environmental goods (known as the Environmental Goods Agreement, or EGA). Once final, this agreement would bind at zero[19] import tariffs on several products deemed environmental, many of which help prevent or reduce GHG emissions.

Figure 5.5. **The past decade has seen growth in global trade in environmental goods outpace overall merchandise trade**

(2007 = 100)

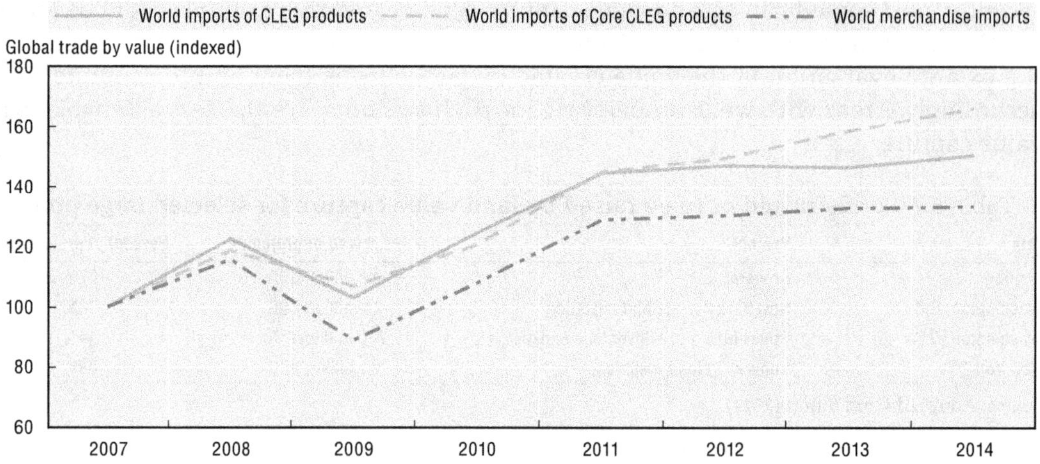

Note: Environmental products are here taken to refer to the OECD's Combined List of Environmental Goods (CLEG), which contains about 250 six-digit HS lines. See Annex 1 in Sauvage (2014) for more information on the CLEG.

Source: OECD calculations based on UN Comtrade and WTO data.

StatLink http://dx.doi.org/10.1787/888933486424

Other trade obstacles can also hamper the diffusion of low-carbon technologies, for example non-tariff measures like local-content obligations attached to feed-in tariffs. Such measures tie the award of public support to the use by solar- or wind-energy investors of locally made inputs, which can increase the costs of renewable-energy projects and slow technological diffusion when foreign-made inputs exist that are cheaper, of better quality, and more technologically advanced (OECD, 2015o; Bahar et al., 2013). Where governments want to create jobs in low-carbon sectors, it is probably more effective to foster the deployment of renewable-energy capacity than to require the use of locally made equipment. This is because the majority of jobs along the renewable-energy value chain are in installation, construction (e.g. erecting wind turbines) and maintenance services, as opposed to upstream and midstream segments (e.g. the manufacturing of solar PVs), which tend to be capital-intensive and automated (OECD, 2017c).

The mutually supportive relationship between trade and investment in low-carbon infrastructure is most apparent in the area of environmental and related services. From the design and construction of a geothermal power plant to the repair and maintenance of a wind turbine, numerous services are essential to the uptake of cleaner technologies, such as customising, installing and maintaining equipment (OECD, 2017d). This underscores the need for governments to accelerate their efforts to liberalise trade in environmentally related services, for instance by shortening work permit processing time for people providing such services, given the environmental and economic gains that would result. These efforts would not only complement ongoing negotiations to liberalise trade in environmental goods, but also reinforce current initiatives for mitigating pollution in all its forms.

Tax incentives, corporate income tax and accounting rules

Aside from taxes on GHG emissions or other externalities, policies related to tax and financial accounting may favour or disadvantage investment in low-carbon technologies. These include tax incentives not motivated by environmental concerns; provisions of the corporate income tax system that may inadvertently skew investment incentives; and accounting rules that may hinder energy efficiency investments.

Tax incentives within the tax code

Tax incentives should be designed with care and used only in particular circumstances. Those aimed specifically at encouraging low-emission investment were covered above under core climate policies. Tax incentives that are not motivated on environmental grounds may hinder the low-carbon transition. For example, providing company cars for private use for tax reasons and enabling commuting costs to be deducted from personal income taxes have unintended environmental effects. Employee compensation is effectively taxed more lightly than cash wages when the compensation comes in form of personal company car use (Harding, 2014). Such tax incentives encourage the use of cars over other modes of transport, increasing air pollution, GHG emissions and noise. Deductibility of car commuting costs encourages consumers to live further from work, increasing environmental impacts. Deductibility of public transport commuting costs can have similar effects but it also may reduce the proportion of commuters using cars, and hence related impacts.

Corporate income tax and technology bias for or against low-emission technologies

Features of the corporate income tax code may favour or disadvantage low-carbon technologies for producing electricity. Unintended technology bias from corporate income tax provisions can arise when the cost profiles of substitute technologies differ. Electricity generation technologies using renewable sources of electricity tend to feature upfront investment costs (including financing costs) that are higher than the variable and fixed costs during production. Technologies relying on conventional sources of energy, such as coal or gas, exhibit a more evenly spread cost profile, with operating costs reflecting the market price of the fuel. When capital cost profiles of technologies differ, at least three features of corporate tax systems may favour or disadvantage low carbon technologies: immediate expensing of variable costs but not investment costs, accelerated depreciation, and restrictions to loss carryovers.

Immediate expensing of variable costs favours investment in carbon-intensive electricity generation technologies. Corporate income tax is levied on profits that are calculated as the difference between revenues from goods sold and expenses. Variable costs are expenses and therefore fully deductible in the period in which they occur, thereby decreasing taxable profit. Investment costs, on the other hand, are depreciated over the asset's lifetime and not immediately deducted from taxable profits. Technologies with a high share of variable costs will benefit more from immediate expensing of variable costs than technologies having a high proportion of costs upfront during investment, such as most renewable electricity technologies.

Fiscal depreciation can favour investment in renewable electricity projects to the extent that the applicable tax depreciation rate does not reflect the actual economic depreciation rate of the capital asset. The majority of G20 countries provide accelerated depreciation to both carbon-neutral and carbon-intensive technologies, with often higher levels of acceleration for carbon-neutral technologies, suggesting there is no inadvertent misalignment with the low-carbon transition on average.[20]

Restrictions to loss carryovers limit the ability of businesses to carry over losses to subsequent (or preceding) fiscal years in order to offset profits. Such provisions therefore disfavour technologies that require a higher share of costs in the investment stage, such as renewables. This effect may be particularly relevant in the case of young, innovative firms, which often have limited access to external finance and are more likely to incur losses during the start-up phase. Loss carryovers are restricted in all but four G20 countries, which suggests there is misalignment.[21]

Accounting rules and energy efficiency investments

Accounting rules can deter investments in energy efficiency, especially in the public sector, in the context of energy performance contracts (EPCs). EPCs can take different forms but usually involve a service provider who is paid from the energy savings made possible on the client's premises. EPCs accounted for about 70% of energy service companies' (ESCOs) revenue in the United States, and are also considered the main business model of Chinese ESCOs (IEA, 2016b). Nevertheless, the accounting treatment of the costs and expected benefits arising from EPCs poses specific challenges, especially if the contract involves governments and local authorities.

Local authorities are often required to record future payments to the service provider as future liabilities on their accounting books. Consequently, as public accounting rules record the cost but not the economic benefit of investments in terms of lower future energy bills, local authorities who face particular constraints on their debt levels are discouraged from engaging in energy savings through EPCs. This is the case in Europe, where EUROSTAT rules (guidance note of 07/08/15) state that investments in energy saving measures have to be attributed to public building owners (EPC client), even if financed through an EPC (unless certain specific requirements are met). This can be problematic as public debt levels are constrained by the Maastricht Treaty. Private firms with limited ability to raise debt may face similar hurdles as, even if their debt levels are not regulated, they are exposed to market discipline and the energy efficiency investment may worsen their debt to equity ratio (IPEEC, 2016; EFFIG, 2015). The need to review accounting rules to identify possible barriers to investment in energy efficiency was recognised by G20 Energy Ministers in the Voluntary Energy Efficiency Investment Principles for G20 participating countries. Progress in implementing these principles and relevant best practice was reviewed in the 2017 G20 Energy Efficiency Investment Toolkit (EEFTG, 2017).

Policies influencing business conduct

Policies that shape business conduct can encourage companies, investors and consumers to make less carbon-intensive choices. The absence of such policies can be considered a misalignment, given that business engagement is crucial to the successful design, financing and implementation of measures to address climate change and is an integral part of responsible business conduct.[22] Policies to influence business conduct include the information policies discussed under climate policies above, such as labelling schemes requiring information on the energy consumption or carbon footprint of products. Others are specific to influencing business behaviour, including voluntary and mandatory disclosure and reporting, mandatory energy audits and consumer protection policies to limit company "greenwashing" (OECD, 2010; Klintman, 2016).

Governments can influence company behaviour, as well as investor and consumer choices, by requiring that financial and non-financial businesses disclose accurate, high quality, and comparable information on how companies are exposed to climate change and how they are addressing related risks and opportunities. Currently, 16 G20 governments require companies to disclose some type of climate-related information. Disclosure requirements vary from country to country, however – for example, in their scope and measurement methodology – thus providing insufficient and non-comparable information (OECD/CDSB, 2015). Often the scope of required disclosure does not extend beyond GHG emissions, so it does not cover the company's exposure to climate-related risks, or wider strategies to limit emissions or address risks.

The need for private sector investment to drive the low-carbon transition has prompted policymakers, including finance ministries, to encourage greater alignment between market behaviour and the Paris Agreement and the Sustainable Development Goals. This in turn has exposed limitations of corporate reporting schemes on climate issues, and in the use of information by investors and others as a basis for decision-making. The way in which companies manage climate change risks has rarely been integrated into reporting on risks, opportunities, strategy and governance practice. Corporate climate risk management targets have been developed by individual companies internally rather than by reference to sectoral, national or global goals, thus making it difficult to assess corporate contributions to wider policy goals. Reporting tends to focus only on historical information (such as GHG emissions) rather than future-oriented assessments of climate risk (OECD/CDSB, 2017, forthcoming). Disclosure of climate risks by financial institutions and institutional investors is a fast-moving area, however (see Chapter 7).

Public infrastructure choices and procurement for low-carbon and resilient pathways

All G20 governments are important economic actors in their respective territories, in particular through procurement of basic infrastructure – such as roads, bridges, harbours, sanitation systems, electricity networks – and the public purchase of good and services. Although varying between countries, public procurement expenditures[23] amount to 13% of GDP on average in OECD countries, and an often higher share of developing economies' GDP (Figure 5.6).

Figure 5.6. **Public procurement as a share of GDP**

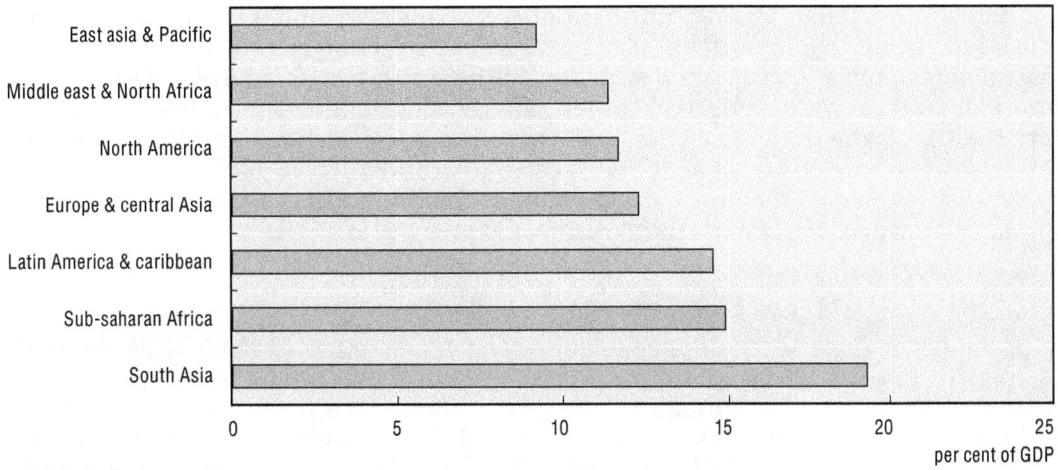

Source: blog/World Bank (2016).

In all countries, it is crucial to integrate climate factors into the methods and policies that governments use to plan for and procure infrastructure – even where significant private sector investment has already been made in infrastructure, whether through public-private partnerships (PPPs; Box 5.9) or other means. At the planning and project selection level, the infrastructure selection process in the next decade may lock in carbon-intensive and climate-change-vulnerable infrastructure (Chapter 3), or conversely create momentum towards sustainable and climate-resilient solutions. Public procurement can also support low-carbon and climate-resilient innovation through the creation of lead markets, ultimately lowering the risk for suppliers of these innovations.

Box 5.9. **PPPs and low-emission infrastructure investment**

A growing proportion of infrastructure services has been delivered through public-private partnerships (PPPs), though with significant differences across countries. In 2013, the PPP capital stock represented around 1% of GDP in advanced economies and around 5% of GDP – with peaks of up to 9% – in emerging and low-income countries. Evidence of whether they provide infrastructure more efficiently than traditional public procurement is mixed (IMF, 2015). Numerous low-carbon infrastructure projects have been developed through PPP frameworks. Examples include mass rapid transit systems in India, bike-sharing schemes in European cities, building-energy-efficiency projects in Berlin, and a forest and watershed restoration programme in São Paulo. Public-private partnerships have also proven to be efficient instruments for food-waste reduction and prevention, thus helping to reduce unnecessary GHG emissions from agriculture (OECD, 2016h).

A practical question is whether PPPs are particularly suited for the procurement of low-carbon infrastructure. If governments decide to leverage the framework of PPPs, then government PPP units may be the appropriate administrative unit for managing the delivery of low-carbon performance as an integral part of infrastructure projects (OECD, 2012b). While subnational governments – jointly with publicly owned companies – are the major contributors to public investment in advanced and large emerging economies (IMF, 2015), they often lack the adequate capacity for managing PPPs (OECD, 2013b). For this reason, mandating dedicated government PPP units to work with local authorities and creating guidelines for their staff can help to spread best practices in designing PPPs (OECD, 2014e).

Sources: OECD (2008); Kennedy and Corfee-Morlot (2012); and as cited.

Reflecting climate factors in public infrastructure decisions

Integrating climate change considerations into government infrastructure decisions is crucial to ensure that investment plans and project selection are consistent with climate change mitigation and adaptation ambitions. Where cost-benefit analysis (CBA) is used to prioritise infrastructure projects,[24] climate change considerations need to be incorporated into the CBA framework – both for GHG emissions and for climate resilience (Box 5.10). To internalise the value of GHG emissions in infrastructure, some OECD countries systematically use a "shadow" carbon value in their appraisal of proposed policies and infrastructure decisions (Smith and Braathen, 2015).

Box 5.10. **Cost-benefit analysis in transport**

In most G20 countries, CBA is compulsory and standardised for large-scale transport infrastructure investments, but its scope is very limited. CBA is mainly used to detect "white elephants" – projects that are far too costly given the social benefits they provide. To ensure optimal allocation of investment from the point of view of climate objectives, CBA should remain one of the major tools for programming and prioritising projects under a given budget allocation. Although national infrastructure plans exist, they are usually assessed using some form of multi-criteria analysis, making the underlying trade-offs between costs and benefits unclear.

The scope of appraisal should include not only transport infrastructure but also all transport policies. Achieving transport decarbonisation requires complete policy packages that go beyond infrastructure investments. Even simple transport policies can involve complex trade-offs, and significant money streams and economic costs. For instance, many national low-carbon strategies support and subsidise electric cars. In Norway, as a result of generous policies to increase the use of electric vehicles, sales are rapidly increasing, reaching a market share of 20% in 2015 (IEA, 2016c). Yet this policy has side effects: thanks to the subsidies, driving an electric vehicle implies very low costs, potentially leading to more driving at the expense of public transport and cycling. It is also costly: the electric vehicle subsidy package costs up to 13 500 USD/tCO_2 (Holtsmark and Skonhoft, 2014).

National methodologies usually include a single carbon value that is applied to every project, thus guaranteeing comparability in the decision-making process and ensuring that investments are aligned toward climate targets. This allows decision-makers to weigh trade-offs between climate change mitigation and other policy objectives such as travel time savings, economic competitiveness and air quality. As long as carbon values follow pathways that are consistent with the mitigation efforts required globally, they ensure that carbon-intensive investments are avoided – except if such investments provide very significant benefit to the transport system and no credible alternative is available.

Incorporating climate-resilience factors into CBA is, however, more challenging due to uncertainty of climate-change impacts that have clear implications for the way networks should be designed (see section "Promoting resilient infrastructure"). First, ensuring continued infrastructure and services performance may decrease the present value of networks or increase maintenance and refurbishment costs. Second, authorities or private operators must design and build new infrastructure in the context of these same changing climate variables. Uncertainty regarding these variables presents the risk of over- or under-specification of infrastructure design standards. Over-specification of design standards results in stranded investments whereas under-specification may lead to network service degradation (OECD/ITF, 2016).

Fostering low-carbon innovation in infrastructure: the role of public procurement

For public procurement to drive low-carbon innovation in the construction value chain, most countries need to change practices. Criteria for appraising public procurement bids can evolve to improve economic efficiency and facilitate the low-carbon transition. An important barrier to the penetration of innovation through public procurement is continued reliance on a single lowest-price criterion for the appraisal of bids, even though this does not guarantee the best value for money for clients, or indeed the lowest cost. Countries should be encouraged to use the Most Economically Advantageous Tender methodology (Box 5.11), which relies on attributes beyond the price.

For instance, governments have been using the total cost of ownership of the procured goods or services, which takes into account the cost of operation and maintenance. Some jurisdictions have introduced life-cycle analysis and the monetisation of externalities, including CO_2, in the evaluation of bids, with measurable effects on life-cycle CO_2 emissions (Box 5.11). This has been backed with a transparent framework, helping both bidders and procuring teams. Generally, robust methods are required to ensure that non-financial metrics do not introduce more subjectivity than price-only auctions.

Several changes can be made to improve the low-carbon performance of public procurement, in infrastructure and beyond:

- Public procurement teams must be adequately resourced to be able to introduce this dimension in their procedures. In particular, the procurement of functionality (e.g. mobility needs vs. number of cars) should be favoured, to avoid the over-specification of bids that might hinder innovation.
- Procurement teams must be also provided with methodologies that allow objective evaluation of bids that include non-price attributes, such as GHG performance. Several standards exist for measuring the GHG performance of products, plants or companies. Relying on well-established methodologies helps avoid risk of bid appraisals not being conducted properly.
- Budget and accounting rules should accommodate the choice of projects whose economic superiority is measured over several years, to facilitate the use of public procurement criteria beyond the lowest price.

- Market dialogues can be established before procurement processes to allow the private-sector suppliers and public clients to exchange information about available innovations and the future direction of demand. This can be done in a transparent way, without harming the competitive nature of procurement.

Identifying activities where public procurement can mitigate carbon lock-in and trigger climate-friendly innovation should be made a priority. Upgrading public-procurement practices in these activities will take time and resources (including professionalisation of the work force and development of new procurement methods) and should be considered an important soft investment for the transition. It is an opportunity to create lead markets and to introduce a competitiveness and innovation agenda into the process.

Public-procurement authorities from different jurisdictions should be encouraged to co-operate. In some instances, jurisdictions have joined forces internationally to tip the supply side of the market towards sustainable solutions. For example, 11 European cities recently launched joint procurement of compressed natural gas garbage trucks (Baron, 2016). There is a significant community of practice in the area of sustainable public procurement that could be used to share best practice. These changes ought to be envisioned in the broader context of public procurement challenges, as presented in the OECD Council Recommendation on Public Procurement (OECD, 2015p). The Recommendation promotes transparency, integrity, open competition, stakeholder participation, risk management, appropriate integration in overall public finance management, and specific measures to ensure accountability throughout the procurement cycle.

Box 5.11. **Encouraging low-carbon performance through infrastructure procurement**

The Department of Public Works of the Dutch Ministry of Infrastructure and the Environment (Rijkswaterstaat, or RWS) has developed an approach to encourage the minimisation of environmental impacts related to infrastructure building. The policy direction was given by the House of Commons, asking that public procurement be 100% sustainable by 2015, with green criteria included in all tenders.

RWS uses the Most Economically Advantageous Tender (MEAT) methodology, which includes both price and quality attributes. In RWS tenders, however, quality attributes are fully monetised and translate into discounts on the bid price; the contract is awarded to the bidder with the lowest adjusted price. RWS tenders combine two sustainability criteria in the quality attributes of the tender:[25]

- The CO_2 Performance Ladder rates a company on a scale from one to five on the basis of energy savings, resource efficiency and renewable energy use. Companies must monitor their performance, formulate ambition and evaluate effects. More ambitious contractors, as rated by the ladder, benefit from a discount applied to their tendering price of 1% to 5%. The ladder is backed by an auditing system.
- A Sustainable Building Calculator (DuboCalc), provided to tenderers, assesses the environmental impacts of material use in the proposed project. DuboCalc provides a transparent assessment of environmental impacts, and helps contractors "optimise" on the basis of environmental costs, rather than mandating specific levels of performance.[26]

DuboCalc has led to significant reductions in projects' CO_2 footprint whenever bids included the design as well as the construction of the infrastructure. This gives more opportunities to innovate and to reduce the use of materials. It does not necessarily entail higher financial costs – even before the positive impacts of innovation are taken into account.

Box 5.11. **Encouraging low-carbon performance through infrastructure procurement** *(cont.)*

As an example, a recent tender for the reconstruction of a motorway was won by a company operating at the highest rung of the performance ladder, for a bid price of less than EUR 200 million, with a EUR 10 million discount to account for high performance on the environmental cost indicator. This high performance was triggered by the use of recycled and innovative materials and lower use of asphalt with certified life-cycle analysis. This illustrates the role of this. The winning project offered a 50% saving in CO_2 emissions compared with the last tender in this bid, showing how procurement procedure can create lead markets for new low-carbon products.

Multi-level governance of climate investment

Subnational governments are among the major contributors to public investment in G20 countries, accounting for 59% of public investment on average in the G20, but with a wide range across countries (Figure 5.7). In addition, they often have authority across several policy domains crucial to the transition, including spatial planning, housing, local roads and city public transport.

Figure 5.7. **Subnational government investment as a share of public investment in the G20, 2013**

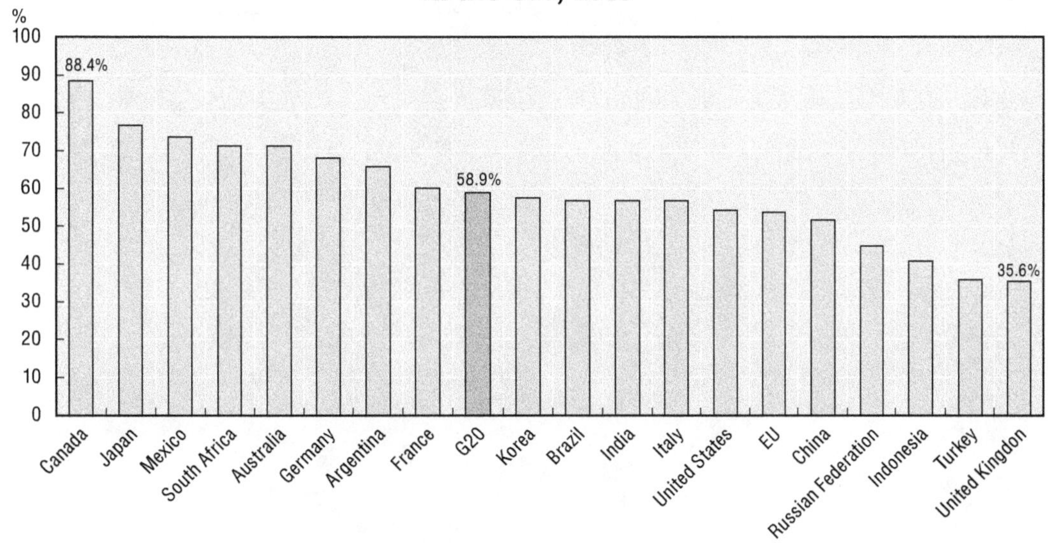

Source: OECD elaboration based on OECD/UCLG (2016), *Subnational around the world: structure and finance.* Note: no data for Saudi Arabia. G20 average is unweighted.

StatLink http://dx.doi.org/10.1787/888933486432

For these reasons, effective multi-level governance is critical to implement the Paris Agreement, as national and subnational governments are also mutually dependent for their climate investment policy. Cities and regions are well situated to identify complementary local infrastructure or services tailored to local contexts, and to co-ordinate investment needs and priorities (OECD, 2014f). For example, to address a 'mega-city malaise' China has adopted a strategy for co-ordinated development of its Jing-Jin-Ji region, consisting of Beijing and Tianjin municipalities and Hebei province. The guideline for the region's integrated development adopted in 2015 envisages transfer of some administrative functions away from Beijing, and prioritises environmental protection, energy security, traffic management and industrial upgrading (China Daily, 2017a and 2017b).

The OECD Council Recommendation on Effective Public Investment Across Levels of Government (Box 5.12) can help governments assess the strengths and weaknesses of their public investment capacity from a multi-level governance perspective. It provides practical guidance on the use of various policy instruments (Allain-Dupré et al., 2017 forthcoming), and a set of indicators to support their implementation.

Box 5.12. Council Recommendation on Effective Public Investment Across Levels of Government (2014)

The OECD adopted in 2014 a Recommendation on Effective Public Investment Across Levels of Government. A Recommendation is an OECD instrument adopted by the Council. Recommendations are not legally binding, but practice accords them great moral force as representing the political will of member states.

The OECD Recommendation underline how member countries should take steps to ensure that national and sub-national levels of government effectively utilise resources dedicated to public investment for territorial development in accordance with 12 key principles, including adoption of effective instruments for co-ordinating across national and sub-national levels of government, reinforcement of the expertise of public officials and institutions involved in public investment, and ensuring proper framework conditions for public investment at all levels of government.

Source: www.oecd.org/effective-public-investment-toolkit/.

Notes

1. This chapter benefited also from the helpful comments and insights provided by the participants of three OECD-led workshops (OECD workshop on financing green infrastructure, Paris, November 2016; OECD-ORF workshop Growth, Investment and the Low-Carbon Transition in India, Delhi, March 2017, hosted by the Observer Research Foundation; and OECD-DRC workshop Growth, Investment and the Low-Carbon Transition in China and the World', Beijing, April 2017, hosted by the Development Research Center of the State Council).
2. Adopting a lower-bound of EUR 30 can be seen as a weak test, in the sense that carbon prices should be *at least* at this level to reflect damages.
3. VAT is not included in the ECR as it applies to all products, at least approximately, and generally leaves the relative prices of energy products unaffected (cf. Chapter 3 in OECD, 2016a)
4. These numbers include emissions from biomass in the emission base. OECD (2016h) discusses the rationale for and the impacts of including or not including emissions from biomass in the emissions base.
5. Where there is perfect competition, permit allocation rules do not affect marginal mitigation incentives as permits carry opportunity costs (Goulder and Schein, 2013). However, when rents occur, incentives for mutually exclusive investments are affected by them (Devereux and Griffith, 1998a; Devereux and Griffith, 1998b). Free allocation of permits is a source of economic rents.
6. In a monopolistic environment such a tax would act a tax on importers and producers allowing the government to capture parts of their monopoly rent.
7. www.umweltbundesamt.de/publikationen/umweltschaedliche-subventionen-in-deutschland-2016; www.bundesfinanzministerium.de/Content/DE/Standardartikel/Themen/Oeffentliche_Finanzen/Subventionspolitik/2015-08-26-subventionsbericht-25.html.
8. Phase-out clauses are necessary not only for financial support measures but also for those aiming at increasing "utility of ownership", commonly applied in the case of electric vehicles. Additional policies like waivers on access restrictions, possibility to drive on bus lanes and free parking should be only temporary: as the number of electric vehicles grows, increased congestion of bus lanes or city centres will lead to decreased benefits for EV owners and to negative impact on public transportation.
9. See, for example, IMF, OECD, UN and World Bank (2015).
10. Further, Borenstein and Davis (2015) find that higher income households receive a disproportionate share of tax credits under various tax incentives in the United States.
11. The difference mechanism provide a variable top-up on the market price up to a pre-agreed strike price determined through tendering. Interestingly, at times when the market price exceeds the strike price, the generator is required to pay back the difference, thus protecting consumers from over-payment.
12. This realisation of the importance of climate change's importance for the sector is broader than just OECD countries; in January 2017, G20 agriculture ministers recognised "the need for agriculture and forestry to adapt to mitigation and emphasised their role in its mitigation" (G20, 2017).
13. Other relevant OECD tools and reports are also used here, including OECD, (2015l) and OECD/IEA/NEA/ITF (2015).
14. The FDI Regulatory Restrictiveness Index (FDI Index) measures statutory restrictions on foreign direct investment. It gauges the restrictiveness of a country's FDI rules by looking at the four main types of restrictions on FDI: 1) Foreign equity limitations; 2) Discriminatory screening or approval mechanisms; 3) Restrictions on the employment of foreigners as key personnel and 4) Other operational restrictions,
15. The production of wind- and solar-based electricity increases 15-fold between 2015 and 2050 in the IEA 66% 2°C scenario (IEA, 2017).
16. To date, the market pilots have focused almost entirely on implementation of direct trading. Within the pilot market, generators that are approved to participate in these programs can sign contracts directly with eligible industrial customers, with contracted output no longer counted in the operating-hour planning process (RAP, 2016a).
17. The regulated rigid pricing and production mechanism may also prevent some of the efficiency gains expected from the Chinese Emission Trading Scheme (Baron et al., 2012; OECD/IEA/NEA/ITF, 2015).
18. Reflecting the growing linkages that exist between trade and investment, the Chinese Presidency of the G20 established in 2016 a working group dedicated to that question (i.e. the Trade and Investment Working Group, or TIWG).

19. That is, the agreement would impose a legal obligation on all Parties to maintain zero import tariffs on the selected environmental goods.
20. Data sources: Hanappi, T. (2017a, forthcoming) and International Bureau of Fiscal Documentation (IBFD).
21. In 2015, companies are allowed to claim unlimited loss carryovers only in four G20 countries: Australia, Germany, South Africa and the United Kingdom. Data sources: Hanappi, T. (2017b, forthcoming) and International Bureau of Fiscal Documentation (IBFD).
22. The OECD Guidelines for Multinational Enterprises are one of the leading international standards on responsible business conduct. Many of their recommendations (including on the chapters on environment, consumer protection, disclosure, etc.) promote business practices relevant to addressing climate related impacts and risks. http://mneguidelines.oecd.org/guidelines.
23. Public procurement is defined as the process of purchasing goods, services or works by the public sector from the private sector, following the definition used in World Bank, 2016.
24. Trends in the use of CBA are worrying, however. A The use of CBAs dropped considerably in small-scale infrastructure projects in recent decades (World Bank, 2010). In addition, decisions on large EU transport infrastructure projects are often made without support from CBA results (Proost et al., 2011). Other evidence suggests CBA results hardly affect actual decision-making (Nellthorp and Mackie, 2000; Odeck, 2010).
25. Other quality attributes include: public-oriented approach, project management, design, and risk management.
26. The costs are derived from an authoritative life-cycle analysis of materials (from extraction to demolition and recycling), including CO_2 emissions and ten other externalities. Penalties are applied if performance is less than promised, for both the CO_2 performance ladder and the environmental cost evaluated with DuboCalc (OECD, 2015f).

References

Acemoglu, D. et al. (2012), "The Environment and Directed Technical Change," *American Economic Review, Vol. 102/1, American Economic Association, Pittsburgh, pp. 131–66,* http://doi.org/10.1257/aer.102.1.131.

Alberta (2016), Budget 2016: The Alberta Jobs Plan – Fiscal Plan, http://finance.alberta.ca/publications/budget/budget2016/fiscal-plan-complete.pdf (accessed 16 January 2017).

Albrizio, S., et al. (2014), "Do Environmental Policies Matter for Productivity Growth?: Insights from New Cross-Country Measures of Environmental Policies", OECD Publishing, Paris, http://dx.doi.org/10.1787/5jxrjncjrcxp-en.

Allain-Dupré, D., T. Matsumoto and A. Robert (2017, forthcoming), "Climate Investment and Multi-Level Governance: Applying the OECD Principles to Climate Mitigation Policies", OECD Publishing, Paris.

Alvarez, F. E., F. J. Buera and R. E. Lucas, Jr. (2013), *Idea Flows, Economic Growth, and Trade,* National Bureau of Economic Research, Cambridge, Massachusetts, www.nber.org/papers/w19667.pdf.

Ameli, N. and N. Brandt (2015), "What Impedes Household Investment in Energy Efficiency and Renewable Energy?", OECD Publishing, Paris, http://dx.doi.org/10.1787/5js1j15g2f8n-en.

Amon, A.Holmes, I. (2015), *What is holding back energy efficiency financing in g20 countries?* E3G, November 2015. Available at: https://www.e3g.org/docs/E3G-G20-EE_as_infra-FULLreport-final.pdf.

Andrews, D. and C. Criscuolo (2013), "Knowledge Based Capital, Innovation and Resource Allocation", OECD Publishing, Paris, http://dx.doi.org/10.1787/5k46bj546kzs-en.

Andrews, D. and F. Cingano (2014), "Public policy and resource allocation: Evidence from firms in OECD countries", *Economic Policy,* Vol. 29/78, Oxford University Press, Oxford, pp. 253-296, https://doi.org/10.1111/1468-0327.12028.

Andrews, D., A. Caldera Sánchez and Å. Johansson (2011), "Housing Markets and Structural Policies in OECD Countries", OECD Publishing, Paris, http://dx.doi.org/10.1787/5kgk8t2k9vf3-en.

Andrews, D., C. Criscuolo and C. Menon (2014), "Do Resources Flow to Patenting Firms?: Cross-Country Evidence from Firm Level Data", OECD Publishing, Paris, http://dx.doi.org/10.1787/5jz2lpmk0gs6-en.

Ang, Röttgers and Burli (2017, forthcoming), "The Empirics of Enabling Investment and Innovation in Renewable Energy", OECD Publishing, Paris.

Angelo, C. (2012), "Brazil's fund for low-carbon agriculture lies fallow", Nature News, Springer Nature, London, http://doi.org/10.1038/nature.2012.11111.

Appelt, S. et al. (2016), "R&D Tax Incentives: Evidence on design, incidence and impacts", OECD Publishing, Paris, http://dx.doi.org/10.1787/5jlr8fldqk7j-en.

Arlinghaus, J. (2015), "Impacts of Carbon Prices on Indicators of Competitiveness: A Review of Empirical Findings", OECD Publishing, Paris, http://dx.doi.org/10.1787/5js37p21grzq-en.

AsDB (2015), *Fossil Fuel Subsidies in Indonesia: Trends, Impacts, and Reforms, Asian Development Bank, Manila,* www.adb.org/sites/default/files/publication/175444/fossil-fuel-subsidies-indonesia.pdf.

AsDB/World Bank (2015), *Unlocking Indonesia's Geothermal Potential,* Asian Development Bank, Manila, and World Bank, Washington DC, www.adb.org/sites/default/files/publication/157824/unlocking-indonesias-geothermal-potential.pdf.

Bahar, H., J. Egeland and R. Steenblik (2013), "Domestic Incentive Measures for Renewable Energy with Possible Trade Implications", OECD Publishing, Paris, http://dx.doi.org/10.1787/5k44srlksr6f-en.

Baron R. (2016), *The Role of Public Procurement in Low-carbon Innovation, background paper for the 33rd* Round Table on Sustainable Development, 12-13 April 2016, OECD, Paris, www.oecd.org/sd-roundtable/papersandpublications/The%20Role%20of%20Public%20Procurement%20in%20Low-carbon%20Innovation.pdf.

Baron, R. et al. (2012), *Policy Options for Low-Carbon Power Generation in China,* International Energy Agency/OECD Publishing, Paris, www.iea.org/publications/insights/insightpublications/Insight_PolicyOptions_LowCarbon_China.pdf.

BDI (2016), "G20 should enact carbon pricing", Federation of German Industries (BDI), Berlin.

Bento, A.M., M.R. Jacobsen and A.A. Liu (2014), "Environmental policy in presence of an informal sector", paper presented at the World Congress of Environmental and Resource Economists 2014, Istanbul, http://works.bepress.com/antonio_bento/25.

Bidegaray, M. (2015), "Subsidios Eléctricos: Habrá Dos Tipos De Tarifas Sociales", Clarín, www.clarin.com/ieco/economia/Subsidios-electricos-tipos-tarifas-sociales_0_ry_2qA_wmg.html (accessed 28 March 2017).

Bloomberg (2016a), "Mexico Attracts $2.6 Billion in Investments in Power Sale Rerun", www.bloomberg.com/news/articles/2016-03-31/mexico-attracts-2-6-billion-in-investments-in-power-sale-rerun (accessed 28 February 2017).

Bloomberg (2016b), "Obama Rule Could Take Wind Out of Renewable Power on Public Land", Bloomberg, 22 August 2016.

Borenstein, S. and L. Davis (2015), The Distributional Effects Of U.S. Clean Energy Tax Credits, National Bureau of Economic Research, Cambridge, Massachusetts, www.nber.org/papers/w21437.pdf.

Bowen, A. (2015), "Carbon pricing: how best to use the revenue?", Grantham Research Institute on Climate Change and the Environment, London, and Global Green Growth Institute, Seoul, www.lse.ac.uk/GranthamInstitute/wp-content/uploads/2015/11/Bowen-policy-brief-2015.pdf.

Buenos Aires Herald (2017), "Electricity tariffs soar between 61 and 148 percent in BA area",

Bureau of the Environment, Tokyo Metropolitan Government (2010), *Tokyo Cap-and-Trade Program: Japan's first mandatory emissions trading scheme*, Tokyo Metropolitan Government, Tokyo, www.kankyo.metro.tokyo.jp/en/attachement/Tokyo-cap_and_trade_program-march_2010_TMG.pdf.

Burniaux, J.M. and J. Château (2011), "Mitigation Potential of Removing Subsidies to Energy Consumption: A General Equilibrium Assessment, OECD Publishing, Paris, http://dx.doi.org/10.1787/5kgdx1jr2plp-en.

Burniaux, J.M., J. Château and J. Sauvage (2011), "The Trade Effects of Phasing Out Fossil-Fuel Consumption Subsidies", OECD Publishing, Paris. http://dx.doi.org/10.1787/5kg6lql8wk7b-en.

Caprotti, F. (2014), "Eco-urbanism and the Eco-city, or, Denying the Right to the City?" *Antipode*, Vol. 46/5, Wiley, Hoboken, pp. 1285-1303, http://doi.org/10.1111/anti.12087.

CARB (2013), "Cap-and-Trade Program Summary of Vintage 2013 Electrical Distribution Utility Allocated Allowance Value Reports", website, California Environmental Protection Agency Air Resources Board, www.arb.ca.gov/cc/capandtrade/allowanceallocation/edu-v2013-allowance-value-report.pdf. (accessed 12 January 2017).

CARB (2017), "Auction Proceeds Funded Programs and Events", website, California Environmental Protection Agency Air Resources Board, www.arb.ca.gov/cc/capandtrade/auctionproceeds/ggrfprogrampage.htm (accessed 12 January 2017).

Carl, J. Fedor, D. (2016), "Tracking global carbon revenues: A survey of carbon taxes versus cap-and-trade in the real world", *Energy Policy*, Vo.96, pp. 50-77. DOI: https://doi.org/10.1016/j.enpol.2016.05.023.

Casullo, L. (2015), "The Efficiency Impact of Open Access Competition in Rail Markets", ITF discussion paper, www.itf-oecd.org/sites/default/files/docs/dp201607.pdf.

Causa, O. and Å. Johansson (2009), "Intergenerational Social Mobility", OECD Publishing, Paris. http://dx.doi.org/10.1787/223106258208.

CEFF (2016), *Report of The Clean Energy Finance Forum, December 16, 2016, Clean Energy Finance Forum, New Delhi*, www.yesbank.in/pdf/clean_energy_finance_forum.pdf.

CERCIND (2016), *Report on Short-term Power Market in India: 2015-16*, Economics Division, Central Electricity Regulatory Commission, New Delhi, www.cercind.gov.in/2016/MMC/AnnualReport15-16.pdf.

Cervero, R., et al. (2004), *Transit-Oriented Development in the United States: Experiences, Challenges, and Prospects*, Transportation Research Board, Washington DC, http://doi.org/10.17226/23360.

Chattopadhyay, D. (2014), "Modelling renewable energy impact on the electricity market in India", *Renewable and Sustainable Energy Reviews*, Vol. 31, Elsevier, Amsterdam, pp. 9-22, https://doi.org/10.1016/j.rser.2013.11.035.

Chen, G., & Kauppila, J. (2017). *Global Urban Passenger Travel Demand and CO_2 Emission to 2050: A New Model*, presented at the 96th TRB Annual Meeting. Available at: https://trid.trb.org/view.aspx?id=1438189.

China Daily, (2017a). PBOC official stresses financial support for Beijing-Tianjin-Hebei. Available at: http://www.chinadaily.com.cn/business/2017-04/11/content_28869717.htm.

China daily, (2017b). Available at: http://www.chinadaily.com.cn/china/2017-02/28/content_28377361.htm.

Commonwealth of Australia (2012) "Securing a Clean Energy Future: Implementing the Australian Government's Climate Change Plan", Statement by Greg Combet, Minister For Climate Change and Energy Efficiency, 8 May 2012, Commonwealth of Australia, Canberra, www.budget.gov.au/2012-13/content/ministerial_statements/climate/download/climate_change.pdf.

Couture, T. et al. (2010), *A Policymaker's Guide to Feed-in Tariff Policy Design*, National Renewable Energy Laboratory, Golden, Colorado, www.nrel.gov/docs/fy10osti/44849.pdf.

Coysh, D. et al. (2017, forthcoming), "Effects of vintage-differentiated environmental regulations: Evidence from survival analysis of coal-fired power plants", OECD Publishing, Paris.

Cramton, P. et al. (eds.) (2017), *Global Carbon Pricing: The Path to Climate Cooperation*, MIT Press, Cambridge, Massachusetts.

Cramton, P., A. Ockenfels and S. Stoft (2017), "An International Carbon-Price Commitment Promotes Cooperation", *in Global Carbon Pricing: The Path to Climate Cooperation*, Cramton, P. et al. (eds.), MIT Press, Cambridge Massachusetts.

de Serres, A., F. Murtin and G. Nicoletti (2010), "A Framework for Assessing Green Growth Policies", OECD Publishing, Paris, http://dx.doi.org/10.1787/5kmfj2xvcmkf-en.

Dechezleprêtre A., R. Martin and S. Bassi (2016), "Climate change policy, innovation and growth", Grantham Research Institute on Climate Change and the Environment, London School of Economics and Political Science, London, www.lse.ac.uk/GranthamInstitute/wp-content/uploads/2016/01/Dechezlepretre-et-al-policy-brief-Jan-2016.pdf.

Dechezleprêtre, A., R. Martin and M. Mohnen (2014), "Knowledge spillovers fromdirty and clean technologies", Centre for Economic Performance, London School of Economics and Political Science, London, http://cep.lse.ac.uk/pubs/download/dp1300.pdf.

Devereux, M. P. and Griffith, R. (1998b) "Taxes and the location of production: evidence from a panel of US multinationals", *Journal of Public Economics, Vol. 68/3, Elsevier, Amsterdam, pp.* 335-367, http://doi.org/10.1016/S0047-2727(98)00014-0.

Devereux, M. P. and R. Griffith (1998a), "The taxation of discrete investment choices", Institute for Fiscal Studies, London, www.ifs.org.uk/wps/wp9816.pdf.

Di Bella, G., et al. (2015), "Energy subsidies in Latin America and the Caribbean: Stocktaking and policy challenges", International Monetary Fund, Washington DC, www19.iadb.org/intal/intalcdi/PE/2015/15526.pdf.

DOE South Africa (2015) *State of Renewable Energy in South Africa, Department of Energy, Pretoria*, www.gov.za/sites/www.gov.za/files/State%20of%20Renewable%20Energy%20in%20South%20Africa_s.pdf.

Eberhard, A., J. Kolker, and J. Leigland (2014), *South Africa's Renewable Energy IPP Procurement Programme: Success Factors and Lessons, World Bank, Washington DC*, www.gsb.uct.ac.za/files/ppiafreport.pdf.

EDF and IETA (2016), Carbon Pricing: The Paris Agreement's Key Ingredient, Environmental Defense Fund, New York, and International Emissions Trading Association, Geneva, www.ieta.org/resources/Resources/Reports/Carbon_Pricing_The_Paris_Agreements_Key_Ingredient.pdf.

EEFTG (2017), *G20 Energy Efficiency Investment Toolkit, G20 Energy Efficiency Finance Task Group/IEA/International Partnership for Energy Efficiency Collaboration*, https://ipeec.org/upload/publication_related_language/pdf/626.pdf.

EFFIG (2015), *Energy Efficiency Financial Institutions Group: Final report covering Buildings, Industry and SMEs*, Energy Efficiency Financial Institutions Group, Brussels, http://ec.europa.eu/energy/sites/ener/files/documents/Final%20Report%20EEFIG%20v%209.1%2024022015%20clean%20FINAL%20sent.pdf.

ENS (2015), *Energy Policy Toolkit on Physical Planning of Wind Power: Experiences from Denmark, Danish Energy Agency (ENS)*, Copenhagen, https://ens.dk/sites/ens.dk/files/Globalcooperation/physical_planning_of_wind_power.pdf.

European Commission (2016), "Auctioning", website, European Commission Climate Action, https://ec.europa.eu/clima/policies/ets/auctioning_en, (accessed 12 January 2017).

FAO (2016), *The State of Food and Agriculture: Climate Change, Agriculture and Food Security, Food and Agriculture Organization of the United Nations, Rome*, www.fao.org/3/a-i6030e.pdf#=44.

Fattouh, B. and L. El-Katiri (2013), "Energy subsidies in the Middle East and North Africa", *Energy Strategy Reviews, Vol. 2/1, Elsevier, Amsterdam, pp.* 108-115, http://dx.doi.org/10.1016/j.esr.2012.11.004.

Financial Times (2016), "Court overturns Argentine gas price rises", *Financial Times, 18 August 2016*, www.ft.com/content/d903f5e2-6575-11e6-8310-ecf0bddad227.

G20 (2017), "Towards food and water security: Fostering sustainability, advancing innovation", G20 Agriculture Ministers' Declaration, 22 January 2017, Berlin, www.bmel.de/SharedDocs/Downloads/EN/Agriculture/GlobalFoodSituation/G20_Declaration2017_EN.pdf;jsessionid=48AF37D05D7C0366B55297B7217593BE.2_cid376?__blob=publicationFile.

GGGI (2014), *A Guide to Innovation System Analysis for Green Growth, Global Green Growth Institute, Seoul,* www.greengrowthknowledge.org/sites/default/files/downloads/resource/A_Guide_to_Innovation_System_Analysis_for_Green_Growth_GGGI.pdf.

GIZ and ICLEI (2014), *Operationalizing the Urban NEXUS: Towards resource-efficient and integrated cities and metropolitan regions, Deutsche Gesellschaft für Internationale Zusammenarbeit GmbH (GIZ) and ICLEI – Local Governments for Sustainability, Eschborn and Bonn, Germany,* www.iclei.org/fileadmin/PUBLICATIONS/Papers/Urban_NEXUS_Publication_ICLEI-GIZ_2014_web.pdf.

Goulder, L.H. and A.R. Schein (2013), "Carbon taxes versus cap and trade: A critical review", *Climate Change Economics, Vol. 4/3, World Scientific Publishing, Singapore,* http://dx.doi.org/10.1142/S2010007813500103.

Government of Brazil (2016), *Third National Communication of Brazil to the United Nations Framework Convention on Climate Change,* Government of Brazil, Brasilia.

Greene, J. and N.A. Braathen (2014), "Tax Preferences for Environmental Goals: Use, Limitations and Preferred Practices", OECD Publishing, Paris, http://dx.doi.org/10.1787/5jxwrr4hkd6l-en.

Haas, R. et al. (2011), "Efficiency and effectiveness of promotion systems for electricity generation from renewable energy sources: Lessons from EU countries", *Energy, Vol. 36/4, Elsevier, Amsterdam, pp. 2186–2193,* https://doi.org/10.1016/j.energy.2010.06.028.

Hanappi, T. (2017a, forthcoming), "Effective Tax Rates: Model Application and Policy Analysis", OECD Publishing, Paris.

Hanappi, T. (2017b, forthcoming), "Loss Carryover Provisions: Measuring Effects on Tax Symmetry and Automatic Stabilisation", OECD Publishing, Paris.

Harding, M. (2014), "Personal Tax Treatment of Company Cars and Commuting Expenses: Estimating the Fiscal and Environmental Costs", OECD Publishing, Paris, http://dx.doi.org/10.1787/5jz14cg1s7vl-en.

Holtsmark, B. and A. Skonhoft (2014), "The Norwegian support and subsidy policy of electric cars. Should it be adopted by other countries?", *Environmental Science and Policy, Vol. 42, Elsevier, Amsterdam, pp. 160-168,* https://doi.org/10.1016/j.envsci.2014.06.006.

Hood, C. (2013), "Managing interactions between carbon pricing and existing energy policies", IEA/OECD Publishing, Paris, www.iea.org/publications/insights/insightpublications/managinginteractionscarbonpricing_final.pdf.

Hood, C (2011), "Summing up the parts: Combining policy instruments for least-cost climate mitigation strategies", IEA/OECD Publishing, Paris, www.iea.org/publications/freepublications/publication/Summing_Up.pdf.

http://english.bdi.eu/#/article/news/g20-should-enact-carbon-pricing (accessed 6 February 2017).

IASS (2016), "Sustainable Energy in the G20: Prospects for a Global Energy Transition", Institute for Advanced Sustainability Studies, Potsdam, Germany, www.iass-potsdam.de/sites/default/files/files/iass_study_dec2016_en_sustainableenergyg20_0.pdf.

ICMM (2013), "Options in recycling revenues generated through carbon pricing: How 16 governments invest their carbon revenues", International Council on Mining and Metals, London, www.icmm.com/website/publications/pdfs/5362.pdf.

IDB and World Bank (2017, forthcoming), *Energy Sector Subsidies and Mitigation to Climate Change Policies in Lating America and the Caribbean, Inter-American Development Bank, Washington DC.*

IEA (2015), *World Energy Outlook – Fossil fuel subsidies database, IEA/OECD Publishing, Paris,* www.worldenergyoutlook.org/media/weowebsite/2015/Subsidies20122014.xlsx.

IEA (2016a), *Repowering Markets,* IEA/OECD Publishing. https://www.iea.org/publications/freepublications/publication/REPOWERINGMARKETS.PDF.

IEA (2016b), *Energy Efficiency Market Report 2016,* IEA/OECD Publishing, Paris, www.iea.org/eemr16/files/medium-term-energy-efficiency-2016_WEB.PDF.

IEA (2016c), *Global EV Outlook 2016, IEA/OECD Publishing, Paris,* www.iea.org/publications/freepublications/publication/global-ev-outlook-2016.html.

IEA (2017), *Chapter 2 in Perspectives for the Energy Transition: Investment Needs for a Low-Carbon Energy System,* IEA/OECD Publishing, Paris, www.energiewende2017.com/wp-content/uploads/2017/03/Perspectives-for-the-Energy-Transition_WEB.pdf.

IEA, OPEC, OECD, and World Bank (2010), *Analysis of the Scope of Energy Subsidies and Suggestions for the G-20 Initiative,* joint report prepared for submission to the G-20 Leaders' Summit (Toronto, June 2010), IEA/ OECD Publishing, Paris, www.oecd.org/env/45575666.pdf.

IEA, OPEC, OECD, and World Bank (2011), *Joint report by IEA, OPEC, OECD and World Bank on fossil-fuel and other energy subsidies: An update of the G20 Pittsburgh and Toronto Commitments,* Prepared for the G20 Meeting of Finance Ministers and Central Bank Governors (Paris, October 2011) and the G20 Summit (Cannes, November 2011), IEA/OECD Publishing, Paris, www.oecd.org/env/49090716.pdf.

Ignaciuk, A. (2015) "Adapting agriculture to climate change: A role for public policies", OECD Publishing, Paris, http://dx.doi.org/10.1787/5js08hwvfnr4-en.

IMF (2015), *Making Public Investment More Efficient,* International Monetary Fund, Washington, DC, www.imf.org/external/np/pp/eng/2015/061115.pdf.

IMF, OECD, UN and World Bank (2015), Options for Low Income Countries' Effective and Efficient Use of Tax Incentives for Investment: A report to the G20 Development Working Group by the IMF, OECD, UN and World Bank, OECD Publishing, Paris, www.oecd.org/tax/tax-global/options-for-low-income-countries-effective-and-efficient-use-of-tax-incentives-for-investment.pdf.

INPE (2017), Satellite Monitoring System of the Brazilian Amazon Forest Project – PRODES, National Institute for Space Research (INPE), Brasilia, www.obt.inpe.br/prodes/index.php.

IPEEC (2016), *The 2016 Activity Report of the G20 Energy Efficiency Finance Task Group,* International Partnership for Energy Efficiency Collaboration, Paris, https://ipeec.org/upload/publication_related_language/pdf/541.pdf.

IRENA (2016), "Unlocking Renewable Energy Investment: The Role of Risk Mitigation and Structured Finance", IRENA, Abu Dhabi, www.irena.org/DocumentDownloads/Publications/IRENA_Risk_Mitigation_and_Structured_Finance_2016.pdf.

IRENA and CEM (2015), *Renewable Energy Auctions: A Guide to Design,* International Renewable Energy Agency, Abu Dhabi, and Clean Energy Ministerial, www.irena.org/DocumentDownloads/Publications/Renewable_Energy_Auctions_A_Guide_to_Design.pdf.

Johnstone, N., I. Haščič and M. Kalamova (2010), "Environmental Policy Design Characteristics and Technological Innovation: Evidence from Patent Data", OECD Publishing, Paris, http://dx.doi.org/10.1787/5kmjstwtqwhd-en.

KAPSARC (2017), *Growth, investment and the low-carbon transition: A view from Saudi Arabia,* King Abdullah Petroleum Studies and Research Centre, Riyadh.

Keller, W. (2004), "International Technology Diffusion", *Journal of Economic Literature,* Vol. 42/3, American Economic Association, Pittsburgh, pp. 752-782, http://doi.org/10.1257/0022051042177685.

Kennedy, C. and J. Corfee-Morlot (2012), "Mobilising Investment in Low Carbon, Climate Resilient Infrastructure", OECD Publishing, Paris, http://dx.doi.org/10.1787/5k8zm3gxxmnq-en.

Klintman, M. (2016), "A Review of Public Policies relating to the Use of Environmental Labelling and Information Schemes (ELIS)", OECD Publishing, Paris, http://dx.doi.org/10.1787/5jm0p34bk7hb-en.

La Rovere et al. (2017), "IES Brasil 2050", Centro Clima, Rio de Janeiro, http://www.centroclima.coppe.ufrj.br/index.php/br/estudos-e-projetos/encerrados/33-ies-brasil-2050.

Lakner, C. et al. (2016), "The Incidence of Subsidies to Residential Public Services in Argentina: The Subsidy System in 2014 and Some Alternatives", Centro de Estudios Distributivos Laborales y Sociales, La Plata, Argentina, http://EconPapers.repec.org/RePEc:dls:wpaper:0201.

Lanzi, E., et al. (2013), *Addressing Competitiveness and Carbon Leakage Impacts Arising from Multiple Carbon Markets: A Modelling Assessment,* OECD Publishing, Paris, http://dx.doi.org/10.1787/5k40ggjj7z8v-en.

Markandya, A., M. Gonzalez-Eguino and M. Escapa (2013), "From shadow to green: Linking environmental fiscal reforms to the informal economy", *Energy Economics,* Vol. 40/Sup. 1, Elsevier, Amsterdam, pp. 108-118, http://doi.org/10.1016/j.eneco.2013.09.014.

Martin, R., et al. (2014), "Industry Compensation under Relocation Risk: A Firm-Level Analysis of the EU Emissions Trading Scheme", *American Economic Review,* Vol. 104/8, American Economic Association, Pittsburgh, pp. 2482-2508, http://doi.org/10.1257/aer.104.8.2482.

MEEDM (2017), "Fiscalité Carbone", website, Ministry of the Environment, Energy and the Sea, Paris, www.developpement-durable.gouv.fr/fiscalite-carbone (accessed on 10 February 2017).

Ministero Dell' Ambiente (2016), *Catalogo dei Sussidi Ambientalmente Favorevoli e dei Sussidi Ambientalmente Dannosi*, Ministry for the Environment, Rome, www.minambiente.it/sites/default/files/archivio/allegati/sviluppo_sostenibile/catalogo_sussidi_ambientali.pdf.

NDRC (2015), *Catalogue for the Guidance of Foreign Investment Industries (Amended in 2015)*, National Development and Reform Commission, Beijing, www.fdi.gov.cn/1800000121_39_4830_0_7.html.

Nellthorp, J., and P. Mackie (2000), "The UK Roads Review: A hedonic model of decision making", *Transport Policy*, Vol. 7/2, Elsevier, Amsterdam, pp. 127–138, https://doi.org/10.1016/S0967-070X(00)00002-0.

Newton, P. et al. (2016), "Overcoming barriers to low carbon agriculture and forest restoration in Brazil: The Rural Sustentável project", *World Development Perspectives*, Vol. 4, Elsevier, Amsterdam, pp. 5-7, https://doi.org/10.1016/j.wdp.2016.11.011.

OCDE/ITF (2017), *ITF Transport Outlook 2017, OECD Publishing, Paris,* http://doi.org/10.1787/9789282108000-en.

Odeck, J. (2010), "What Determines Decision-Makers' Preferences for Road Investments? Evidence from the Norwegian Road Sector", *Transport Reviews*, Vol. 30/4, Taylor and Francis, Abingdon, United Kingdom, pp. 473-494, http://dx.doi.org/10.1080/01441640903138640.

OECD (2017a), *Economic Policy Reforms 2017: Going for Growth*, OECD Publishing, Paris, http://dx.doi.org/10.1787/growth-2017-en.

OECD (2017b forthcoming) Evaluating Environmental Tax Policies – Case Studies, OECD Publishing, Paris.

OECD (2017c forthcoming), "Global Value Chains for Environmental Goods and Services", OECD Publishing, Paris.

OECD (2017d forthcoming), "Trade in Services Related to the Environment", OECD Publishing, Paris.

OECD (2016a), *Effective Carbon Rates: Pricing CO_2 through Taxes and Emissions Trading Systems*, OECD Publishing, Paris, http://dx.doi.org/10.1787/9789264260115-en.

OECD (2016b), *Agricultural Policy Monitoring and Evaluation 2016*, OECD Publishing, Paris, See also the associated database at www.oecd.org/tad/agricultural-policies/producerandconsumersupportestimatesdatabase.htm.

OECD (2016c), *Innovation, Agricultural Productivity and Sustainability in Turkey*, OECD Publishing, Paris, http://dx.doi.org/10.1787/9789264261198-en.

OECD (2016d), *Innovation, Agricultural Productivity and Sustainability in the United States*, OECD Publishing, Paris, http://dx.doi.org/10.1787/9789264264120-en.

OECD (2016e), "Mitigating Droughts and Floods in Agriculture: Policy Lessons and Approaches", OECD Publishing, Paris, http://dx.doi.org/10.1787/9789264246744-en.

OECD (2016f), *OECD Clean Energy Investment Policy Review of Jordan*, OECD Publishing, Paris, http://dx.doi.org/10.1787/9789264266551-en.

OECD (2016g), *2016 OECD Recommendation of the Council for Development Co-operation Actors on Managing the Risk of Corruption*, OECD Publishing, Paris, www.oecd.org/corruption/anti-bribery/Recommendation-Development-Cooperation-Corruption.pdf.

OECD (2016h), *"Synergies and trade-offs between agricultural productivity and climate change mitigation and adaptation: French case study", OECD Publishing, Paris,* www.oecd.org/officialdocuments/publicdisplaydocumentpdf/?cote=COM/TAD/CA/ENV/EPOC(2016)6/FINAL&docLanguage=En.

OECD (2016i), "Synergies and trade-offs between agricultural productivity and climate change mitigation and adaptation: Dutch case study", OECD Publishing, Paris, www.oecd.org/officialdocuments/publicdisplaydocumentpdf/?cote=COM/TAD/CA/ENV/EPOC(2016)7/FINAL&docLanguage=En.

OECD (2015a), *Policy Framework for Investment, 2015 Edition*, OECD Publishing, Paris, http://dx.doi.org/10.1787/9789264208667-en.

OECD (2015b), *Innovation, Agricultural Productivity and Sustainability in Brazil*, OECD Publishing, Paris, http://dx.doi.org/10.1787/9789264237056-en.

OECD (2015c), *The Future of Productivity*, OECD Publishing, Paris, http://doi.org/10.1787/9789264248533-en.

OECD (2015d), *OECD Companion to the Inventory of Support Measures for Fossil Fuels 2015*, OECD Publishing, Paris. See the associated database at www.oecd.org/site/tadffss/data/.

OECD (2015e), "Executive Summary of the Hearings on the Use of Auctions and Tenders", Working Party no. 2 on Competition and Regulation, OECD Publishing, Paris, www.oecd.org/officialdocuments/publicdisplaydocumentpdf/?cote=DAF/COMP/WP2/M(2015)1/ANN7&docLanguage=En.

OECD (2015f), *Innovation, Agricultural Productivity and Sustainability in the Netherlands*, OECD Publishing, Paris, http://dx.doi.org/10.1787/9789264238473-en.

OECD (2015g), *Innovation, Agricultural Productivity and Sustainability in Canada*, OECD Publishing, Paris, http://dx.doi.org/10.1787/9789264238541-en.

OECD (2015h), *OECD Environmental Performance Reviews: Brazil 2015*, OECD Publishing, Paris, http://dx.doi.org/10.1787/9789264240094-en.

OECD (2015i), *Climate Change Risks and Adaptation: Linking Policy and Economics*, OECD Publishing, Paris, http://dx.doi.org/10.1787/9789264234611-en.

OECD (2015j), *National Climate Change Adaptation: Emerging Practices in Monitoring and Evaluation*, OECD Publishing, Paris, http://dx.doi.org/10.1787/9789264229679-en.

OECD (2015k) "Agriculture and Climate Change Policy Note", OECD Publishing, Paris, www.oecd.org/tad/sustainable-agriculture/agriculture-climate-change-september-2015.pdf.

OECD (2015l), "Policy Guidance for Investment in Clean Energy Infrastructure: Expanding Access to Clean Energy for Green Growth and Development", OECD Publishing, Paris, http://dx.doi.org/10.1787/9789264212664-en.

OECD (2015m), "Impact of investment restrictive measures on investment", OECD Publishing, Paris, www.oecd.org/g20/topics/trade-and-investment/Impact-of-investment-restrictive-measures-on-investment.pdf.

OECD (2015n), *OECD Guidelines on Corporate Governance of State-Owned Enterprises, 2015 Edition*, OECD Publishing, Paris, http://dx.doi.org/10.1787/9789264244160-en.

OECD (2015o), *Overcoming Barriers to International Investment in Clean Energy*, OECD Publishing, Paris, http://dx.doi.org/10.1787/9789264227064-en.

OECD (2015p), *OECD Recommendation of the Council on Public Procurement*, OECD Publishing, Paris, www.oecd.org/gov/ethics/OECD-Recommendation-on-Public-Procurement.pdf.

OECD (2014a), *Addressing social implications of green growth*, OECD Publishing, Paris, www.oecd.org/greengrowth/Issue%20Note%20Session%20One%20GGSD%20Forum.pdf.

OECD (2014b), *Global Implications of Fossil-Fuel Consumption Subsidies Reform and Nuclear Phase-Out: An Economic Analysis*, OECD Publishing, Paris.

OECD (2014c), *Green Growth Indicators for Agriculture: A Preliminary Assessment*, OECD Publishing, Paris, http://dx.doi.org/10.1787/9789264223202-en.

OECD (2014d), *Climate Change, Water and Agriculture: Towards Resilient Systems*, OECD Publishing, Paris, http://dx.doi/10.1787/9789264209138-en.

OECD (2014e), "A national strategy for cities: Taking ownership of urban policy", in *OECD Regional Outlook 2014: Regions and Cities: Where Policies and People Meet*, OECD Publishing, Paris.

OECD (2014f), *OECD Regional Outlook 2014: Regions and Cities: Where Policies and People Meet*, OECD Publishing, Paris, http://dx.doi.org/10.1787/9789264201415-en.

OECD (2013a), *OECD Economic Surveys: Brazil 2013*, OECD Publishing, Paris, http://dx.doi.org/10.1787/eco_surveys-bra-2013-en.

OECD (2013b), *Investing Together: Working Effectively across Levels of Government*, OECD Publishing, Paris, http://dx.doi.org/10.1787/9789264197022-en.

OECD (2012a), *OECD Employment Outlook 2012*, OECD Publishing, Paris, http://dx.doi.org/10.1787/empl_outlook-2012-en.

OECD (2012b), *Recommendation of the Council on Principles for Public Governance of Public-Private Partnerships*, OECD Publishing, Paris, www.oecd.org/governance/budgeting/PPP-Recommendation.pdf.

OECD (2011), *Convention on Combating Bribery of Foreign Public Officials in International Business Transactions and Related Documents*, OECD Publishing, Paris, www.oecd.org/daf/anti-bribery/ConvCombatBribery_ENG.pdf.

OECD (2010) *Transition to a Low-carbon Economy: Public Goals and Corporate Practices*, OECD Publishing, Paris, http://dx.doi.org/10.1787/9789264090231-en.

OECD (2009), *Recommendation of the Council for Further Combating Bribery of Foreign Public Officials in International Business Transactions, Working Group on Bribery in International Business Transactions*, OECD Publishing, Paris, www.oecd.org/daf/anti-bribery/44176910.pdf.

OECD (2008), Public-Private Partnerships: In Pursuit of Risk Sharing and Value for Money, *OECD Publishing, Paris.*

OECD (2006), *OECD Council Recommendation on Bribery and Officially Supported Export Credits*, Working Party on Export Credits and Credit Guarantees, OECD Publishing, Paris, www.oecd.org/officialdocuments/publicdisplaydocumentpdf/?cote=td/ecg(2006)24&doclanguage=en.

OECD/CDSB (2015,) *Climate change disclosure in G20 countries: Stocktaking of corporate reporting schemes*, OECD Publishing, Paris, www.oecd.org/daf/inv/mne/Report-on-Climate-change-disclosure-in-G20-countries.pdf.

OECD/CDSB (2017, forthcoming), *Climate change disclosure in G20 countries: Supporting the Paris Agenda*, OECD Publishing, Paris.

OECD/IEA/NEA/ITF (2015), *Aligning Policies for a Low-carbon Economy*, OECD Publishing, Paris. http://dx.doi.org/10.1787/9789264233294-en.

OECD/ITF (2016), *Adapting Transport Infrastructure to Climate Change and Extreme Weather: Decisionmaking under Uncertainty*, OECD Publishing, Paris, http://doi.org/10.1787/9789282108079-en.

OECD/ITF (2017), ITF Transport Outlook 2017, OECD Publishing, Paris. DOI: http://dx.doi.org/10.1787/9789282108000-enOntario (2017), "Cap and Trade in Ontario", website, Ministry of the Environment and Climate Change, Ontario, www.ontario.ca/page/cap-and-trade-ontario (accessed 9 February 2017).

Oueslati, W., S. Alvanides and G. Garrod (2014), "Determinants of urban sprawl in European cities", *Urban Studies*, Vol. 52/ 9, Sage, Thousand Oaks, California, pp. 1594-1614, http://doi.org/10.1177/0042098015577773.

Pagani, F. (2002), "Peer review: A tool for global co-operation and change", *OECD Observer*, No. 235, OECD Publishing, Paris, http://doi.org/10.1787/observer-v2002-6-en.

Partnership for Market Readiness (2015), *Carbon Leakage; Theory, Evidence and Policy Design*, World Bank Group, Washington, DC. Available online at: http://hdl.handle.net/10986/22785.

Prag A., I. Scherrer and D. Röttgers (2017 forthcoming), *State-owned Enterprises and the Low-carbon Transition*, OECD Publishing, Paris.

Proost, S. et al. (2011), "When are subsidies to trans-European network projects justified?", *Transportation Research Part A: Policy and Practice*, Vol. 45/3, Elsevier, Amsterdam, pp. 161–170, https://doi.org/10.1016/j.tra.2010.12.001.

Quebec (2012), *Climate Change Action Plan 2013-2020*, Government of Quebec, Quebec City, www.mddelcc.gouv.qc.ca/changements/plan_action/pacc2020-en.pdf.

RAP (2016a), "Issues in China Power Sector Reform: Generator Dispatch Discussion Paper", Regulatory Assistance Project, Beijing, www.raponline.org/wp-content/uploads/2016/07/rap-kahrl-dupuy-wang-china-generator-dispatch-reform-july-2016.pdf (accessed on 28 February 2017).

RAP (2016b), "Wholesale Electricity Markets and Pricing in China: How is Reform Going?", Regulatory Assistance Project, Beijing, www.raponline.org/wholesale-electricity-markets-pricing-china-reform-going/ (accessed on 28 February 2017).

REN21 (2016), Renewables 2016 Global Status Report, REN21, Paris, http://www.ren21.net/status-of-renewables/global-status-report/.

REN21 (2014), *Mini-Grid Policy Toolkit, Africa-EU* Renewable Energy Cooperation Programme, www.ren21.net/Portals/0/documents/Resources/MGT/MinigridPolicyToolkit_Sep2014_EN.pdf.

RGGI (2016), *The Investment of RGGI Proceeds through 2014*, Regional Greenhouse Gas Initiative, New York, www.rggi.org/docs/ProceedsReport/RGGI_Proceeds_Report_2014.pdf.

Robalino, J. and L.D. Herrera (2010), *Trade and Deforestation: What Have We Found?* World Trade Organisation, Geneva, www.wto.org/english/res_e/publications_e/wtr10_robalino_herrera_e.htm.

Rydge, J. (2015), *Implementing Effective Carbon Pricing*, New Climate Economy, London and Washington DC, http://newclimateeconomy.report/2015/wp-content/uploads/sites/3/2015/10/Implementing-Effective-Carbon-Pricing.pdf.

Salon, D. (2014), "Location Value Capture Opportunities for Urban Public Transport Finance: A whitepaper prepared for the Transit Leadership Summit, London", Regional Plan Association, New York, http://library.rpa.org/pdf/TLS-2014-Research-Paper-Value-Capture.pdf.

Sauvage, J. (2014), "The Stringency of Environmental Regulations and Trade in Environmental Goods", OECD Publishing, Paris, http://dx.doi.org/10.1787/5jxrjn7xsnmq-en.

Shrimali, G. and S. Tirumalachetty (2013), "Renewable energy certificate markets in India: A review", *Renewable and Sustainable Energy Reviews*, Vol. 26, Elsevier, Amsterdam, pp. 702–716, https://doi.org/10.1016/j.rser.2013.06.034.

Siemiatycki, M. (2006), "Message in a Metro: Building Urban Rail Infrastructure and Image in Delhi, India", *International Journal of Urban and Regional Research*, Vol. 30/2, Wiley, Hoboken, pp. 277-292, http://doi.org/10.1111/j.1468-2427.2006.00664.x.

Smith, S. and N.A. Braathen (2015), "Monetary Carbon Values in Policy Appraisal: An overview of Current Practice and Key Issues", OECD Publishing, Paris, http://dx.doi.org/10.1787/5jrs8st3ngvh-en.

Solar PEIS (2017), "Priority Development Area (Solar Energy Zone) and Variance Area Maps", website, http://solareis.anl.gov/maps/alternatives/index.cfm (accessed 28 February 2017).

Srinivasan, S. (2001), "Land Use Change as a Tool: A Framework to Link Transportation and the Environment in New Delhi, India," Workshop on Transportation, Land Use and the Environment, Pune, India, 3-4 December 2001, www.deas.harvard.edu/TransportAsia/workshop_papers/Srinivasan.pdf.

STEP up (2017), "STEP up! – More power efficiency in companies", website, www.stepup-energieeffizienz.de (accessed on 28 February 2017).

Stiglitz, J.E. (2017), "Overcoming the Copenhagen Failure with Flexible Commitments" in *Global Carbon Pricing: The Path to Climate Cooperation*, Cramton et al. (eds.), 2017, MIT Press, Cambridge Massachusetts.

The Davis Tax Committee (2015), First Interim Report on Carbon Tax for the Minister of Finance, The Davis Tax Committee, Pretoria, www.taxcom.org.za/docs/20151116%20DTC%20Carbon%20Tax%20First%20Interim%20Report.pdf.

UNCTAD (2013), "Promoting low-carbon investment, Investment Advisory Series Series A, No. 7", United Nations Conference on Trade and Development, Geneva, http://unctad.org/en/PublicationsLibrary/diaepcb2013d2_en.pdf.

Vallejo, L. and M. Mullan (2017), "Climate-resilient infrastructure: Getting the policies right", OECD Publishing, Paris, http://dx.doi.org/10.1787/02f74d61-en.

Weitzman, M.L. (1974), "Prices vs. quantities", *Review of Economic Studies*, Vol. 41/4, Oxford University press, Oxford and New York, pp. 477-491, http://doi.org/10.2307/2296698.

World Bank (2010), Cost-Benefit Analysis in World Bank Projects, World Bank, Washington DC, http://doi.org/10.1596/978-0-8213-8578-4.

World Bank (2012), "The Real Costs of Fossil Fuel Subsidies", website, www.worldbank.org/en/news/feature/2012/05/09/real-costs-fossil-fuel-subsidies (accessed 28 February 2017).

World Bank (2016), "Is public procurement a rich country's policy?", blog, World Bank, Washington DC, http://blogs.worldbank.org/governance/public-procurement-rich-country-s-policy (accessed on 23 February 2017).

World Bank, Ecofys and Vivid Economics (2016), *State and Trends of Carbon Pricing 2016*, World Bank, Washington, DC, http://hdl.handle.net/10986/25160.

Chapter 6

Towards an inclusive transition

The significant structural change implied by the transition of whole economies to balance greenhouse gas emissions and carbon sinks will inevitably create tensions among those affected – from central and local governments, to the private sector, the labour force and citizens. This chapter examines the social and economic factors that affect the ability of governments to envision and implement the long-term policy choices needed to stabilise the global climate. It considers how governments might draw on experience with industrial restructuring; the potential impacts of climate policy on households; and case studies that illustrate the need for a just transition for workers and communities. The chapter concludes by exploring how best to take political economy dimensions into account in preparing robust, long-term, low-emission development strategies.

G20 countries vary widely in the speed of change they pledged under the Paris Agreement, reflecting their differing perceptions of the challenges and opportunities presented by the global response to climate change. Transforming whole economies to balance greenhouse gas (GHG) emissions and carbon sinks will inevitably create tensions among those affected by the required changes, from central and local governments, to the private sector, the labour force and citizens.

Regardless of each country's starting point, significant structural change will be necessary. Activities with high GHG emissions need to change technologies or business models, or face decline. Some assets will be stranded. Jobs will be lost, even if the shift to low-emission, climate-resilient economies could result in net job creation. So a "just transition" is needed (as recognised in the Paris Agreement) that creates jobs in low-emission sectors, anticipates changes in employment patterns and fosters business plans that help workers find new jobs and opportunities.

Since climate change emerged as a planetary problem three decades ago, vested interests and incumbent actors in the high-emission economy have made it hard for governments to implement the long-sighted policy choices needed to stabilise the global climate. Policy-makers must take these circumstances into account early on to facilitate the transition while meeting other pressing policy agendas, such as poverty alleviation, job creation, ageing and inclusiveness.

This chapter examines the social and economic factors that make it easier or harder for governments to envision and implement an effective climate response. It draws on current observations and experience with industrial restructuring, analyses the impact of climate mitigation policy on household income, and considers case studies that reflect the need to ensure a just transition for workers and communities. The chapter concludes by exploring how best to take political economy dimensions into account in preparing robust, long-term, low-emission development strategies.

The political economy dimensions of the transition

Setting the scene

Political economy can be defined as the interaction of political and economic processes in a society, including the distribution of power and wealth among groups and individuals, and the processes that create, sustain and transform these relationships (OECD-DAC, 2005). Taking a political economy approach can answer the basic questions of who wins, who loses, how and why (Newell, Phillips and Mulvaney, 2011).

A number of political economy dimensions of climate policy arise at the **macro-economic** level:

- As indicated in Chapter 4, low-emission, climate-resilient policies will have various effects on the economy, welfare, and jobs. This will also have repercussions for inclusiveness.
- Such policies will also impact royalties or tax revenues from fossil-fuel production and consumption, and hence government revenue streams.
- A country's innovation and technological capability, its position as leader or follower, will affect how readily its economic and industrial strategy can adjust to and support its climate policies.
- A country's exposure to immediate and future climate risk, and the vulnerability of its assets and infrastructure, can be powerful drivers of action at national and international level.

A number of **sector-specific** aspects and stakeholders will also influence the politics of climate policies:

- High- and low-emission industries have different weights in a country's economy and trade balance; the share of GHG-intensive activities in total employment is another important factor.
- The marked impact of the transition on fossil-fuel demand has clear implications for dependency on fossil-fuel exports and imports and hence the energy security of countries. The other visible dimension of energy security is the reliability of the electricity system as low-carbon variable renewable energy sources challenge the flexibility of electricity grids.
- Existing infrastructure (electricity networks, pipelines, roads, railways, ports, city plans and buildings) has locked countries into development paths that could be difficult to change. Such change will be influenced by the balance of ownership of key infrastructure between private and public sectors.
- As governments introduce regulatory changes for the transition, incumbent players are likely to try to influence them in order to seek new rents. This will matter especially when critical technology choices require regulatory foresight and intervention – for example, whether electricity or gas infrastructures should be prioritised for the transition.

Local factors can also be critical for the success of climate policy efforts:

- As policy incentives favour low-emission choices, communities specialised in high-emission activities will be affected, as local jobs may decline or be eliminated.
- When assets are stranded or divested, the impact will be felt mostly locally. Bankruptcy law and vulture funds may prolong the lifetime of assets, undermining climate mitigation policy.
- Climate-friendly policy measures may produce other benefits, such as reducing local pollution or energy poverty.

Overall, the effectiveness of climate policy also hinges on a country's **general political conditions**:

- Political accountability and stability of the executive and of key supporting institutions, such as the civil service and the judiciary, are crucial.
- Institutions are necessary that enable consultation in public policy settings and in driving investment strategies with key stakeholders, including social dialogue between governments, business and trade unions.
- Civil society groups have a key role in influencing the debate and holding governments and other interests to account.
- The nature and role of the news media, and its positioning on climate change, can also shape the debate.
- Public voice, awareness and perceptions of climate change play a vital role.
- Business leaders, vested interests and lobby groups can wield significant influence.
- The balance of power between national and local authorities needs to be taken into account.

As shown above, numerous stakeholders, interests, capabilities and interactions come into play when countries design and implement low-emission, climate-resilient policies. The rest of this chapter provides examples of just how important these political economy dimensions can be.

The public sector can be entangled in the high-emission economy

Government budgets, which hinge on the health of economic activity, may rely on fossil fuels and GHG-intensive sectors, for example by collecting royalties on the extraction of oil, natural gas and coal. The share of such revenue in government revenues is typically below 5%, although it is above two-thirds in most countries that belong to the Organization of the Petroleum-Exporting Countries (OPEC) and some fossil-resource-rich countries (Table 6.1). The same considerations are likely to apply, but at a far smaller scale relative to GDP, to the production of palm oil (it is Indonesia's second export in value and contributes significantly to deforestation), and to other GHG-intensive facilities that generate income and tax revenues at local and national level.

Table 6.1. **Estimated rents from extraction of oil, natural gas and coal resources**

	Billion USD			% GDP			% Total government revenue		
Period average	**2001-05**	**2006-10**	**2011-15**	**2001-05**	**2006-10**	**2011-15**	**2001-05**	**2006-10**	**2011-14**
G20 (excl. EU)	**485**	**1 020**	**1 130**	**1.5**	**2.2**	**2.0**	**4.7**	**6.8**	**6.8**
Argentina	6	12	12	3.8	3.7	2.3	17.5	14.4	8.2
Australia	8	25	26	1.6	2.6	1.8	4.3	7.5	6.2
Brazil	11	29	41	1.7	1.9	1.7	4.8	5.2	5.4
Canada	35	46	27	3.7	3.1	1.5	9.1	8.0	4.7
China	51	180	207	2.7	4.1	2.4	16.7	19.6	10.1
France	<1	<1	<1	<1	<1	<1	<1	<1	<1
Germany	2	5	3	<1	<1	<1	<1	<1	<1
India	14	37	44	2.0	2.9	2.3	11.2	14.8	13.4
Indonesia	12	29	34	5.2	5.7	3.7	29.8	32.2	25.3
Italy	1	2	3	<1	<1	<1	<1	<1	<1
Japan	<1	<1	<1	<1	<1	<1	<1	<1	<1
Korea	<1	<1	<1	<1	<1	<1	<1	<1	<1
Mexico	27	49	52	3.4	4.8	4.3	17.1	20.9	20.9
Russia	82	194	249	16.9	14.6	12.5	45.1	38.6	35.6
Saudi Arabia	94	213	293	38.3	46.8	41.0	>90	>90	>90
South Africa	5	12	10	2.7	4.0	2.5	11.0	14.3	10.3
Turkey	<1	2	1	<1	<1	<1	<1	<1	<1
United Kingdom	17	26	20	<1	<1	<1	2.3	2.6	2.3
United States	119	162	106	1.0	1.1	<1	3.3	3.7	2.5
Rest of the World	**453**	**950**	**1 000**	**2.2**	**3.2**	**2.9**	**16.3**	**20.0**	**20.5**
OPEC (excl. Saudi Arabia)	133	336	363	26.6	29.8	20.8	80.9	85.9	81.1
World	***938***	***1 970***	***2 130***	***1.8***	***2.6***	***2.3***	***7.1***	***10.0***	***9.9***

Notes: Estimated oil, natural gas and coal rents are the difference between the production value at world prices and total costs of production, based on national sources and methods described in "The Changing Wealth of Nations: Measuring Sustainable Development in the New Millennium" (World Bank, 2011). Mexico indicated that less than 20% of government revenues were oil-related in 2016 (16.3%).

Sources: IMF (2017) and World Bank (2017).

The decisive transition outlined in Chapter 4 shows the importance of fossil-fuel exports in some countries' ability to navigate the transition. Other countries may be subject to similar situations due to the importance of land-use activities or GHG-intensive industrial sectors, whose evolution in a low-emission, climate-resilient scenario is much harder to project. The consumption of fossil fuels also generates tax revenues, mostly from transport. The prospect of the market for fossil fuels progressively declining as low-emission technologies and practices are deployed is therefore a significant structural issue for governments, asset owners and the workforces involved and may prompt resistance to policies that seek to constrain emissions. This situation presents two important issues for the transition:

- It may be difficult to initiate a constructive domestic dialogue on climate policy when immediate budgetary, economic and financial interests work against decisive climate change action that will drive investment away from GHG-intensive activities.

- Economies that rely on GHG-intensive exports are exposed to an external risk of declining demand and prices as the rest of the world embarks resolutely on emissions reductions, and have therefore limited, if any, control over the pace and moment of a decrease in revenues. Reliance on depleting fossil-fuel resources and the volatility of international market prices is already a major concern in fossil resource-rich countries[1]. External factors may in this way force more proactive domestic change in fossil-fuel exporters, despite entrenched interests.

As the International Energy Agency (IEA) shows in its scenario consistent with a 66% likelihood of keeping the global average surface temperature increase to below 2°C throughout the century (IEA 66% 2°C scenario, IEA 2017), the world will still need coal, oil and gas for some time. As such, a progressive yet timely exit and diversification strategy is feasible if it is well planned. The challenge for economies rich in fossil fuels is how to best build on today's revenues, workforce skills, education and training institutions, infrastructure and other capacities to engage in diversification.[2] Saudi Arabia, for example, has identified such risks and opportunities in its Saudi Vision 2030 transition plan to a less oil-dependent economy (see Box 5.4 in Chapter 5). Diversifying government income through energy price reforms and other public revenue measures can help to align diversification goals with GHG mitigation. Such policies are always best undertaken from a position of strength, when international energy prices are high. However, most often, the fiscal pressures that drive structural reforms come only when commodity prices fall. Indonesia and Russia, for example, were also set for austerity with public spending cuts following the recent decline in global oil prices, although some cuts have been held back by a more recent upswing in oil prices.[3] Countries with fewer reserves will need to get fossil-fuel subsidy removal and economic reforms under way even more quickly.

The policy and social challenge facing countries rich in fossil fuels is multi-dimensional, hinging on elements such as their ability to spur new activities and innovate; the education levels of their workforces; and their financial infrastructure. Norway, for example, manages its oil wealth effectively through a sovereign wealth fund, of which the government is allowed to spend no more than 4% a year.[4] This has helped to achieve high living standards while reducing exposure to risks related to the oil price.

The fossil-fuel industry also matters for public and broader institutions in other economies, through its footprint in financial markets and pension funds (Box 6.1).

Box 6.1. **Pension funds rely on energy companies' stocks**

Stranded assets will sooner or later affect the capital value of energy companies. Chapter 3 provides latest estimates of stranded assets under the IEA 66% 2°C scenario, assuming an orderly energy transition (IEA, 2017).

Energy companies can be an important share of a stock market's capital value or of the indices that it proposes to fund managers. Pension funds need to be aware of the possible risk caused by policies that directly target the revenues of companies that rely on fossil fuels. For that reason, France introduced legislation requiring institutional investors to evaluate and report on their exposure to risks related to climate change, including the effect of climate change policies on their portfolios (Journal Officiel, 2015; see also Chapter 7).

In 2014, the total of equity, bond and credit exposures of EU financial institutions to the fossil-fuel industry were EUR 260-330 billion for EU pension funds, EUR 460-480 billion for banks and EUR 300-400 billion for insurance companies (Weyzig et al., 2014). This represents approximately 5% of total assets for pension funds, 4% for insurance companies and 1.4% for banks.

Box 6.1. **Pension funds rely on energy companies' stocks** *(cont.)*

De Nederlandsche Bank (2016) has analysed the exposure of Dutch financial institutions to commodity price risks and loans to carbon-intensive producers (Figure 6.1). Dutch banks have EUR 38 billion in outstanding loans to emerging economies that are vulnerable to falling commodity prices. Their largest exposure is to Brazil and Russia. Pension funds' exposure to these countries amounts to EUR 30 billion. Dutch financial institutions' exposure to producers of fossil fuels, including oil and gas companies, amount to EUR 40 billion for the three largest Dutch banks, EUR 38 billion for the three largest pension funds, and EUR 9 billion for the three largest insurance groups. Overall, the Dutch bank association concludes that the costs of a gradual transition will probably be manageable, but a rapid shift could see carbon-intensive companies written off abruptly. A bursting "carbon bubble" may damage not only producers of fossil fuels, but also other GHG-intensive sectors, such as energy generation, transport and agriculture. Exposure to these sectors accounts for a large part of the balance sheets of Dutch financial institutions.

Figure 6.1. **Fossil fuels on the balance sheet of the three largest Dutch banks, the five largest insurance groups, and the three largest pension fund administrators**

EUR billion and as a share of balance sheet

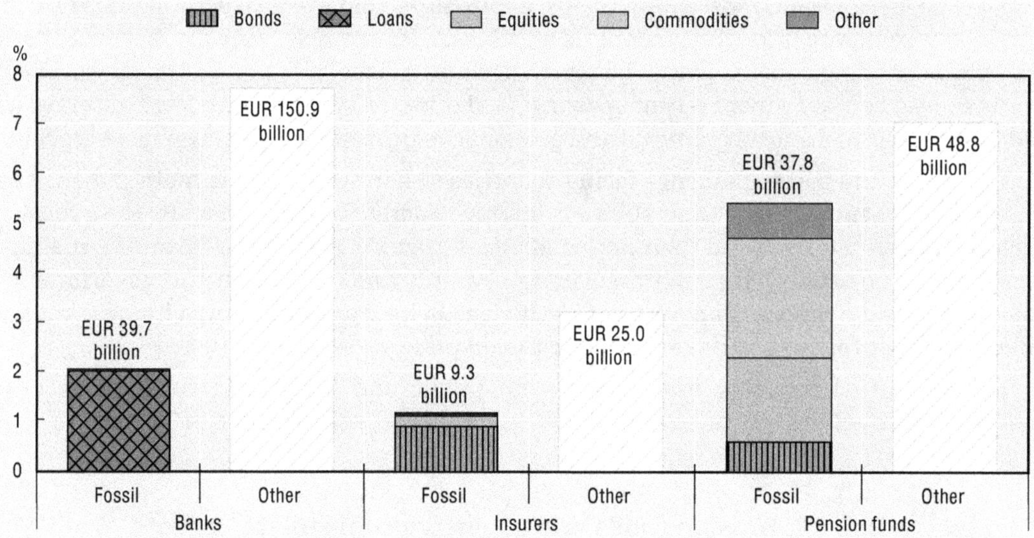

Source: De Nederlandsche Bank (2016).

Revenues from taxes on fossil-fuel use will also be eroded. If a carbon tax were the only instrument for the transition, estimates of demand-price elasticities indicate that carbon tax revenues would rise faster, given constraints on emissions, than energy tax revenues would if no further policies on climate were implemented. In fact, however, several other instruments have already been mobilised to reduce GHGs, including support schemes for low-carbon alternatives, performance standards and other regulations, representing tax *expenditures* rather than *revenues* through a carbon tax. The rising constraints on emissions are also being anticipated by energy-using equipment manufacturers, which are rapidly introducing low-carbon technology such as electric vehicles.

As a result, new tax revenues will need to be mobilised as fossil-fuel consumption drops. In Portugal, for example, the Green Tax Reform led to an increase in the vehicle tax of around 3%. The scope for broadening the base and for increasing rates is considerable, so revenue erosion is not an immediate concern.[5] While abrupt changes are unlikely in this area, detailed country-level modelling could help to anticipate when revenue erosion may arise and what alternatives should be envisioned. These could include taxation of transport and other energy services. In the United Kingdom, for instance, GBP 28 billion per annum of tax income comes from fuel duty, which is tied to the use of combustion engines in the transport sector (Office for National Statistics, 2017).

The scale of the fiscal challenge underlines the fact that moving to a clean economy concerns not just environment, energy or transport ministries. Long-term planning also needs to involve finance ministries to ensure a successful and fiscally sustainable transition. In particular, there is a tension between setting taxation levels that maximise revenue while minimising deadweight loss (the principle of fiscal efficiency), and the use of tax policy to change behaviour. For some countries with high energy consumption taxes, CO_2 reduction policies will impose fiscal policy shifts towards other steady bases for taxation.

Why political economy matters in climate policy: lessons from technology deployment and industrial transitions

Technological change vs. stakeholder interests: carbon capture and storage in the EU

From the early 2000s, it was recognised that any credible 2°C pathway had to deal with the scale of coal use in power generation. Carbon capture and storage (CCS) was seen as offering a way to do so while meeting climate and energy security requirements simultaneously. Global emission scenarios consistent with the Paris Agreement objective of well below 2°C also assume an important role for CCS (Chapter 3). Experience in Europe shows how important it is to ensure that different interests align in this area.

Between 2005 and 2009, Europe led global efforts to avoid dangerous climate change and advance CCS. In 2007, a programme was launched that aimed to develop 12 CCS projects by 2015. The programme was supported by a funding mechanism (NER300) that was linked to auction revenues from the EU Emissions Trading System (ETS) and economic stimulus support for six leading projects.

This political impulse clashed with underlying political economy interests at the sectoral level, however. Policy makers had assumed that incumbent players would want to deploy CCS as a means of protecting high-carbon assets and business models. In reality, utility companies and the coal sector perceived CCS as a threat to their assets and in conflict with their lobbying positions: the coal sector remained reticent about strong climate action for coal, and utilities favoured a single policy instrument, the emissions trading scheme. Meanwhile, the economic crisis combined with the spread of renewables was challenging their business models. Their response was to prioritise the pursuit of capacity payments for existing power plants, and to call for a stronger carbon price under the EU ETS.

At the same time, policy makers charged with advancing CCS focused on coal rather than on climate change. They paid insufficient attention to the application of CCS to energy-intensive industries, even though two of the most advanced CCS projects in Europe were proposed for steel and hydrogen production. By tying CCS to coal power, policy makers failed to provide a clear public interest case for deploying CCS to clean up and modernise "old" industries that are under continuous pressure to restructure.

What could have been done differently? Engaging the gas sector and industrial emitters from the cement, steel, biofuels and chemicals sectors earlier would have been more productive, as their long-term interests and skill sets are more closely aligned with the deployment of CCS. More emphasis should also have been given to developing new business models that provided CO_2 transport and storage infrastructure alongside targeted deployment incentives.

Such an approach requires greater government willingness to create "market maker" infrastructure providers, as has often been done to accelerate the deployment of public interest infrastructure as diverse as sewerage systems and gas pipelines. This is not a question of "picking technology winners", but does mean identifying key geographies and geologies where CCS can be deployed. The broad alignment of interests and incentives for CCS require careful attention by all stakeholders if this option is to deliver the sizeable emission reductions projected by long-term scenarios to stay well below 2°C.

Divestment: the need for exit strategies

Several financial actors, companies and individuals are lowering the carbon footprints of assets as policy interventions start to reduce profits and market shares, and increase the economic and reputation risks these assets carry. Some are also acting on ethical grounds (Baron and Fischer, 2015).

Vattenfall, a Swedish energy company owned by the Swedish government, announced its intention to divest its lignite mines and associated power plants in eastern Germany in October 2014. During the sales process, prospective buyers lost interest as market conditions for lignite power deteriorated and political pressure to reduce German power sector emissions grew. The portfolio, initially valued at EUR 2-3 billion, was eventually sold for a "symbolic price" to the Czech power producer Energetický a průmyslový holding (EPH), with Vattenfall additionally having to make a cash transfer of EUR 1.7 billion to EPH to cover high expected land reclamation costs (Reuters, 2016). EPH expects that rising European power prices can restore the profitability of these assets.[6] The company also indicated its interest in assets that can benefit from capacity payments, as found in the company's 2015 annual report (EPH, 2016) although this was not mentioned in the course of the transaction. Two of the acquired units are in fact part of the German electricity system's emergency reserve and will generate revenues outside the energy market as a result.

When it seemed that Vattenfall might not find a buyer, several proposals for a managed closure of the plants through the creation of a charitable foundation were discussed. These became irrelevant when the Swedish government approved the sale in July 2016, despite widespread concerns about the sustainability and environmental responsibility of the buyer.

The possibility of closure raised social concerns as the lignite industry provides around 15 000 direct jobs – 3.7% of local employment – in Lusatia, where the plants are located (Schwarzkopff and Schulz, 2015). The social implications of such job losses cannot be ignored. The sale created serious difficulties for the German federal government and for the state governments of Brandenburg and Saxony, which wanted to keep the lignite assets open, whether this meant selling or having Vattenfall maintain ownership.

While Vattenfall reduced its carbon footprint, the mid-term impact may be a net increase in GHG emissions, as EPH will seek to maximise revenues from the newly acquired assets. The mines and power plants could have been wound down in a way that helped the region's workers – including economic diversification, pension bridging, reskilling and redeployment of workers – but this option was not considered given the sale.

This case illustrates several pitfalls that may arise during the transition, as governments need to make politically unpopular but necessary decisions to facilitate the exit of carbon-intensive activities. It also demonstrates the need for a comprehensive climate mitigation policy package: a stronger price on CO_2 emissions through the EU ETS would have clarified the not-so-favourable economics of lignite-based power and probably facilitated the elaboration of an exit strategy.

Governments that want to avoid the social fallout of mining closures will have to either stop such sales or create an enabling environment for buyers. They should take a more active approach in negotiating divestment deals with companies or in managing future phase-outs. In Germany, the Climate Action Plan 2050 finalised in November 2016 has opened discussions on coal phase-out, and options should be presented in 2018 (Weiss, 2017). If a political solution is needed, it should be elaborated sooner rather than later. The following section covers some of the policy instruments and approaches that can be used to handle these situations.

Lessons from earlier industrial transitions

The pathways described in Chapter 2 show that a significant shift in the global energy profile is necessary to mitigate climate change, with a decline in the use of coal, oil and gas, in that order. In the power generation sector, decarbonisation implies the stranding of production assets. The best way to avoid massive stranding is to align energy infrastructure decisions with the Paris Agreement climate objective (Chapter 3), but some of the productive capacity that exists today will not recoup its investment cost in a low-carbon scenario.

Beyond fossil fuels, other activities that are very GHG-intensive, such as cement, steel, chemicals, paper and pulp, and glass – face similar risks, until technological change makes their decarbonisation possible.[7] They will also be challenged by alternative solutions, improved resource efficiency or recycling of their outputs. New business models could emerge, undermining or reinforcing incumbent companies. Today's GHG-intensive companies also have considerable human capital that will be essential for successful shifts in technology and business models. Some sectors have already shown that they can undergo rapid changes as new technologies appear. The steel sector, for example, adopted electric arc furnaces that enabled scrap to account for 25% of global crude steel output.

The policy challenge is to manage the reduction of emissions from today's GHG-intensive sectors while minimising the destruction of assets' value. This first requires sending clear signals to investors and decision-makers in companies, starting with coherent climate policy instruments, including public procurement that spurs innovation in the right direction (see Chapter 5). It also requires corporate disclosures that reflect firms' positions in the face of climate risks (see Chapter 7).

It is impossible to predict how efficient these heavy industries will be in their shift to a low-emission, climate-resilient economy; although policy makers should aim for an orderly evolution, disruptive changes may happen. Disruptions can also be exacerbated by business cycles and other factors, such as the global excess capacity of the iron and steel industry, or the current pressure on government budgets in oil-rich economies. These challenging times for every industry also open a window of opportunity for governments to prepare industry for competition in a world consistent with the objectives of the Paris Agreement. Failure to do so is likely to result in further carbon lock-in or stranded assets down the line.

Governments sometimes intervene to facilitate large-scale industrial restructuring, with mixed results. Numerous firms also enter and exit markets without government intervention. Within any given economy and any given sector, managers of incumbent companies will adopt different strategies and new entrants may drive incumbents out, so an overall picture of stranded assets and new opportunities cannot be obtained. Divergent corporate strategies are already visible in the oil and gas sector: some multinationals are expanding renewable energy activities while others remain focused on conventional activities.

Overall, it is unlikely that all high-emission companies will manage to re-allocate capital (financial and human) without problems. Governments should start thinking about policy packages to facilitate and not prolong the exit of less successful plants.

During recent sectoral crises, G20 governments have intervened to facilitate restructuring. Their policy approaches, particularly in the shipbuilding and the iron and steel sectors, offer examples for the low-emission transition.

- Investment aid, strictly related to restructuring: loan guarantees to support companies closing facilities (in the EU and Japan) have been effective when accompanied by commitments to reduce capacity, e.g. in Japan's shipbuilding industry in the 1980s (OECD, 2005).
- Closure aid, to cover social costs (severance pay-outs, counselling services, vocational training), which are addressed below.
- Diversification and modernisation: financing instruments and other measures to promote energy efficiency, process improvements, and environmental protection in the Japanese steel industry in the 1970s (Tamura, 2015); public support to promote job creation in depressed areas (Europe, steel) (Davignon, 2016); site transformation, from shipbuilding to ship-repair and offshore wind (Denmark), or in Gdansk, from container ship-building to specialised hulls and industrial steel structure (Mazurkiewicz-Gorgol and Bomhoff, 2009).

Governments freed up considerable resources to facilitate the restructuring of these sectors. In the 1980-85 restructuring of European steel, for example, EU producers received ECU 38 billion of state aid (23 to support continued operations, 12 to improve operations, 2 for closure and one to support R&D) (Davignon, 2016).

In China, Liu and Xu (2015) consider overcapacity in coal and heavy industry and advocate for a "Fund for output reduction and transition of industries with excess overcapacity" to finance special bonds used by local governments, with penalties for non-compliance with output reduction plans. In early 2016, China established a Special Fund for Excess Capacity Reduction to finance lay-offs and reemployment in the coal and steel industry.[8] Funds will be raised mainly from power grid enterprises and the amounts will be based on the energy generation of these enterprises. The special fund will co-finance worker layoff and re-employment, alongside local funds.

Box 6.2. **Barriers to "green" restructuring**

Several problems may hamper industrial restructuring, including the exit of inefficient plants and firms:

- General conditions may sometimes create inadvertent barriers to exit, including the administrative costs of closing plants, such as the costs incurred by companies when going through bankruptcy procedures, the cost of decommissioning and rehabilitating industrial sites, or severance payments and relocation of workers.
- Industries with high capital and sunk costs and low salvage values may be able to run relatively unproductive (and polluting) plants for a long time.
- The financial sector may discourage exit as the sudden closure of firms facing overcapacity could create market risks.
- Some heavy and fossil-fuel industries are geographically concentrated and account for an important share of local economic activity and jobs. Local governments have a direct interest in sustaining economic activity and may offer support to firms that should otherwise exit the market.
- In certain G20 economies, state-owned enterprises are often active in heavy and fossil-fuel industries, which may have financial and social implications for structural change.

A rational approach for government interventions is to first facilitate the exit of less profitable, less energy- and resource-efficient firms. Support for the modernisation of the industry should be targeted to the remaining companies, subject of course to the country's state aid rules. If broad subsidies to upgrade the sector are introduced first, they risk being wasted on companies that are likely to shut down; they also undermine the main effort to cut capacity, essential to restore the viability of remaining companies. Box 6.2 highlights some challenges with industrial restructuring.

Addressing the impacts of mitigation-related policies on household income

The success of the transition, and public support for it, depends on equitable and transparent distribution of its costs and benefits across society. Energy and transport affordability is central to the discussion on how changes in energy prices would affect development and welfare of households in different income groups (Box 6.3). Equity outcomes and public support could be undermined by concerns that instruments such as carbon pricing and reform of fossil-fuel subsidies may lead to energy price increases that disproportionately affect poorer households. Economic simulations and policy observations demonstrate, however, that adverse impacts of energy prices on the affordability of energy and on inter-household equity can be alleviated without harming the environmental effectiveness of policies. The following sections give examples of measures to that effect, in the context of carbon and energy taxes and fossil-fuel subsidies.

Solving distributive impacts of carbon pricing

To ensure that public debate is well informed and that effective policies are designed, it is vital to gauge how the transition affects societies and economies. It is also crucial to understand how to prevent or reduce impacts that are regressive – i.e. that fall disproportionately on the poor. Despite a common perception, climate policy instruments are not always regressive. In a sample of 21 mainly European OECD economies, taxes on transport fuels were often proportional to income in high-income countries and progressive in middle-income countries, with limited regressive effects for heating fuels and stronger regressive effects for electricity (Flues and Thomas, 2015). Analysis for low-income countries also shows progressive effects of taxes on transport fuels (Sterner, 2012). A review of relevant work from over two dozen countries concludes that while environmental taxes can be regressive in some high-income countries, transport-fuel taxation is generally progressive, particularly in low-income countries (Kosonen, 2012). In low-income countries, such as India, taxes on transport fuels are generally considered to be highly progressive in both urban and rural areas (Datta, 2008).

Where higher energy prices have a regressive effect, well-designed social policies can improve access to energy and reduce energy affordability risks. An appropriate use of revenues, such as income-tested transfers, can ensure that higher carbon prices have progressive impacts (Klenert and Mattauch, 2016). Simulations for 20 high- and middle-income countries show that transferring a third of the additional revenues from higher energy taxes can improve energy affordability (Flues and Van Dender, 2017). A review of more than 120 studies notes that if a sufficient, but small, part of revenues from energy taxes is handed back to households, it is possible to avoid negative distributional effects, and also reduce poverty and deprivation (OECD, 2014). More broadly, concerns about the post-tax distribution of income should be addressed through personal income tax reform rather than through differentiated carbon or energy taxes, which can frustrate or hinder the underlying environmental objective of reform.

Overall, policy makers can take full account of the expected impacts of proposed policies on income distribution and embed solutions that ensure their progressivity. There need not be a trade-off between the effectiveness of a carbon pricing mechanism or other measures to lower emissions and their impact on household equity. There is now much policy experience in this area, which can be replicated as countries decide to introduce mitigation policies targeted to the general public.

Box 6.3. Energy access and affordability within the climate policy context

The success of climate policies partly hinges on how well they guarantee energy access and affordability. Households need access to modern energy services such as lighting, cooking, heating and transport. Lack of energy access is sometimes called energy poverty. Households should not face difficulties in paying for necessary levels of energy use. A lack of such energy affordability is referred to as energy affordability risk, or fuel poverty, and may compromise health and participation in society.

In the case of energy access, energy use remains flat for the lowest income groups until the household is well clear of the poverty line. This means that even though a household may be earning more, it will not spend the additional income on energy at first: there are therefore many more households that lack access to energy than there are households below the poverty line. In the case of energy affordability, there is a clear link between energy affordability risk and low disposable income.

For G20 countries, lack of energy access is largely a rural problem in emerging economies, notably in India where 44% of households lack access to electricity (IEA, 2016). The use of traditional fuels is widespread in Indonesia (72% of the population) and Brazil (13%), and just under half of Chinese households use solid fuels for cooking (and often for heating as well) (Tang and Liao, 2014). Energy affordability risk can be high in both rural and urban areas, including in advanced economies where it affects 2% to 30% of the population, depending on the indicator used (Flues and Van Dender, 2017). These are important considerations when thinking about policy instruments that aim to transform energy use, including by phasing out fossil fuels (see below).

It is more important to address the causes of insufficient access to energy and energy affordability risks (through access to cleaner and more efficient fuels, new and more targeted social transfers, more efficient electric appliances and housing, and improved public transport) than the symptoms (such as increased costs and health problems). Energy efficiency measures can help deliver similar or higher end-use service at lower overall cost. Rather than increasing product prices, for example, energy efficiency labelling schemes significantly lower lifecycle costs.

Such energy efficiency measures can produce a "triple dividend". A positive environmental effect from reduced emissions is accompanied by a positive economic impact: large-scale energy efficiency policies can boost annual economic growth by 0.25% to 1.1% (IEA, 2014). In addition, re-investing carbon revenues in energy efficiency programmes aimed at low-income households would also yield an equity dividend. Low-income households may not always fully benefit, however, because they often use older or second-hand appliances, so energy efficiency programmes specifically targeting the poor tend to be most effective. Through Brazil's Electricity Public Benefit Fund, for example, utilities are required to invest 0.5% of their revenues in energy efficiency programmes, of which 50% must be devoted to low-income households.

Sources: Flues and Van Dender (2017, forthcoming); Tang and Liao (2014); IEA (2016); IEA (2014).

Lessons from fossil-fuel subsidy reforms

When governments start to appreciate the environmental, financial or opportunity costs of fossil-fuel subsidies, and attempt to reform them, they typically face a difficult, politically risky task. Subsidies benefitting consumers of fuels or electricity are usually widely available, and may be regarded by citizens as their rightful share of their nation's natural wealth. Raising the price of a basic commodity is rarely popular. Subsidies benefitting fossil-fuel production flow to powerful companies with considerable resources to defend the status quo.

Resources freed up by reducing fossil-fuel subsidies can also be used to alleviate energy affordability pressures on poorer households. Analysis before subsidy reforms in Indonesia showed that removing fossil-fuel and electricity consumption subsidies would result in real GDP gains of 0.4% to 0.7% and aggregate welfare gains for consumers ranging from 0.8% to 1.6% in 2020 (Durand-Lasserve et al., 2015). Such findings suggest that the choice of redistribution system may determine the impacts on different income groups. Cash transfers, and to lesser extent food subsidies, tend to reduce poverty and hence appeal more to poorer households. Compensation mechanisms proportional to labour income tend to benefit higher-income households and increase poverty, as informal employment is higher among poorer households, which renders them ineligible to receive these payments.

An increasing number of countries have overcome the political obstacles to subsidy reform. Successful reforms generally have several features in common. First, the government will have gathered data on the monetary value of the subsidies, their distribution across beneficiaries, and how energy-related services — and, usually, air quality — could be improved when prices better reflect costs. As countries are also committed to reducing national GHG emissions, they would have looked at the cost per tonne of carbon emissions reduced through fossil-fuel subsidy reform and compared that with alternative policies.

Surprisingly, at the start of reform, governments often have poor information even on how much money they are spending, or losing, and how the subsidies have grown through time. The actual distribution of subsidies across income classes can also come as a surprise to policy makers. Price subsidies may have even been conceived initially as a way of reducing poverty. Yet untargeted subsidies that simply reduce the price of fuel or electricity for everyone are by their nature regressive: the richer the household, the higher their energy consumption, the more they benefit from the subsidy.

An IMF review of fossil-fuel subsidies in 32 developing countries found that for every USD 100 spent on subsidising fuels, only USD 18 goes to the bottom 40% of households. In other words, for each USD of benefit provided to the poorest 40% of households in each country using energy subsidies, governments spend on average USD 5.6 (Coady et al., 2015). In Mexico, for example, 40% of the subsidies for gasoline between 2006 and 2014 went to the top 10% of its population. In India, 87% of electricity consumed by domestic consumers is subsidised, based on 2010 estimates, of which more than half goes to the richest 40% of households, while 25% of households do not even have access to electricity (Pargal and Banerjee, 2014).

These figures compare unfavourably to the cost of direct cash transfer programmes, recognised as one of the most efficient ways to deliver social assistance to poor households. In Bolivia, 58% of the Juancito Pinto cash transfer programme goes to the poorest 40% (Arauco et al., 2014). In Peru, 71% of the Juntos cash transfer programme goes to households below the international poverty line (Jaramillo, 2014). In Brazil, it is estimated that 80% of the Bolsa Familia programme goes to the poorest 32% (Soares et al., 2006).

Communicating information on subsidies is essential in the reform process. In Egypt, 70% of the population did not know the scale of energy subsidies in 2014 (Fay et al., 2015); in Morocco, a 2010 survey found 70% unaware that energy was subsidised at all.

The governments' strategies included explaining that the subsidy absorbed a huge part of government revenues (39% in Egypt and 17% in Morocco) and that the compensation package would address citizens' concerns about "what's in it for me?".

To target energy subsidies better, countries such as India and Egypt have experimented with voucher systems. The United States, which deregulated its petroleum and natural gas markets in the 1970s and 1980s, also targets its energy-related assistance to poorer households through its Low Income Household Energy Program (LIHEAP). Iran introduced the Targeted Subsidy Reform Act, replacing some fossil-fuel subsidies with targeted cash transfers to households (Guillaume, Zytek and Farzin, 2011). Similar targeted assistance to help with heating bills was introduced in 2015 by Ukraine in connection with its subsidy reforms (Ogarenko and Gerasimchuk, 2016). India transfers subsidies directly into the bank accounts of means-tested consumers of liquid petroleum gas (LPG) and kerosene. When the informal economy is large, it can be difficult to target assistance for energy fuels to lower income households. In such situations, electricity subsidies can be targeted to specific households, to a specific amount of energy (e.g. the first 80 kilowatt-hours every month), or both.

Numerous other countries have shifted assistance from fuel or electricity consumption to cash transfers. In 2015, for example, the Kingdom of Saudi Arabia raised retail gasoline prices by about 50% as part of a plan to restructure energy subsidies, pushing down oil consumption by 2% in the first 11 months of 2016 compared with the corresponding period in 2015; it raised prices again at the end of 2016. Families affected by the measures could register for cash transfers from 1 February 2017; the payments are expected to start in June 2017 (Mahdi, Carey and Nereim, 2016; Lee, 2017).

Overall, effective reform of fossil-fuel subsidies requires an accurate understanding of its primary impact on budgets and on targeted consumers. Alternative methods, such as cash transfers, are needed to correct the social distortions that subsidies were intended to rectify in the first place. Governments that need to undergo this delicate process now have much experience to build on.

Towards a "just transition"

"Taking into account the imperatives of a just transition of the workforce and the creation of decent work and quality jobs in accordance with nationally defined development priorities [...]"

– UNFCCC, 2015, preamble to the Paris Agreement.

Alongside households, the other critical group affected is the workforce of companies that are likely to be restructured or closed. Trade unions are fully aware of these challenges and have publicly advocated for a role as active players in a "just transition". For climate action to be successful, workers should have a say in their company's strategy to respond to the climate challenge; and when restructuring and closures are anticipated, proper social measures should be planned.[9]

The UNFCCC estimates that there are nearly 1.5 billion jobs in sectors critical to climate stability (Table 6.2). This is probably more an indication of the magnitude of economic activities that will contribute to mitigation and the need to adapt to ongoing climate changes, than an estimate of total jobs at stake in the transition. For instance, the IEA's 66% 2°C scenario indicates that around 1 million direct jobs would be "lost due to the premature closure of assets, around 20% of current coal mining employment", to be put in perspective with the current 30 million jobs in the energy sector (IEA, 2017).

Although the aggregate effect on jobs may be modest (see Chapter 4), the net number hides significant job losses at local level, with potential for geographical dislocation of affected communities, as well as creation of new jobs, some of which require new skills.

Workers who lack mobility or means to acquire new jobs in different industries and regions may find themselves with skills and expertise no one wants, leaving whole communities vulnerable (Caldecott et al., 2016). Upgrading and diversifying workers' skills is vital to strengthen their resilience to risks and shocks, particularly where access to education is limited and incomes are low, limiting the opportunities to re-skill and relocate.

Table 6.2. **Global direct employment in sectors critical to climate stability**

Sector	Employment (millions of people)
Agriculture	1 000
Forestry	44
Energy	30
Manufacturing (resource-intensive)	200
Buildings	110
Transport	88
TOTAL	1 472

Source: UNFCCC, 2016.

Historically, the term "just transition" was used by North American trade unionists to describe a programme of support for workers who lost their jobs due to environmental protection policies. It is now used globally and understood by trade unions and their partners as an effort to plan and invest in a shift to environmentally and socially sustainable jobs, sectors and economies.

Trade unions have recognised the urgency of acting to mitigate climate change. They point out the need to open negotiations with companies on how to best move to a low-emission economy. The International Trade Union Confederation has set out several demands, including recognising workers in the fossil-fuel industry; supporting innovation in the manufacturing sector; investing in community renewal; guaranteeing social protection and human rights; establishing just transition funds; and pursuing social dialogue backed by collective bargaining (ITUC, 2010). In 2015, the International Labour Organization established *Guidelines for a just transition towards environmentally sustainable economies and for all* (ILO, 2015), which rests on the following principles:

a. Strong social consensus on the goal and pathways to sustainability.

b. Policies that respect rights at work.

c. The recognition of the strong gender dimension of environmental challenges and opportunities, and the consideration of policies to promote equitable outcomes.

d. Policy coherence across economic, environmental, social, education, training and labour portfolios to generate an enabling environment for the transition.

e. The anticipation of impacts on employment, social protection for job losses and displacement, skills development and social dialogue – including the right to organise and bargain collectively.

f. The need to take into account the specific conditions of countries, including their level of development, economic sectors and sizes of enterprises – no "one size fits all" solutions.

g. The importance of fostering international co-operation among countries.

Even if there is a high-level consensus among trade unions on sustainable development and climate protection, however, those whose jobs are directly at stake sometimes oppose climate policies (ETUC, 2016). This reinforces the case for engagement with workers and unions to ensure community ownership.

In the wake of the financial and the economic recessions, it is not enough just to mitigate the effects of climate policy on the workforce: climate policy also has to help to create new jobs, for example in renewable energy and energy efficiency measures.[10] The urgency to act on climate could be challenged by a workforce that does not see new jobs and livelihoods arising in parallel with abrupt industrial and economic changes. Chapter 4 stresses the importance of active labour market policies to guide employment towards new, low-GHG activities or other growth sectors.

The following case studies show how labour force aspects are playing out, positively or negatively, in recent transition processes.

Just transition at the enterprise level: Enel in Italy and Latrobe Valley in Australia

Enel, an Italian electricity multinational, is committed to renewable energy and to researching and developing environmentally friendly technologies. In 2015, approximately half of the group's electricity was from non-fossil sources. Enel has committed to decarbonize its energy mix by 2050.

As part of its decarbonisation plan, Enel will close 13 GW of thermal power stations in Italy, while expanding renewable energy, demand management and other measures. It has entered into social dialogue and a framework agreement with its Italian union partners. The framework covers retention, redeployment, reskilling and early retirement. It is a good example of a just transition agreement for this sector as it envisions the workforce evolving hand in hand with the structural change of the sector from a small number of large plants to a more decentralised model. The employability problem was managed by Enel together with trade unions, using opportunities provided by recent Italian legislation.[11]

The closure of the Latrobe Valley Hazelwood coal power plant in Victoria, Australia, provides another example of worker transition measures. Its owner, the French energy multinational Engie, decided to close the plant on 31 March 2017, a decade after the end of its envisioned technical life. The Victoria government signed an agreement with Engie to transfer some workers to the AGL Loy Yang B station, also operated by Engie. Vacancies at this station will be created through early retirement packages. A funding scheme of AUD 20 million was put in place to support the workers, part of a AUD 266 million package for the Latrobe Valley, AUD 50 million of which is to support business growth in this community (Victoria Government, 2017).

Community level: perspectives for social and ecological industrial policy

Schweinfurt, Germany, has 50 000 inhabitants and a strong automotive industry, so decarbonisation will bring structural changes that affect lives and working conditions. The Bavarian metalworkers' union IG Metall and Friends of the Earth in Bavaria (BUND) are collaborating on a project that aims to identify how to create a social and ecological transition in the area. The project will consider the consequences of climate change and demographic trends on employment as well as on private lives. During the project, the partners will tackle challenges like increasing energy efficiency and reparability of products, as well as the energy and resource efficiency of plant processes. The provision of regional mobility options is another strong focus. Improving quality of life and protecting the environment are held equally important. The organisers also seek to answer the following questions, putting people first: How should the region's industries develop their products, processes and employees' skills? How can new jobs be created at the same time? How can the interests of employees and the environment be brought in line?

Port Augusta, South Australia, hosted coal mining and coal-based power plants for more than six decades until its plants, among Australia's oldest, closed in 2016. A local initiative bringing together citizens, workers and trade unions elaborated a plan to replace coal-based activities with renewable power generation plants (6 concentrated solar plants and 95 wind turbines; Repower Port Augusta, 2017). The first plant was completed in October 2016.[12] The initiative attracted interest from three potential employers and received funding from the Clean Energy Finance Corporation (a Commonwealth entity) and the South Australian government, an example of how central and local governments can collaborate to help communities move away from CO_2-intensive activities. The Port Augusta example shows how important investment can be for community transitions. Low-carbon infrastructure can help communities to avoid stranding but may require public support.

The transition of workers during Canada's phase-out of coal power generation

In 2016, the government of Canada announced plans to phase out coal-fired electricity by 2030, by which time 90% of Canada's electricity will be non-emitting. Coal-fired power produces 8% of Canada's total emissions and almost three-quarters of emissions from its power sector. To support an overall transformation of the economy, the government intends to use CAD 21.9 billion over 11 years for low-carbon infrastructure and commercially viable clean energy, including funds flowing through the Canadian Infrastructure Bank.

The government has also committed to working with provincial governments and organised labour to "ensure workers affected by the accelerated phase-out of traditional coal power are involved in a successful transition to the low-carbon economy of the future" (Government of Canada, 2016). This includes a national Just Transition task force, including organised labour, to guide and support the shift away from coal.

Social dialogue: the closure plan of the Diablo Canyon facility

Diablo Canyon is the last remaining commercial nuclear power plant in California, meeting 8.6% of California's power consumption. Although its low-carbon technology does not make it an obvious case study for this report, its closure plan offers a useful example of a just transition process.

Diablo Canyon has been targeted by environmental groups from its construction onward, because of general opposition to nuclear power and because of concerns about its safety in an area prone to earthquakes. It has been the target of numerous protests and legal challenges. Pacific Gas and Electric (PG&E), one of California's three investor-owned utilities and the largest utility in the United States, owns and operates the facility. As of 2011, PG&E employed 1 200 workers at Diablo Canyon, and 200 workers were employed by subcontractors.

In 2016, the plant faced uncertainty about whether its lease and permits would be renewed. If the state had not extended the lease, the plant would have had to close down as early as 2018. The local trade union (IBEW local 1245) worked to ensure that the plant would stay open to preserve employment, while Friends of the Earth US was campaigning for the plant to phase out and for PG&E to invest instead in renewable energy, energy efficiency and energy storage (Moglen and Peek, 2016). PG&E informed the trade union that it would not seek relicensing of the plant after 2024, based on market conditions, but was seeking a coalition to stave off an abrupt shutdown. The trade union started working on a retention package. The final deal was to operate the plant for eight to nine more years, combined with a package including annual bonuses, severance allowances, and retraining – a positive outcome from the union's perspective. Under the agreement, PG&E will also compensate the community for its loss of property tax revenues with USD 85 million, and will reserve up to USD 62.5 million for plant decommissioning.

Both unions and civil society groups describe the Diablo Canyon closure plan as a good example of a just transition process. It was based on effective social dialogue involving strong unions and a large and well-funded employer who recognises and supports rights at work. For workers, the long lead time to closure (eight years), good retraining and redeployment provisions, and generous retrenchment packages were key. PG&E's ability to offer a Just Transition Fund, compensate the community for lost tax revenues and create new jobs in renewable energy and energy efficiency were also important (see PG&E et al. 2016).

It will nonetheless take several years to see how many workers at the plant are able to retrain and redeploy within the company. Similarly, the surrounding community now needs to try to diversify its economy and attract a broader tax base. The real results of the agreement may not be apparent for a decade after closure.

Retraining and skills development: examples from Belgium, France, Germany and Saudi Arabia

Stronger energy efficiency standards in the EU and Belgium have driven an increase in jobs retrofitting buildings as well as building new structures to a higher standard of energy efficiency. These standards, along with technological innovation in the building sector, have developed so quickly that it is challenging for workers to keep up to date (see also OECD/IEA/NEA/ITF, 2015).

To address this issue, the three main trade unions in **Belgium** and sector employers have collaborated on developing courses for construction workers linked to green buildings and energy efficiency measures. Workers and employers assess the need for skills training together and develop proposals, which lead to very concrete training programmes executed by the government organisations.

Saudi Arabia started the development of its energy efficiency policy in 2010, leading to the Saudi Energy Efficiency Program (SEEP) in 2012, focusing on three sectors (buildings, transport and industry) totalling 90% of the country's energy consumption. The programme contains standards, labels and regulations covering a range of equipment (vehicles, white goods, building codes, etc.) and accompanying measures (e.g. a national energy-service company). Human capital development is an important component, with the development of energy efficiency curricula in five engineering schools, the development of an energy efficiency technician degree, and professional training for energy managers, based on the ISO 50001 energy management certification system (Alabbadi, 2016). Energy efficiency retrofit programmes could generate up to 247,000 jobs in Saudi Arabia over 10 years depending on the ambition of implementation (Dubey, Howarth and Krarti, 2016).

In **France**, workers from a car factory in Le Mans that was closing were retrained in mechanics and electromechanics to work in the new wind turbine industry, which helped to maintain jobs and create a new local activity. In Bielefeld, **Germany**, the project "Meine Energie hat Zukunft" (My Energy has a Future) brought together the public employment services, 120 companies, higher education institutions and vocational training providers, among others, to attract students to renewable energy-related activities (ILO, 2016).

Informal workers in the transition: Bus Rapid Transit in Bogotá, Colombia

Transmilenio, the bus rapid transit (BRT) system of Bogotá, Colombia, is the most used BRT system in the world, with 1.6 million passengers a day. Before BRT, informal workers driving minibuses provided the bulk of Bogotá's transport. Their work involved long days, leading to health problems and insecure conditions; these workers also had little ability to organise themselves.

BRT has been praised as a sustainable transport solution, reducing pollution and congestion. In theory, BRT should also be good for workers, replacing insecure informal jobs with formal work with better conditions. But each newly created formal job in BRT replaces seven jobs driving minibuses, according to some estimates. Despite this, there has been no effort by the government, the donors (the World Bank) or the bus company to assess the impact of this restructuring on the existing workforce, and to ensure that affected workers are supported to gain employment in the new system, to retrain in order to find alternative work, or to take early retirement. "The experience suggests that even a progressive restructuring programme that results in the creation of formal jobs can have significant negative labour impacts, and exclusion of workers and their representatives from dialogue about the reforms undermines the possibility of tackling those problems fairly" (Porter, 2010).

Such examples support the case for prior assessment of social impacts. ILO argues that "mitigation actions with potential social impacts, should only be taken once the potential affected population are protected and compensation measures through adequate social protection measures are in place" (ILO, 2016).

The above case studies illustrate several points that echo the ILO's just transition guidelines:

- Local communities and unions have a key role to play in the shift to a low-emission, climate-resilient economy, including identifying activities that can substitute the declining high-carbon ones.
- An active social dialogue is necessary between unions, employers, and local or central government.
- The transition needs to be anticipated years in advance in order to facilitate retraining and mobility plans.
- High-level policy and corporate commitments are vital, including funding commitments.
- Overall coordination, co-operation and trust among stakeholders is crucial.

Managing the transition: towards robust, low-emission, climate-resilient, development strategies

The call of the Paris Agreement to formulate and communicate long-term low GHG-emission development strategies offers governments an opportunity to start a discussion among all stakeholders on the key domestic ingredients of a successful shift to a low-carbon, climate-resilient economy. A few countries, including G20 members, have built long-term GHG emission pathways to guide their strategies (see Chapters 2 and 3).

This report's chapters on pathways (Chapter 2), infrastructure needs (Chapter 3) and financing (Chapter 7) all show how important it is for each country to have a long-term vision of a low-emission economy. Such a vision is necessary to drive a set of near- and medium-term decisions on policies and infrastructure, and generate common expectations among public and private stakeholders.

A purely technical approach that leaves out political economy aspects can miss opportunities and challenges, limiting buy-in for implementation. In countries where they have been developed, 2050 planning processes have initiated debate and deepened understanding. Experience with domestic and international initiatives to design long-term GHG strategies provides the following lessons:[13]

- Long-term strategies help to make it clear to stakeholders that economic and political transformations are necessary. They can offer stakeholders opportunities for competing visions of the low-emission transition, which can expose vested interests and help ensure the robustness of the agreed outcome. In the Netherlands, a detailed Energy Agreement for Sustainable Growth was reached by the government with employers, trade unions, environmental organisations and others, containing provisions on energy conservation, renewable energy and job creation, and associated policies and funding measures. The involvement of non-governmental stakeholders in the process ensures broad support (Government of the Netherlands, 2017).

- In their long-term economic plans, countries should take into account actions by the rest of the world in the context of the Paris Agreement – especially by export-oriented economies, whether in high- or low-carbon products. Global linkages among economies that are starting to take climate mitigation actions are likely to exacerbate disruptive effects.
- Climate change impacts should be part of the conversation on long-term strategies, informed by the latest available science. Even if projected changes are gradual in the medium term, they can affect the viability of certain technology or infrastructure options.
- The modelling of domestic economic pathways may not reflect the possible emergence of new activities, infrastructure, business models or disruptive technologies. As an example, a cost-based analysis may indicate a growing role for CCS, but will say little about the need to align various stakeholders for the provision of critical infrastructure for CO_2 transport and storage.
- Sectoral change is most effective when aligned with investment cycles and trends. As infrastructure reaches the end of its technical life, windows of opportunity open and close rapidly. In the United Kingdom, for example, the government had recognised by 2009 that there could be no new coal plant without CCS, due to lifetime emissions associated with any new coal power plant. The quantitative tools used for long-term projections sometimes miss this dimension of capital stocks, giving a misleading sense of progressive change when step changes are more probable.
- Sectoral targets help to provide guidance to investors and planners about future trends, supporting a more orderly transition. Transparency is important to avoid capture of sectoral goals by vested interests.
- Low-emission development strategies can be regularly revised – as is the case of the UK carbon budgets and the German 2050 plan. Numerous examples in recent climate policy history show why revisions will be necessary. These include: unexpected levels of effectiveness of policy instruments (EU ETS, renewable support policies); significant cost changes of low-carbon options (onshore and offshore wind, solar, nuclear, hybrid and electric vehicles); technology disruptions (LED, compact fluorescent and incandescent lighting); the emergence of new business models; changes in macro-economic conditions; and new information on climate change impacts and vulnerabilities.
- Strategies need to take into account constraints and opportunities related to the workforce, such as the availability of skills and training needs.

The process of elaborating low-emission development strategies can be critical in a country's response to the climate challenge. Nationally determined contributions pledged under the Paris Agreement are only the first step towards a long-term strategy that brings together various business, public sector and civil society interests. Long-term strategies should create an opportunity to gather the views of different ministries and levels of government. Governments can choose from a number of approaches to organise co-operation across ministries. For instance, there could be a systematic regulatory review of proposed legislative changes, testing their impact on GHG emissions and climate resilience – i.e. an extended climate impact assessment.[14]

Notes

1. This is illustrated by some of the main themes of the 2016 Arab Fiscal Forum: "Revenue Diversification in the Arab World: Challenges and Opportunities"; "Macro-Fiscal Challenges – Learning to live with Cheaper Oil"; and "Managing Oil Wealth – A Fiscal Framework for Uncertain Times" (Arab Fiscal Forum, 2016a, 2016b)
2. IMF (2016) underlines the differences across oil-exporting Arab countries and lays out policy recommendations to promote economic diversification contingent on country circumstances and capacities. Macroeconomic stability and supportive institutional and regulatory frameworks are prerequisites to promote diversification of government revenues and unlock the private sector potential in non-oil sectors.
3. See: https://www.bloomberg.com/news/articles/2016-06-06/saudi-arabia-approves-plan-to-bolster-non-oil-economy; https://www.bloomberg.com/news/articles/2016-06-21/for-jokowi-real-test-on-indonesia-fuel-subsidy-may-be-starting.
4. In February 2017, the Norwegian government announced that it would reduce the cap on spending from 4% to 3%. See https://www.bloomberg.com/news/articles/2017-02-16/norway-central-bank-chief-warns-of-sharp-drop-in-wealth-fund
5. If transport became fossil-fuel-free, today's main source of environmentally related tax revenue in most countries would disappear. However, it is possible to tax other, closer proxies to transport services, for example, the distance driven. The technology for doing so is available and getting cheaper, leading to growing adoption of distance-based charges.
6. "...we are convinced that lignite is in a position to contribute successfully to the rapidly evolving German power mix". Jan Springl, Board member of EPH, Reuters, 2016.
7. Wyns and Axelson (2016) give an overview of decarbonisation options for these industrial actors, from technology to business models, highlighting also the cross-sectoral linkages that the low-emission transition could generate as industries look for improved resource efficiency.
8. See http://szs.mof.gov.cn/bgtZaiXianFuWu_1_1_11/mlqd/201601/t20160122_1655180.html; http://english.gov.cn/state_council/ministries/2016/05/19/content_281475352712538.htm
9. See Rosemberg (2010) for a history of the emergence of "just transition" in the climate policy context and an early overview of climate and jobs-related issues.
10. The US Energy and Employment Report finds that 6.4 million Americans work in the energy and energy efficiency sector, with a 5% increase in 2016. Some 800 000 individuals work in the low-emission generation industries and 2.2. million "are employed, in whole or in part, in the design, installation and manufacture of Energy Efficiency products and services, adding 133 000 jobs in 2016 [...] Almost 1.4 million Energy Efficiency jobs are in the construction industry" (US DOE, 2017). Other studies try to project the job impacts of the low-emission energy transition; see Saussay et al. (2016), in the case of France, and Bivens (2015) for estimates of employment impacts of the U.S. EPA Clean Power Plan.
11. The measures included: early retirement incentives for older workers; apprenticeships to ensure transfer of knowledge from older to younger workers; company agreements preventing laid-offs through "solidarity agreements"; dedicated training to ensure qualification and employability of resources both during the recruitment phase and in professional mobility, including acquisition of new skills for the development of new businesses.
12. "World-first solar tower powered tomato farm opens in Port Augusta" RenewEconomy.com.au website, accessed on 12 February 2017.
13. These lessons are drawn from national processes for the elaboration of long-term emission pathways in France, Germany, South Africa and the United Kingdom.
14. In the United Kingdom, the Carbon Plan lists contributions of policies towards the overall carbon budget, although these are not legally binding on departments. France established legally binding sectoral carbon budgets aligned with its long-term goal.

References

Alabbadi, N.M. (2016), "Cross-sectoral perspective: Strategies for improving energy efficiency in Saudi Arabia", *G20 Workshop, Energy Policy after the Paris Agreement, Tuesday 13 December 2016, Munich, Germany*, www.bmwi-registrierung.de/g20-workshop-munich/pdf/G20_Workshop_all_Sessions.pdf.

Arab Fiscal Forum (2016a), *Revenue Diversification in the Arab World: Challenges and Opportunities, Arab Fiscal Forum, February 12, 2017*, www.imf.org/en/News/Seminars/Articles/2017/02/06/Second-Arab-Fiscal-Forum.

Arab Fiscal Forum (2016b), *Fiscal Policy and Growth, Arab Fiscal Forum, jointly organised by the Arab Monetary Fund and the International Monetary Fund, February 22-23, 2016*, www.imf.org/external/np/seminars/eng/2016/fiscalforum/pdf/agenda.pdf.

Arauco, V.P. et al. (2014), "Explaining Low Redistributive Impact in Bolivia", *Public Finance Review, Vol. 42/3, Sage, Los Angeles, pp. 326-45*, https://doi.org/10.1177/1091142113496133.

Baron, R. and D. Fischer (2015), *Divestment and Stranded Assets in the Low-Carbon Transition*, background paper for the 32nd Round Table on Sustainable Development, 28 October 2015, OECD Headquarters, Paris, www.oecd.org/sd-roundtable/papersandpublications/Divestment%20and%20Stranded%20Assets%20in%20the%20Low-carbon%20Economy%2032nd%20OECD%20RTSD.pdf.

Bivens, J. (2015), "A comprehensive analysis of the employment impacts of the EPA's proposed Clean Power Plan", Economic Policy Institute, Washington DC, www.epi.org/files/pdf/79246.pdf.

Caldecott, B. et al. (2016), "Stranded Assets: A Climate Risk Challenge", Inter-American Development Bank, Washington DC, http://dx.doi.org/10.18235/0000517.

Coady, D., V. Flamini and L. Sears (2015), "The Unequal Benefits of Fuel Subsidies Revisited: Evidence for Developing Countries", International Monetary Fund, Washington DC, www.imf.org/external/pubs/ft/wp/2015/wp15250.pdf.

Coady, D. et al. (2015), "How large are global energy subsidies?", International Monetary Fund, Washington DC, www.imf.org/~/media/Websites/IMF/imported-full-text-pdf/external/pubs/ft/wp/2015/wp15105.ashx. See also the associated database at www.imf.org/external/np/fad/subsidies/data/codata.xlsx.

Datta, A. (2008), "The Incidence of Fuel Taxation in India", Indian Statistical Institute, New Delhi, www.webmeets.com/files/papers/EAERE/2008/512/Incidence.pdf.

Davignon, E. (2016), "Restructuring of the European steel industry", presentation at the High-Level Meeting: Excess Capacity and Structural Adjustment in the Steel Sector, Palais d'Egmont, Brussels, 18 April 2016, www.oecd.org/industry/ind/Presentation-1-Davignon-Europe.pdf.

De Nederlandsche Bank (2016), "Time for transition: towards a carbon-neutral economy", De Nederlandsche Bank, Amsterdam, www.dnb.nl/en/binaries/tt_tcm47-338545.pdf?2017050504.

Dubey, K., N. Howarth and M. Krarti (2016) "Evaluating building energy efficiency investment options for Saudi Arabia", King Abdullah Petroleum Studies and Research Center (KAPSARC), Riyadh, www.kapsarc.org/wp-content/uploads/2016/10/KS-1655-DP049A-Evaluating-Building-Energy-Efficiency-Investment-Options-for-SA_web.pdf.

Durand-Lasserve, O. et al. (2015), "Modelling of distributional impacts of energy subsidy reforms: an illustration with Indonesia", OECD Publishing, Paris. http://dx.doi.org/10.1787/5js4k0scrqq5-en

EPH (2016), *Annual Report 2015, Energetický a průmyslový holding (EPH), Prague*, www.epholding.cz/wp-content/uploads/EPH_Vyrocni_Zprava_2015_repre.pdf.

ETUC (2016), "Industrial Regions and Climate Policies : Towards a Just Transition ?" European Trade Union Confederation, www.etuc.org/issue/project-climate-change.

Fay, Marianne, et al. (2015). *Decarbonizing development: three steps to a zero-carbon future. World Bank Publications, Washington D.C.* https://openknowledge.worldbank.org/bitstream/handle/10986/21842/Decarbonizing_Development_Overview.pdf?sequence=4&isAllowed=y.

Flues, F. and A. Thomas (2015), "The distributional effects of energy taxes", OECD Publishing, Paris, http://dx.doi.org/10.1787/5js1qwkqqrbv-en.

Flues, F. and K. Van Dender (2017, forthcoming), "The impact of energy taxes on the affordability of domestic energy", OECD Publishing, Paris.

Government of Canada (2016), "The Government of Canada accelerates investment in clean electricity", 21 November 2016, website consulted on 5 May 2017, http://news.gc.ca/web/article-en.do?nid=1157989.

Government of the Netherlands (2017), Energy Agreement for Sustainable Growth, *website*, www.government.nl/topics/energy-policy/contents/energy-agreement-for-sustainable-growth. (accessed on 28 March 2017).

Guillaume, D., R. Zytek and M. R. Farzin (2011), "Iran: The Chronicles of the Subsidy Reform", International Monetary Fund, Washington DC, www.imf.org/external/pubs/ft/wp/2011/wp11167.pdf.

IEA (2017), Chapter 2 in *Perspectives for the Energy Transition – Investment Needs for a Low-Carbon Energy System, IEA/OECD Publishing, Paris, and IRENA, Abu Dhabi,* www.energiewende2017.com/wp-content/uploads/2017/03/Perspectives-for-the-Energy-Transition_WEB.pdf.

IEA (2016), *World Energy Outlook 2016, International Energy Agency/OECD Publishing, Paris. See also the associated database at* www.worldenergyoutlook.org/media/weowebsite/2015/Subsidies20122014.xlsx.

IEA (2014), *Capturing the Multiple Benefits of Energy Efficiency, IEA/OECD Publishing, Paris.*

ILO (2016), *A just transition to climate-resilient economies and societies: Issues and perspectives for the world of work, International Labour Organisation, Geneva,* www.ilo.org/wcmsp5/groups/public/---ed_emp/---gjp/documents/publication/wcms_536552.pdf.

ILO (2015), *Guidelines for a just transition towards environmentally sustainable economies and for all, International Labour Organisation, Geneva,* www.ilo.org/wcmsp5/groups/public/---ed_emp/---emp_ent/documents/publication/wcms_432859.pdf.

IMF (2017), IMF Worldwide Government Revenue Database, International Monetary Fund, Washington DC, http://data.imf.org/revenues, (accessed on 27 March 2017).IMF (2016), *Economic Diversification in Oil-Exporting Arab Countries, Annual Meeting of Arab Ministers of Finance, April 2016, Manama, Bahrein, International Monetary Fund, Washington DC,* www.imf.org/external/np/pp/eng/2016/042916.pdf.

ITUC (2010), "Resolution on combating climate change through sustainable development and just transition", International Trade Union Confederation, Brussels, www.ituc-csi.org/resolution-on-combating-climate?lang=en.

Jaramillo, M. (2014), "The Incidence of Social Spending and Taxes in Peru", *Public Finance Review, Vol. 42/3, pp. 391-412, Sage, Los Angeles,* https://doi.org/10.1177/1091142113496134.

Journal Officiel (2015), "Décret no 2015-1850 du 29 décembre 2015 pris en application de l'article L. 533-22-1 du code monétaire et financier", Direction de l'information légale et administrative, Paris, www.legifrance.gouv.fr/eli/decret/2015/12/29/2015-1850/jo/texte.

Klenert, D. and L. Mattauch (2016), "How to make a carbon tax reform progressive: The role of subsistence consumption", *Economics Letters*, Vol. 138, Elsevier, Amsterdam, pp. 100-103, https://doi.org/10.1016/j.econlet.2015.11.019.

Kosonen, K. (2012), "Regressivity of environmental taxation: myth or reality?", European Commission, Brussels, https://ec.europa.eu/taxation_customs/sites/taxation/files/docs/body/taxation_paper_32_en.pdf.

Lee, J. (2017), "Why Saudi Arabia may unravel OPEC's big deal", Bloomberg News, 22 January 2017, www.bloomberg.com/gadfly/articles/2017-01-22/why-saudi-arabia-may-walk-away-from-opec-deal-by-june.

Liu, S. and W. Xu (2015), "Approaches and Suggestions on Output Reduction, Transformation, and Improvement of Quality and Efficiency for Industries with Excess Overcapacity", Development Research Center of the State Council, http://en.drc.gov.cn/2015-08/17/content_21625495.htm.

Mahdi, M., G. Carey and V. Nereim (2017), "Saudi Arabia plans payouts to offset subsidy cuts in 2017", *Bloomberg News, 21 December 2016,* www.bloomberg.com/news/articles/2016-12-21/saudi-arabia-said-to-plan-payouts-to-offset-subsidy-cuts-in-2017-iwysm4vh.

Mazurkiewicz-Gorgol, A. and A. Bomhoff (2009), "Restructuring of Gdansk Shipyard — outlook good for return to viability after four-year investigation", *Competition Policy Newsletter, 2009, Vol. 3, Competition Directorate-General of the European Commission, Brussels,* http://ec.europa.eu/competition/publications/cpn/2009_3_13.pdf.

Moglen D., J. Peek (2016), "Diablo Canyon nuclear plant to be shut down, power replaced by renewables, efficiency, storage". IWW Environmental Unionism Caucus, 21 June 2016, website consulted on 5 May 2017. https://ecology.iww.org/taxonomy/term/189.

Newell, P., J. Phillips and D. Mulvaney (2011), "Pursuing Clean Energy Equitably", United Nations Development Programme, New York, http://hdr.undp.org/sites/default/files/hdrp_2011_03.pdf.

OECD (2014), "Addressing social implications of green growth: Energy sector reform and its impacts on households", Issue Note for 2014 Green Growth and Sustainable Development Forum, OECD, Paris. www.oecd.org/greengrowth/ggsd-2014.htm.

OECD (2005), "Sectoral Case Studies", in *Trade and Structural Adjustment: Embracing Globalisation, OECD Publishing, Paris,* http://dx.doi.org/10.1787/9789264010970-3-en.

OECD-DAC (2005), *Lessons Learned on the Use of Power and Drivers of Change Analyses in Development Co-operation, OECD Development Assistance Committee (OECD-DAC) Network on Governance (GOVNET), Paris,* www.gsdrc.org/docs/open/doc82.pdf.

OECD/IEA/NEA/ITF (2015), *Aligning Policies for a Low-carbon Economy, OECD Publishing, Paris,* http://oe.cd/lowcarbon.

Office for National Statistics (2017), *UK public sector finances: December 2016, ONS, London,* www.ons.gov.uk/releases/ukpublicsectorfinancesdec2016.

Ogarenko, I. and I. Gerasimchuk (2016), "Winter approaches: The real test for Ukraine's energy subsidy reforms", Global Subsidies Initiative, Geneva, www.iisd.org/gsi/news/winter-approaches-ukraine-subsidy-reforms.

Pargal, S. and S.G. Banerjee (2014), *More Power to India: The Challenge of Electricity Distribution, World Bank, Washington DC,* http://dx.doi.org/10.1596/978-1-4648-0233-1.

PG&E et. al. (2016), *Joint Proposal of Pacific Gas and Electric Company, Friends of the Earth,Natural Resources Defence Council, Environment California, International Brotherhood of Electrical Workers Local 1245, Coalition of California Utility Employees and Alliance for Nuclear Responsibility to retire Diablo Canyon nuclear power plant at expiration of the current operating licenses and replace it with a portfolio of GHG free resources.* PG&E Website, consulted on May 5, 2017. https://www.pge.com/includes/docs/pdfs/safety/dcpp/JointProposal.pdf.

Porter, J. (2010), "Trade union responses to World Bank restructuring projects: The case of Transmilenio in Colombia", Public World, London, www.publicworld.org/files/colombiabrtenglish.pdf.

Repower Port Augusta (2017), "Solar and wind rising to the challenge", Repower Port Augusta, website, https://repowerportaugusta.org/thefacts, (accessed on 28 March 2017).

Reuters (2016), "Vattenfall sells German lignite assets to Czech EPH", Monday April 18, 2016, www.reuters.com/article/us-vattenfall-germany-lignite-idUSKCN0XF1DV (accessed 11 February 2017).

Rosemberg, A. (2010), "Building a Just Transition: The linkages between climate change and employment", *International Journal of Labour Research, Vol. 2/2, International Labour Office, Geneva, pp. 125-162.*

Saussay, A. et al. (2016), « Changer de *mix* : urgence et opportunité de la transition énergétique en France », Observatoire français des conjonctures économiques (OFCE), Paris, www.ofce.sciences-po.fr/pdf/pbrief/2016/pbrief08.pdf.

Schwarzkopff J. and S. Schulz (2015), "Structural change in Lusatia: What will come after lignite?", E3G, London, https://www.e3g.org/docs/E3G_Lausitzstudie_FINAL_EN.pdf.

Soares, F. V. et al. (2006). "Cash Transfer Programmes in Brazil: Impacts on Inequality and Poverty", *UNDP, New York,* www.ipc-undp.org/pub/IPCWorkingPaper21.pdf.

Sterner T. (ed.) (2012), *Fuel taxes and the poor: The distributional effects of gasoline taxation and their implications for climate policy, RFF Press, Washington DC.*

Tamura, H. (2015), "Structural adjustment in the steel sector: Japan's experience", presentation at the 79th Session of the Steel Committee, Paris, 30 November-1 December 2015, www.oecd.org/sti/ind/Item8_JapanExperienceOnStructuralAdjustment_2015.pdf.

Tang, X. and H. Liao (2014), "Energy poverty and solid fuels use in rural China: Analysis based on national population census", Centre for Energy and Environmental Policy Research, Beijing institute of Technology, Beijing.

UNFCCC (2016), *Just transition of the workforce, and the creation of decent work and quality jobs, United Nations Framework Convention on Climate Change, Bonn,* http://unfccc.int/resource/docs/2016/tp/07.pdf.

US DOE (2017), *2017 U.S. Energy and Employment Report, United States Department of Energy, Washington DC,* https://energy.gov/downloads/2017-us-energy-and-employment-report.

Victoria Government (2017), "Landmark Scheme Gives Hazelwood Workers a Fresh Start", Premier of Victoria webside, www.premier.vic.gov.au/landmark-scheme-gives-hazelwood-workers-a-fresh-start (accessed on 27 March 2017).

Weiss, M. (2017), "Lessons learnt from German climate policy including the Climate Action Plan 2050", presentation at the Ad Hoc Seminar of Senior Officials on Climate Change, OECD, Paris, 10 March 2017.

Weyzig, F. et al. (2014), "The price of doing too little too late; the impact of the carbon bubble on the European financial system", Green New Deal Series, Greens/EFA Group in the European Parliament, Brussels, http://greennewdeal.eu/green-economy/publications/2011/carbon-bubble-the-price-of-doing-too-little-too-late.html.

World Bank (2017), World Bank Natural Resource Rents database, World Bank, Washington DC, http://data.worldbank.org/topic/energy-and-mining (accessed on 27 March 2017).

World Bank (2011), *The Changing Wealth of Nations: Measuring Sustainable Development in the New Millennium, World Bank, Washington DC*, http://hdl.handle.net/10986/2252.

Wyns T. and M. Axelson (2016), *Decarbonising Europe's energy intensive industries – The Final Frontier, Institute for European Studies, Vrije Universiteit Brussel, Belgium*, www.ies.be/node/3698.

Investing in Climate, Investing in Growth

Chapter 7

Mobilising financing for the transition

Meeting the objectives of the Paris Agreement will require reallocation of investment away from carbon-intensive assets and rapid scale-up of private investment in low-emission, climate-resilient infrastructure and technologies. This chapter describes major trends in private financing for infrastructure and the roles of different private actors and sources of finance. It then explores what is needed to mobilise private finance for the transition, including how to address factors hindering private investment, and the types of instruments and transaction enablers governments have at their disposal. It considers the role of specialised development banks and development finance institutions, and how greater transparency and signaling in the global financial system might improve its capacity to respond to opportunities arising from the transition, while strengthening resilience to climate risks.

Establishing pathways to reach the goals of the Paris Agreement will require large-scale private investment in low-emission infrastructure and innovative technologies, and reallocation of investment away from carbon-intensive assets. Infrastructure, new or old, will also need to be made resilient to climate impacts. Investment in emerging economies and developing countries also needs to be aligned with the Sustainable Development Goals (SDGs). It is crucial to ensure that the global financial system is able to support the necessary rapid scaling up of such investment.

While infrastructure development is ultimately funded by taxpayers or users, it may be financed by public authorities (national and/or subnational governments, development banks, or other public financial institutions) and/or the private sector (corporate or project finance). The mix of financing will depend on the country and stage of development, and on the type of infrastructure and investment model. Despite low interest rates, in advanced economies fiscal constraints combined with concerns about the efficiency of public sector investment have led to a reduction in the share of public funds allocated to infrastructure to around 40% compared with 60-65% in developing countries (NCE, 2016; Ahmad, 2015).

This chapter describes what is needed to mobilise private finance for the transition in G20 countries and beyond. It describes the major trends in private financing for infrastructure, and the roles of different private actors and sources of finance that could support low-emission, climate-resilient infrastructure, including innovative clean technologies. It also examines the major factors hindering and helping private investment, and sets out a range of instruments and transaction enablers that can be deployed to mobilise private finance. The potential role of specialised development banks and development finance institutions in financing infrastructure and mobilising private investment is also explored. Finally, the chapter discusses potential misalignments in the global financial system and shows how transparency and signalling can improve its capacity to respond to opportunities arising from the transition, while strengthening the system's resilience to climate risks.

The shifting global landscape of private finance for low-emission infrastructure

Despite the favourable investment environment in global capital markets, access to long-term finance remains constrained for some infrastructure projects, particularly those in developing countries. To scale up private investment in infrastructure, it is vital to reach a sound understanding of the changing roles of actors and sources of finance, and the differences between traditional actors such as utilities and commercial banks, and non-traditional ones such as institutional investors and capital markets. Understanding the appetites and needs of these actors in terms of risk and liquidity in investments, and their capacity to finance potentially complex and large-scale infrastructure assets, is key to delivering the financing required.

Trends in the private financing of infrastructure

Private financing of infrastructure has undergone a major shift towards low-emission energy sources in the last decade.[1, 2] Overall infrastructure financing levels remained stable between 2010 and 2016, supported by ample liquidity in financial markets and high demand from the private sector. This included project-based primary finance (i.e. financing associated with "greenfield projects" – new activity in new assets) and secondary market transactions. Out of total finance from 2010 to 2016 of about USD 2.6 trillion, the lion's share went to the energy sector and finance for renewables accounted for 50% (USD 1.3 trillion) (IEA, 2016). A further 25% consisted of non-renewable power generation and support for transmission and distribution, while 23% went to the transport sector. Water received more limited private finance at 2% of the total, reflecting the predominant use of public financing in this sector. Europe accounted for 35% of the total, North America 25%, and Asia-Pacific 29%.

Despite the stable total volumes of financing for infrastructure, in recent years fewer new projects have secured primary financing. Primary financing for infrastructure declined for all regions from USD 226 billion in 2010 to USD 153 billion in 2016. For renewable energy projects, however, primary financing increased.

In contrast to the overall drop in primary financing, a large secondary market for infrastructure is developing, boosted by mergers and acquisitions (M&A) and refinancing activity (Figure 7.1). Global infrastructure M&A activity doubled from a low of USD 83 billion in 2012 to a record USD 179 billion in 2016, with the United States and China leading in terms of volumes.[3] Meanwhile, low interest rates and the abundance of liquidity in financial markets encouraged refinancing, which more than doubled from USD 43 billion in 2010 to USD 92 billion in 2015, before declining in 2016 due to a slow-down in activity, particularly in Asia.

While refinancing does not lead to additional investments, it can lower overall costs for users and governments, potentially freeing up fiscal space. Secondary markets also provide opportunities for investors, in particular for institutional investors, who represent a growing source of finance. For example, a project moving from construction to operation could refinance both debt and equity investments, accessing capital markets through bond issues, syndicated loans, and direct or indirect equity investment. In fact, much of this secondary market activity is fuelled by increased investor appetite for operational projects. For example, in the renewable energy sector, the increase of equity provision by institutional investors can be traced mainly to the acquisition of operational assets or portfolios for onshore wind deals (OECD, 2016a). Pension funds and insurers are less involved in greenfield onshore wind-power transactions, suggesting that institutional investors look to the onshore wind sector mainly for the acquisition of existing projects in the operational phase.

Figure 7.1. **Infrastructure finance by type, 2010-16**

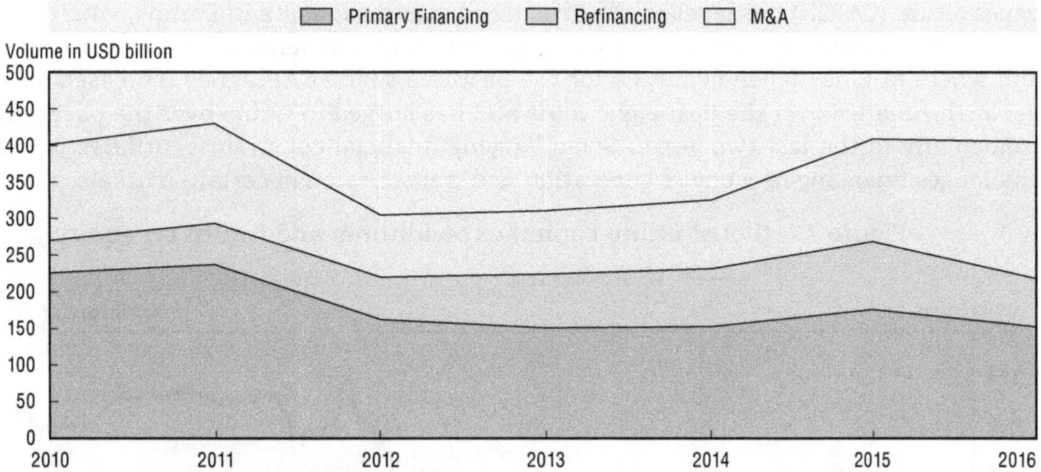

Sources: IJGlobal Transactions and Bloomberg New Energy Finance. OECD calculations.
StatLink http://dx.doi.org/10.1787/888933491468

Improving access to finance and lowering the costs of investment for infrastructure can make a substantial difference to the economic viability of projects and to the affordability of the services provided, in particular for the poor. Investors have a strong preference for operating assets in advanced countries; 76% of the secondary market transactions analysed is located in Europe and North America. This concentration of interest has contributed to competitive debt pricing and to a high valuation of operating assets located in developed economies.

In developing and emerging economies, however, infrastructure investments are often unable to attract long-term financing, as financing conditions such as shorter maturities and higher margins increase refinancing risks and make projects unviable. A recent IMF survey suggests funding and absorptive capacity constraints are a common impediment to scaling up infrastructure investment across low-income countries. Availability of external finance and administrative capacity were seen as key barriers in fragile states, while availability of domestic resources and concerns about debt accumulation were most important for low-income economies (IMF, 2017).

Key sources of finance for low-emission infrastructure

Utilities: pressure for new business models

Utilities, whether state-owned or private, have traditionally been an important source of finance for power generation, transmission and distribution through their balance sheets. Due to disruptions in energy markets, however, utilities' business models are facing challenges and may require adaptation. Across the G20, energy markets are not delivering the right price signals, while emissions markets and carbon-pricing mechanisms need significant adjustment (see Chapter 5). The growth of low marginal cost sources of energy (renewables), which tend to displace high variable cost energy generation (fossil fuels), combined with the rising share of decentralised generation in the electricity generation mix, are disrupting markets. As a result, wholesale electricity prices are flat or declining in Germany, the United Kingdom and the United States (IEA, 2016). In addition, "shadow costs" are mounting in the form of necessary power grid upgrades and mechanisms that balance demand and cope with intermittency.

As a result of these disruptions, the profitability of utilities in some countries and their ability to finance new capital expenditures are facing difficulties. Utility companies saw a global surge in investment in the mid-2000s until the global financial crisis.[4] This coincided with an increase in profitability as measured by return on equity (ROE). Since then, ROE has declined to below long-term averages for an extended period of time, while global capital expenditure (CAPEX) has levelled off. This trend was most acute in Europe, where CAPEX increased dramatically before the sovereign debt crisis and fell substantially afterwards, and where ROE has fallen below 5% for the past two years. CAPEX has increased steadily in North America over the past eight years and has surged in China over the past decade, particularly in the last two years. Given varying financial constraints, utilities may face challenges financing new power generation and transmission in certain markets.

Figure 7.2. **Global utility capital expenditures and return on equity**

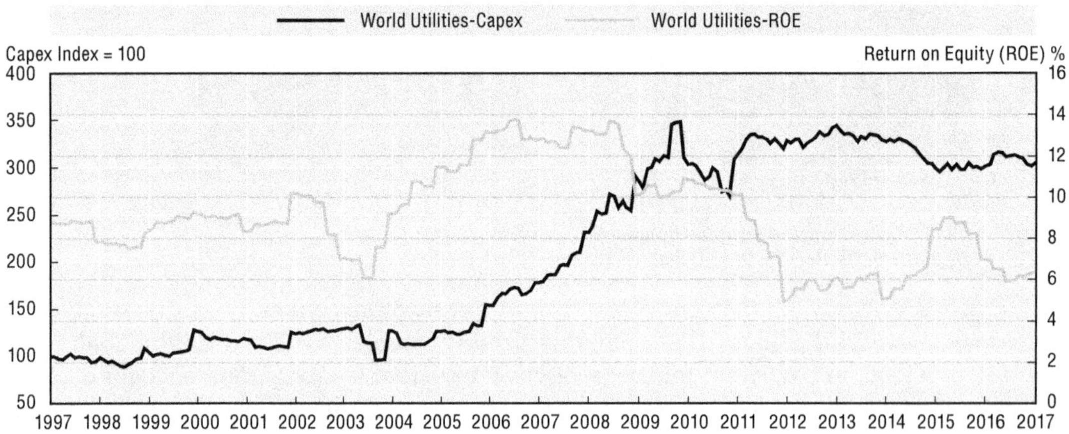

Source: OECD calculations based on Thomson Reuters Datastream.

StatLink http://dx.doi.org/10.1787/888933491479

The changing landscape of competition within electricity markets is driving a shift towards creative business models. Where the ability to self-finance new investment is limited, for example, utilities could act as construction companies, building projects through joint ventures or special purpose vehicles (SPVs) and project finance structures, benefiting from the trend in secondary markets that is opening new channels of finance for renewable energy projects.

Corporations: increasingly active in the transition

Corporations have provided finance in the renewables sector as off-takers through direct investments, power purchase agreements (PPAs), pure equity investments[5] and partnership models. Behind this flow of finance is a trend among large corporations to lower the emissions profile of their core business operations, or to promote renewable energy deployment more generally. Sixty-five global corporations have committed to 100% renewable energy for their future business operations (DLA Piper, 2016). Among firms investing for business reasons, criteria for investment consist mainly of the cost competitiveness of the energy source, but also its proximity to the end-use location and its contribution to the penetration rate of renewables.

In 2015, over 5 GW of renewable capacity was transacted through corporate PPAs in the global market, with the United States representing the largest share. Corporations such as Amazon, Google, Microsoft, Facebook, Norsk Hydro, and Wal-Mart were some of the biggest off-takers in 2016. This trend is also relevant in developing countries: by 2016, India had become the largest corporate procurement market in the Asia Pacific region (BNEF, 2016). As an example of a pure equity investment, Google, along with the Public Investment Corp (PIC) and the Development Bank of Southern Africa, financed the USD 230 million Jasper solar PV project in 2013 in South Africa. Regarding examples of partnerships, Google recently paired up with Eneco in the Netherlands and Duke Energy in the United States, both utilities; and Apple partnered with Goldwind in China, a maker of wind turbines.

Traditional oil and gas companies are also seeking to diversify their revenue sources away from fossil fuels and could be a growing source of finance for renewable energy. Statoil, the Norwegian oil and gas company, has applied its decades of expertise building offshore oil platforms to invest in floating wind turbines, which are cheaper than those fixed to the seafloor.

Banks are a crucial source of primary finance

Bank loans remain a critical source of finance for investment in new infrastructure projects and for the refinancing of operational assets, in both advanced and developing countries. In terms of overall volumes, bank lending remains the largest source of infrastructure finance in global markets. Whether through short-term corporate lending or non-recourse specialised lending, banks provide roughly 80% of green infrastructure finance, including for low-emission, climate-resilient infrastructure (UNEP, 2016a). In lower middle-income and low-income countries, state-owned banks and development banks play a bigger role due to the lack of well-developed commercial banking systems, and support the derisking of investment for greenfield projects.

Loans are flexible in that they can be paired with many different project finance structures, including public and private sponsorship models. Bank financing covers the spectrum of low-emission and climate-resilient infrastructure. In the aftermath of the financial crisis, more stringent banking regulations have caused banks to become more focused in their evaluation of infrastructure projects, leading to projects with higher financial stability over the long run.

Bank lending may be constrained by sizeable exposures in one country, which requires an enhanced focus on country risk, the stability of the fiscal and regulatory framework, and the reliability of new technologies and revenue sources. Lending may also depend on the business model of infrastructure projects; banks prefer models where costs are supported by end-users and stable tariff structures. From a policy perspective, facilitating cross-border transactions would strengthen infrastructure lending, allowing banks to diversify their portfolios and infrastructure projects to raise funds globally. Investing abroad is usually more complicated, however, as larger information asymmetries lead to higher risks, which may hinder cross-border lending.

One form of bank lending that is particularly important in infrastructure finance is the formation of loan syndications, which consist of a group of banks headed by one or more Mandated Lead Arranger that organises the financing package for a single borrower. Global syndicated loan volumes for infrastructure topped USD 1 trillion in 2015 (Figure 7.3), most of which originated in advanced economies; syndicated lending in emerging markets reached its highest level in 2015 in comparison with the earlier six years.

Figure 7.3. **Global syndicated infrastructure loan volumes, 2010-2015**

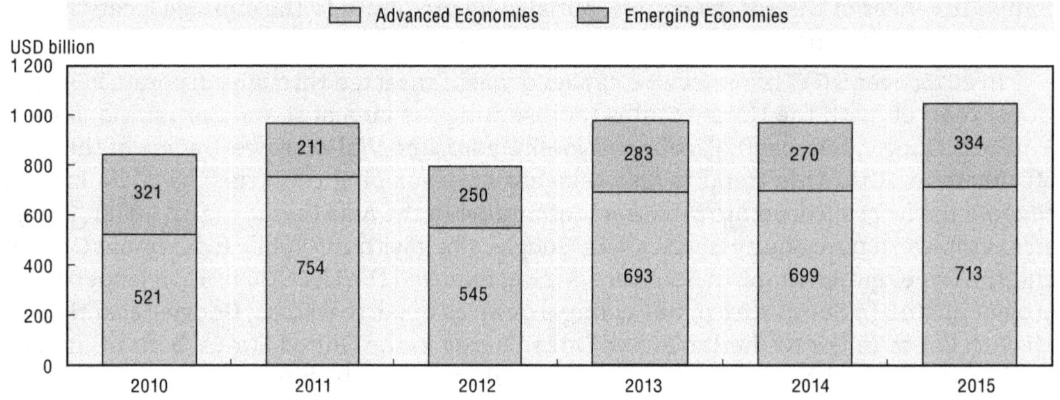

Source: OECD calculations based on Thomson OneBanker, IJGlobal Transactions.

StatLink http://dx.doi.org/10.1787/888933491484

Banks are a particularly crucial source of finance for small-scale investments. For example, bank loans are essential for investment in energy efficiency and renewable energy (such as small-scale solar), both for consumer applications and larger-scale commercial projects. Policy reforms that make energy efficiency profitable and create demand will scale up lending through such channels. On the supply side, governments could work with the banking sector to create specialised financial products designed for energy efficiency and decentralised power generation projects.

Institutional investors: so far, low levels of investment, but immense potential sources of finance

As part of the overall trend to diversify investment portfolios, institutional investors – including pension funds and insurance companies – have been increasingly interested in infrastructure investments. With USD 56.6 trillion in assets under management at the end of 2014 just in pension funds and insurance company portfolios, institutional investors represent a large potential source of investment (OECD, 2015a).

Pension funds and life insurance companies with long-dated liabilities – in particular defined benefit (DB) pension plans and life insurers offering long-term fixed guarantees – seek to match these liabilities with long-term assets, and are thus a potential source of long-term finance for low-emission infrastructure. Despite similar liability profiles, life insurers have less scope than DB pension funds to invest in illiquid assets because policy holders are able to withdraw funds, which can put pressures on insurer liquidity. By contrast, property and casualty insurers tend to have a short liability profile. The differing profiles of institutional investors highlight the need for expanded and diversified channels of financing for low-carbon infrastructure, which can cater to different investment horizons, risk appetites, liquidity needs, and capacities to invest in potentially complex and large-scale projects.

Pension fund demand for investment in illiquid unlisted infrastructure equity markets has increased over the past five years. Despite this strong demand, the pension funds[6] that reported their unlisted infrastructure equity allocation in a recent OECD survey have only increased this allocation slowly, occupying around 3.5% of portfolios, on average, in 2014. At the same time, many funds reported that they were below their investment targets for infrastructure. This suggests that funds have some capacity to increase their investment in unlisted infrastructure equity, including in climate-relevant sectors. These figures only represent those funds that reported investments in infrastructure in the survey – a considerable number of pension funds globally do not currently invest in infrastructure, or do not treat it as an asset class. When considering assets invested in infrastructure across the whole survey population, just 1% is allocated to infrastructure.

By investing in infrastructure projects through funds that invest in infrastructure assets, or indirectly through public equity markets and debt instruments, some pension funds already have exposure to relevant infrastructure investments, depending upon the availability of financial instruments and channels of investment.[7] Notably, some funds reported exposure to renewable energy assets, and most funds reported high investment allocations in transportation and energy sectors (Figure 7.4). Of the 26 pension and reserve funds that reported sector allocations in their infrastructure portfolios, nine reported exposure to renewable electricity. Renewables investments were concentrated in pension funds based in Europe, while funds based in North America and Latin America had low or no reported investments in renewables. Based on these data, evidence is scant that investors have broadly lowered the carbon footprints of their infrastructure portfolios.

Figure 7.4. **Infrastructure allocations by sector at the end of 2014 (%)**

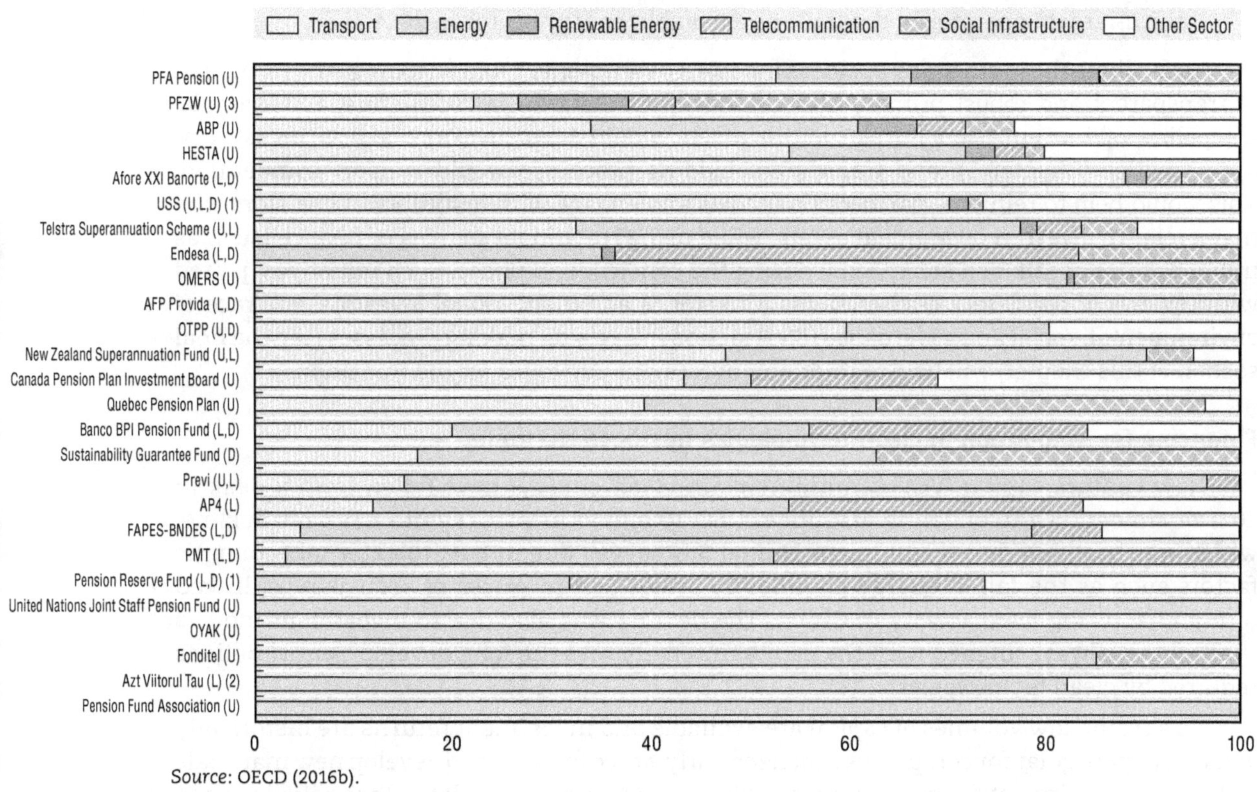

Source: OECD (2016b).

StatLink http://dx.doi.org/10.1787/888933491498

Partnerships combine traditional finance with new business models and non-traditional sources of finance

Engaging institutional investors through the formation of partnerships is a key way for governments to mobilise finance for the transition to a low-carbon economy. As an example, in a recent financing partnership institutional investors invested (pari passu to the lead lender) in infrastructure through project finance loans along with commercial banks, acting as a sponsor and taking advantage of the bank's origination resources and experience. This approach represents an increasingly important financing model, as individual banks may not be able to finance the entire debt portion of a project. Using such a mechanism, construction risk is no longer a barrier for institutional investors in European assets and mature technologies, as investing alongside lead lenders with existing due diligence and monitoring capabilities builds a higher level of comfort.

The asset management industry is crucial to help finance the transition, as not all investors are able to make direct investments in, or perform due diligence on, infrastructure assets. For example, the proliferation of private debt and equity investment strategies geared toward institutional investors such as pension funds and insurers has opened the possibility for many different types of investors to make small allocations while building necessary diversification across a range of projects, geographies and asset risks. There are already examples where asset managers are investing in new projects during the construction phase, and partnering with traditional finance sources such as banks or utilities:

- The French utility EDF and the asset manager Amundi partnered in 2014 to create a joint asset management company that will finance energy-related projects. This partnership sought to raise financing for renewable electricity generation and energy efficiency projects.
- In Italy, the utilities EDF and Edison and the infrastructure fund F2i established the third-largest operator in the Italian renewable energy sector in 2014.

To facilitate these emerging business models, governments could encourage the formation of transparent and robust primary and secondary markets for infrastructure. For example, a pipeline of low-emission infrastructure assets would help investors assess investment opportunities (Chapter 3), and transparent bidding processes would support competition. This would help to reinforce partnership models and engage with institutional investors who may prefer to invest in operational assets, while using traditional sources of finance (such as utility balance sheets and bank loans) during the construction phase. Such initiatives, paired with reviews of regulatory environments (Chapter 5) and institutional investors' embrace of environmental, social and governance (ESG) factors in decision making processes (see the final section of this chapter) will help to re-orient investment portfolios for the transition.

Financing for innovation in clean technologies has been declining

New venture capital (VC)[8] commitments in clean energy[9] have fallen from a peak in 2011 of almost USD 16 billion to just under USD 10 billion in 2016 (Figure 7.5). This decline, which was particularly evident in the United States was due in part to external economic factors such as the fall in energy commodity prices and in prices of solar panels, linked to manufacturing overcapacity in China. The decline was also due to investment models that are not always aligned with the capital intensity and the long development timelines required by clean technologies.

As a result, low volumes of capital are available and investment returns are historically volatile, in particular for companies that need early-stage financing to develop new materials or hardware for industrial or consumer applications,[10] or energy-related technologies such as biofuels and batteries (Gaddy, Sivaram and O'Sullivan, 2016). Attracting financing for clean technology in developing countries can be even more challenging, given a scarcity of long-term capital and country-specific risks. However, USD 2.5 trillion in assets are under management in global private equity and venture capital markets,[11] so there is a significant amount of capital to unlock for the transition.

Figure 7.5. **Global clean energy venture capital and private equity volumes by region, 2009-16**

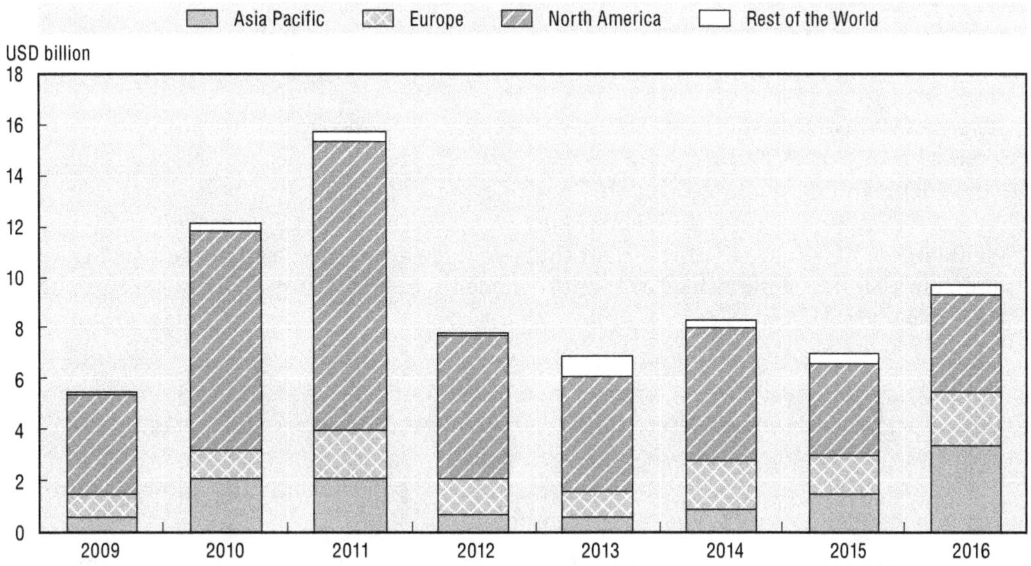

Source: CEP (2016).

As well as reducing their financing of clean technology, VCs have shifted investments from early-stage financing for hardware and materials to later-stage investments and less capital-intensive sectors such as software development and information technology, creating financing gaps within the clean technology value chain. For example, energy efficiency VC and private equity investment progressed from USD 1.2 billion in 2015 to USD 1.3 billion in 2016, whereas energy storage reached just USD 0.6 billion in 2016 (CEP, 2016). Energy storage is a critical area of investment that appears to be underserved by the VC model.

The problems in the existing VC model for the clean technology sector are reflected in the poor returns experienced by investors. Gaddy, Sivaram and O'Sullivan (2016) analysed the recent clean technology VC performance history, comparing the risk/return profile of clean technology investment with those of medical and software technologies. Clean technology companies were more likely to fail and yielded lower returns. Clean technology companies commercialising innovative science and engineering were especially unsuited to the VC investment model for four reasons. They were illiquid, tying up capital for longer than the three- to five-year time horizon preferred by VCs. They were expensive to scale up, often requiring hundreds of millions of dollars to build factories, even while the fundamental technology was still being developed. There was little room for error because these companies competed in commodity markets with thin margins – against cheap silicon solar panels or abundant oil and gas – making it difficult to invest in R&D while operating a lean manufacturing operation. Finally, the likely acquirers – utilities and industrial giants – were unlikely to acquire risky start-ups and were averse to paying a premium for future growth prospects.

New technologies beyond the clean energy sector have the potential to revolutionise the way that energy is generated and delivered to customers, as well as transform transportation (Box 7.1) and energy. In the future, innovative business models in the sharing and on-demand economy – based on services as opposed to hard assets – could make this change possible. In the transport sector, for example, the carbon footprint of urban road and other infrastructure could be much lower with a massive adoption of car-sharing models or driverless car technologies, while artificial intelligence could drastically lower the carbon intensity of the mining industry.

Box 7.1. **Innovation in financing low-carbon transportation**

The provision of transport services has historically relied on public funding and planning, resulting in a heavily regulated sector. However, recent research suggests that large-scale deployment of shared vehicle fleets, perhaps operated by private companies, could be viable. In a case study in Lisbon, Portugal, OECD/ITF (2016) assessed the large-scale deployment of a shared vehicle fleet that provides app-based and on-demand transport, replacing all other motorised transport modes while rail and subway services operate as today.

The results show that shared mobility would decrease congestion and reduce GHG emissions even with current internal combustion engines. At the same time, intensive per-vehicle use would accelerate fleet replacement and thus penetration of newer, cleaner technologies. This could provide an even quicker and more marked reduction of carbon emissions. Other benefits include a decrease in both pollution and the number of transfers, better accessibility and 95% less need for public parking.

The challenges for policy-makers lie in creating the right market conditions and operational frameworks. While a sudden change to a completely shared mobility system is not conceivable, gradual installation is plausible and would yield large benefits from the start. In addition, this deployment scenario would free up significant amounts of space in a city that would need to be managed to ensure the benefits were fully reaped. Management strategies could include allocating space to specified commercial or recreational uses, such as delivery bays, bicycle tracks or enlarged footpaths. Freed-up space in off-street parking could be used for urban logistics purposes, such as distribution centres.

Source: OECD/ITF (2016).

The role of development banks and development finance institutions

Considering the scale of financing required, development banks and development finance institutions will be essential in helping countries to deliver on their Nationally Determined Contributions (NDCs), both within and outside the G20. As publicly owned or controlled institutions with a development mandate, these banks already play a role in infrastructure financing, especially in developing countries. Between 2010 and 2015, development and state-owned banks contributed around 21% of primary financing for privately financed infrastructure projects in developing economies (Figure 7.6). This role can be further strengthened to help countries shift investment for low-emission, climate-resilient infrastructure from "billions to trillions": by developing infrastructure pipelines, by investing in new greenfield projects and by de-risking infrastructure investment and mobilising private investors.

Figure 7.6. **Share of development banks and state-owned banks in privately financed infrastructure, power and transport sectors, 2010-16**

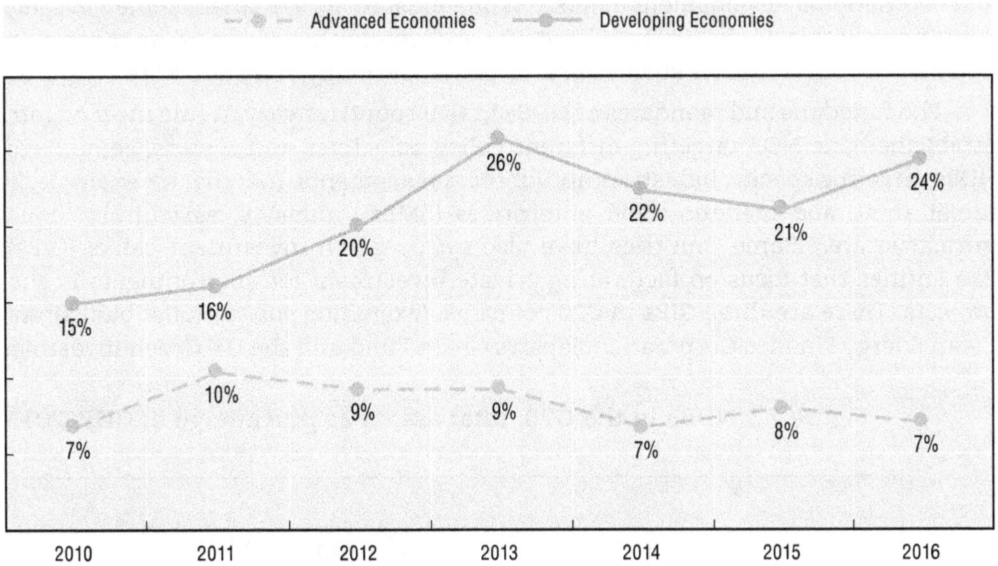

Note: This graph is based on figures for primary financing (i.e. financing associated with "greenfield projects" – new activity in new assets) in power (excluding renewables) and transport sectors.

Source: IJGlobal Transactions. OECD calculations, based on IMF categorisation of advanced and developing economies.

StatLink http://dx.doi.org/10.1787/888933491500

Three types of development finance actors are worthy of attention: national development banks (NDBs), multilateral development banks (MDBs), and bilateral development banks and development finance institutions (DFIs).[12] Each type of institution has a complementary role to play, and their collective potential impact lies in their working collaboratively, building on the comparative advantages of each. Because NDBs work within a domestic context, they can be well integrated into national infrastructure policy and planning frameworks. They also have relationships with private companies operating in the local market and can supply adequate long-term financing in local currency. MDBs and bilateral DFIs are backed by strong credit ratings and the support of their shareholders. They can leverage significant capital and bring knowledge, expertise and innovation, based on broad experience elsewhere, to spur investment in climate-friendly infrastructure.[13]

To fulfil their potential, development banks and finance institutions – national, multilateral and bilateral – will need to scale up efforts to mobilise private capital, and ensure that infrastructure portfolios are aligned with low-emission, climate-resilient development pathways. Governments, which are the major shareholders and clients of development banks and finance institutions, need to encourage and enable them to fulfil this role.

National development banks

NDBs are important actors in financing public policy objectives in G20 countries

National development banks are prevalent across advanced and emerging economies alike. There are more than 250 NDBs worldwide, with assets of over USD 5 trillion, roughly three times the total assets of all major MDBs combined (Studart and Gallagher, 2016). As publicly owned domestic finance institutions with an overarching development mandate, NDBs support the policies and strategies of their governments. Among G20 countries, while only a few NDBs have mandates that focus explicitly on infrastructure, there are at least 21 NDBs with relevance to infrastructure financing,[14] excluding agriculture-related and sub-national development banks.[15] While most NDBs are small, some have substantial operations within their domestic contexts and assets that make up as much as a fifth of national GDP, in countries such as Brazil, China, Germany, Italy and Korea (Figure 7.7).

The functions and mandates of NDBs in G20 countries vary. While most countries have established one NDB targeting different policy objectives and sectors, some have several NDBs targeting specific industries and/or market segments. In India, for example, five NDBs target small and medium-sized enterprises (SMEs), industry, agriculture, housing and infrastructure.[16] Some countries have also set up green investment banks (GIBs),[17] NDB-like entities that focus on facilitating private investment for environmentally sustainable projects. There are three GIBs in G20 countries (excluding sub-national banks): Australia's Clean Energy Finance Corporation, Japan's Green Fund and the UK Green Investment Bank.

Figure 7.7. **NDBs in the G20, total assets as percentage of GDP, 2015**

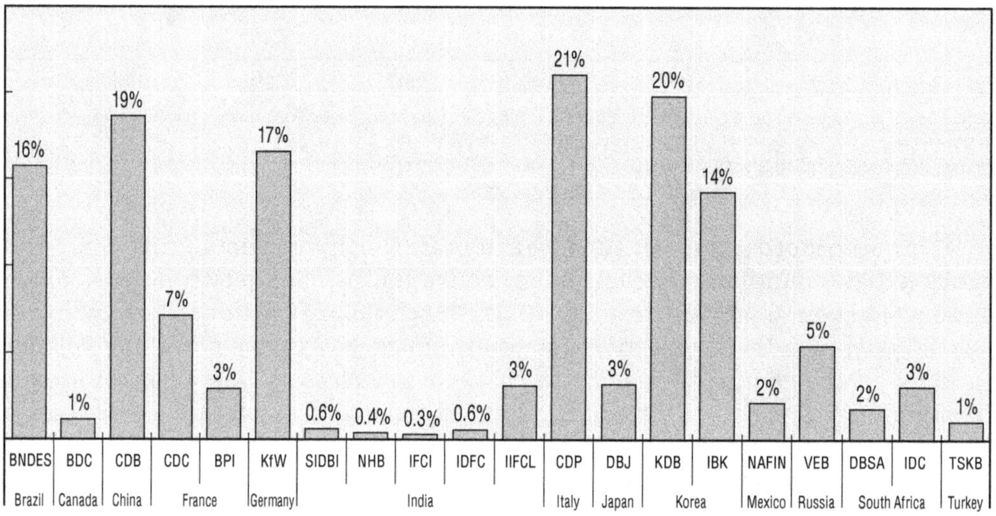

Note: Brazil Development Bank (BNDES); Business Development Bank of Canada (BDC); China Development Bank (CDB); Caisse des dépôts et consignations (CDC); Banque Publique d'Investissement (BPI); Kreditanstalt für Wiederaufbau (KfW); Small Industries Development Bank of India (SIDBI), National Housing Bank (NHB), Industrial Finance Corporation of India (IFCI); Infrastructure Development Finance Company (IDFC); India Infrastructure Finance Company Ltd (IIFCL); Cassa Depositi e Prestiti (CDP), Development Bank of Japan (DBJ); Korea Development Bank (KDB); Industrial Development Bank of Korea (IBK); Nacional Financiera (NAFIN); Vnesheconombank (VEB); Development Bank of Southern Africa (DBSA); South African Industrial Development Corporation (IDC); Industrial Development Bank of Turkey (TSKB). BANOBRAS (Mexico) has been excluded from this figure due to lack of data.

Source: Authors, based on data from 2015 annual reports (from institutions listed) and World Bank data on GDP in current prices and local currency.

StatLink http://dx.doi.org/10.1787/888933484352

It is difficult to determine the share of NDBs' portfolios that targets infrastructure due to the lack of data and information on what these institutions finance. Out of the NDBs in G20 countries that publish sectoral breakdowns of their portfolios, the share of infrastructure financing varies from over 50% to only a minor share. NDBs with larger shares of infrastructure in their portfolios include China Development Bank (CDB), Development Bank of South Africa (DBSA) and India Infrastructure Finance Company Limited (IIFCL). Green investment banks such as Australia's Clean Energy Finance Corporation and the UK Green Investment Bank also focus on financing infrastructure.

Figure 7.8. **Percentage of NDBs' portfolio in infrastructure, selected NDBs (2015)**

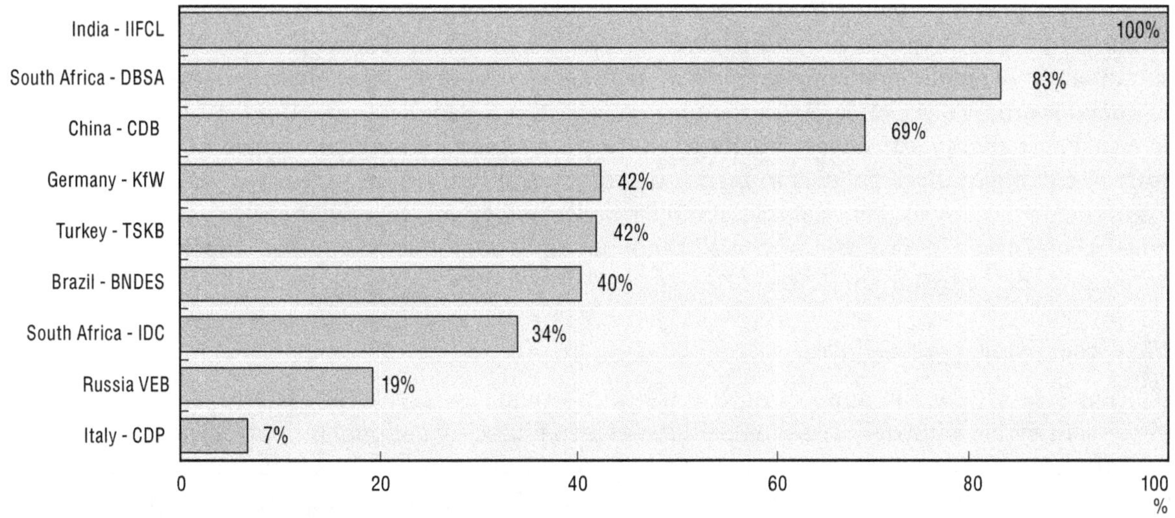

Notes: Comparing the sectoral distribution of NDBs' portfolio is not an exact science, as categories to describe sectors differ from one NDB to another. However, it gives a general sense of priorities. In this figure, portfolio can mean loan portfolio year-to-date, outstanding loan portfolio or disbursements. The following assumptions were used to approximate the infrastructure share of the loan portfolio. For CDB, infrastructure includes railways, public highways, electric power, public infrastructure, petroleum and petrochemical and urban renewal. For TSKB, infrastructure includes electricity production and electricity-gas transmission, but not logistics. For KfW, infrastructure includes housing investment and infrastructure. Finally, for DBSA, infrastructure includes energy, roads and water and sanitation.

Source: Authors, based on data from annual reports of the institutions in 2015.

The extent to which NDBs finance low-emission, climate-resilient infrastructure varies

Some NDBs have made efforts to organise and track climate-related activities through the International Development Finance Club (IDFC), which includes several NDBs, both within and outside G20 countries, as well as some bilateral development finance providers.[18] There is no widely adopted standard for reporting climate-related activities, however, particularly in NDBs' annual and sustainability reports, so it is difficult to assess how much of each NDB's support for infrastructure is low-emission and climate resilient. IDFC members committed USD 98 billion in green finance[19] in 2014, and across members, the shares of green finance among total new commitments ranged from 100% (in one case) to under 5% or even 0% in a few cases (IDFC, 2014).

Several NDBs are taking a leading role, with substantial shares of their portfolios targeting this area. While lack of consistent data limits the extent to which comparisons can be made, a review of annual reports reveals examples of NDB activity related to climate change more broadly, and infrastructure more specifically.[20] Over half of Brazil Development Bank's (BNDES) disbursements for infrastructure in 2015 went to "green" infrastructure, including 27% to renewable energy (excluding hydro) and energy efficiency, and 24% for large hydro plants (above 30MW) (BNDES, 2016). In Germany, just under 40% of KfW's domestic commitments went towards environmental and climate-related goals in 2015, including support for the construction of energy-efficient housing as well as credit lines to help SMEs scale up activities related to environmental protection, renewable energy and energy efficiency (KfW, 2015a).[21]

Many NDBs also support financing for more carbon-intensive infrastructure, in line with national energy policies and priorities that may not yet be aligned with NDCs or the objectives of the Paris Agreement. For example, CDB, TSKB and DBSA support coal power generation. CDB has recently increased its focus on greening its portfolio: in 2014, total outstanding "green credits" were around five times larger than CDB's outstanding loans for coal-related power projects (CDB, 2014).[22] Similarly, DBSA and Industrial Development Bank of Turkey (TSKB) support coal and gas power plants as well as renewables. Nevertheless, sustainability-themed loans make up 50% of TSKB's loan portfolio, excluding loans to the financial sector (TSKB, 2015). As of the end to 2015, TSKB had disbursed USD 487.7 million to renewable energy and energy efficiency, 14% of its total loan portfolio. Some NDBs have started putting in place policies to favour investment in low-emission infrastructure over fossil fuel technologies. For example, BNDES's new strategy for the power sector prioritises renewable energy over coal- and oil-based power, and includes greater subsidies for renewable sources (Climate Home, 2016).

NDBs could play a greater role

The role of NDBs in supporting the transition requires further attention from policy makers. In many countries, these institutions' role in financing national infrastructure could be harnessed for supporting the implementation of NDCs. In South Africa, for example, the Industrial Development Corporation (IDC) and DBSA are financing the development of renewable energy projects as part of the Department of Energy's Renewable Energy Independent Power Producer Procurement Programme.

In addition, partly due to their ability to borrow and lend in local currency, NDBs can mobilise and crowd-in local private finance based on their special status within their countries (Smallridge et al., 2013). In India, for instance, NDBs not only have access to soft funds from the Reserve Bank of India, the country's central bank, they can also mobilise additional capital by issuing securities that qualify as reserves under India's Statutory Liquidity Ratio (Kumar, 2016).

For NDBs to deliver on their potential, governments need to strengthen their mandates to specifically support low-emission infrastructure. As a first step, NDBs should put in place strategies and targets for climate action, including support for low-emission, climate-resilient infrastructure, in line with national low-carbon strategies. Where such cases have occurred, the focus has been on greening existing NDB portfolios and practices (e.g. project screening and preparation), creating green infrastructure windows within NDBs and/or establishing new specialised institutions, such as separate green investment banks (OECD, 2017b). Governments could ensure that NDBs play a role in NDC implementation, and create frameworks to facilitate dialogue and exchange on policy issues between NDBs and government agencies.

To scale up NDBs' role, better reporting and disclosure of information about what NDBs finance is vital. IDFC members use a common methodology to report on "green finance" and efforts have been made to harmonise the tracking methodology with the approach used by MDBs. Despite this, individual NDB annual reports present information in a variety of formats, and there is no standard practice on reporting or disclosure across NDBs more broadly. Governments could encourage the use of comparable and consistent data as well as greater transparency among NDBs and support more detailed disclosure about what they finance, including their support for climate-friendly infrastructure.

Some NDBs have been found to suffer from performance-related problems, which need to be rectified for them to make a full contribution to mobilising investment. These include the risk of market distortions arising from picking winners or from crowding-out effects, as well as corruption, the potentially high opportunity costs associated with subsidised loans, and a lack of efficiency due to the small scale of projects undertaken (Torres and Zeidan, 2016).

International collaboration will enable NDBs to scale up support

The major barriers that NDBs face in scaling up financing for low-emission, climate-resilient infrastructure, particularly in developing countries, include a lack of capacity to mainstream climate objectives into their portfolios, as well as a lack of "readiness" to access international concessional climate finance. In recognition of their potential as well as the challenges, MDBs and bilateral providers work with NDBs to develop and deploy green financing, and support these efforts with capacity building and technical assistance. In Russia, support for "environmental stewardship and energy efficiency" made up 34% of Vnesheconombank's (VEB) portfolio in 2014, and VEB has partnered with the World Bank and the International Finance Corporation (IFC) to support energy efficiency activities. Similarly, many of TSKB's green finance activities have been initiated and continue in partnership with KfW, Agence Française de Développement (AFD), European Bank for Reconstruction and Development (EBRD) and European Investment Bank (EIB). Governments and development partners could encourage further partnerships and collaboration between NDBs, MDBs, and bilateral development banks and DFIs in the financing of low-carbon infrastructure, as well as supporting efforts to shore up institutional and staff capacity. Initiatives such as the NDC Partnership can encourage collaboration aimed at concrete action at national and sub-national levels.

Multilateral development banks

MDBs have made ambitious commitments to increase climate finance and support infrastructure

MDBs are widely recognised as critical providers of financial and technical assistance to developing countries to promote economic and social development. They already play a noteworthy role in mobilising international climate finance, and this role is likely to continue. Within the context of the UNFCCC and the commitment of developed countries to mobilise USD 100 billion to support climate action in developing countries, MDBs supported over a third of estimated flows of public climate finance in 2013-14, on average,[23] and are also estimated to have mobilised roughly 50% of private climate finance (OECD, 2015b). MDBs deliver climate finance by using their own resources and by managing climate finance trust funds for donor governments.

In the run up to COP21 in 2015, MDBs also made ambitious commitments to scale up climate financing in their portfolios (Table 7.1). Based on these commitments, MDBs will provide over 40% of international public climate finance flows in 2020 (OECD, 2016c). Several MDBs have also made efforts to co-ordinate action on climate change, ranging from jointly tracking and reporting climate finance to harmonising measurement of projects' GHG emissions impacts.

Table 7.1. **Climate finance from MDBs to developing countries: Current status and future targets**

	MDB targets to scale up climate action	Climate finance in 2015 (USD '000)	Share of climate finance in MDB portfolios in 2015[24]
AsDB	Double climate finance to USD 6 billion annually by 2020	2 917	15.3%
AfDB	Triple climate financing to reach 40 percent of investments by 2020	1 359	15.6%
EBRD	40% of annual business investment in green finance by 2020	3 217	25.5%
EIB	Global target of greater than 25 percent of all lending. Increased target of 35% of lending in developing countries by 2020	5 137	26.2%
IDB	Double climate finance to 30% of operational approvals by 2020, to an average USD 4 billion per annum	1 744	16.1%
WBG	Increase climate financing by one-third, from 21 percent to 28 percent of annual commitments by 2020.	10 722	17.9%

Note: AsDB: Asian Development Bank; AfDB: African Development bank; EBRD: European Bank for Reconstruction and Development; EIB: European Investment Bank; IDB: Inter-American Development Bank; WBG: World Bank Group.
Source: Adapted from 2015 Joint Report on Multilateral Development Banks.

Alongside climate finance commitments, MDBs have stated in the context of the G20[25] that they intend to increase infrastructure financing, in many cases at scales far greater than the targets shown in Table 7.1. It is important to ensure that infrastructure financing – in the MDB portfolios and more generally across all public finance – aligns with the Paris Agreement by helping developing countries to make the transition; otherwise, there will be a "lock in" of high-carbon development pathways. While it will take time for policy reforms to take hold that mobilise private financing, public financing should be shifted as soon as possible to uphold the legal commitments and long-term goals that countries adopted in the Paris Agreement. Governments – as shareholders and clients of MDBs – can call on MDBs to prepare roadmaps and climate action plans to support implementation of the banks' climate change targets, including support for low-carbon infrastructure. Many MDBs already have climate change strategies that can form the basis of road maps and action plans. The World Bank, for example, prepared an action plan in 2016 for how it will deliver on its climate change commitments.

MDBs could scale up efforts to mainstream low-carbon infrastructure support

While the importance of infrastructure in MDB portfolios has largely decreased since the 1950s[26] (Humphrey, 2015), infrastructure sectors – including energy, transport, water and communications – still remain a key aspect of MDB financing. From 2005 to 2014, MDB commitments to infrastructure doubled, growing at 10% per year on average (Miyamoto and Chiofalo, 2016). In 2014, MDBs' actual spending (i.e. disbursements) on infrastructure was around USD 31 billion. This was roughly half the support provided for infrastructure in developing countries by all bilateral and multilateral development partners reporting to the OECD Development Assistance Committee (OECD-DAC). Infrastructure is a substantial part of the overall portfolios of some MDBs, especially the Islamic Development Bank (IsDB) (85% in 2014), the African Development Bank (AfDB) (63%) and the Asian Development Bank (AsDB) (59%).

Roughly one-third of commitments by the largest MDBs[27] to infrastructure sectors in 2013-15 reported to OECD-DAC were climate-related, on average, ranging from 51.8% of EIB's infrastructure operations to 6.7% of IsDB's (Figure 7.9). Development finance for the energy sector shows the most alignment, while less mainstreaming is noticeable in the transport sector (Figure 7.10). This comparison is enabled by MDB efforts to harmonise their tracking and reporting on climate finance, both among MDBs and with other tracking systems, such as OECD-DAC.

Figure 7.9. **Share of MDB commitments for infrastructure which is climate-related and total MDB commitments for infrastructure, by institution (USD billion), 2013-15 average**

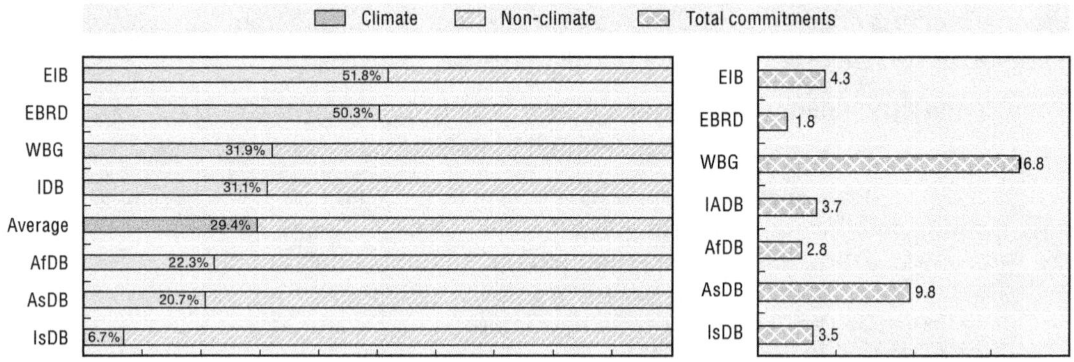

Notes: This graph is based on data reported to the OECD Development Assistance Committee by the following MDBs: the African Development Bank, the Asian Development Bank, the European Bank for Reconstruction and Development, the European Investment Bank, the Inter-American Development Bank, the Islamic Development Bank and the World Bank Group (WBG), which also includes the International Finance Corporation. Climate-related components of projects are those that target mitigation, adaptation, or both mitigation and adaptation, based on the joint MDB Climate Finance Tracking Methodology. MDB commitments include concessional and non-concessional support. Infrastructure sectors include transport, energy, water supply and sanitation, and communications.

Source: OECD-DAC statistical system.

StatLink http://dx.doi.org/10.1787/888933484367

Figure 7.10. **Share of MDB commitments for infrastructure which are climate-related and total MDB commitments for infrastructure, by sector (USD billion), 2013-15 average**

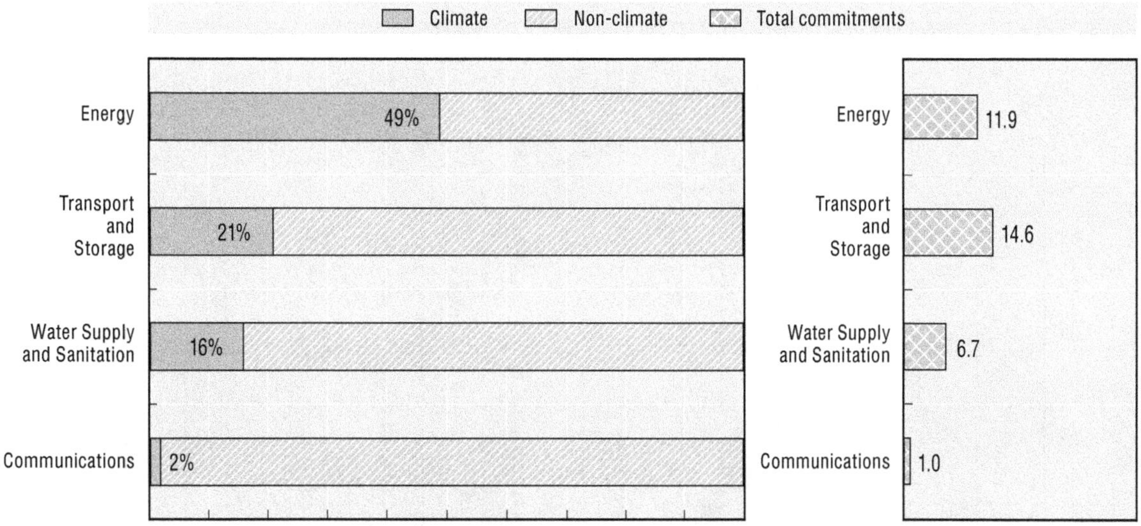

Notes: This graph is based on data reported to the OECD DAC by the following MDBs: African Development Bank (AfDB), the Asian Development Bank (AsDB), the European Bank for Reconstruction and Development (EBRD), the European Investment Bank (EIB), the Inter-American Development Bank (IDB), the Islamic Development Bank (IsDB) and the World Bank Group (WBG), which also includes the International Finance Corporation (IFC). Climate-related components of projects are those that target mitigation, adaptation, or both mitigation and adaptation, based on the joint MDB Climate Finance Tracking Methodology. MDB commitments include concessional and non-concessional support.

Source: OECD-DAC statistical system.

StatLink http://dx.doi.org/10.1787/888933484375

Analysis of MDB support for the energy sector shows there is room to scale up financing for low-emission, climate-resilient technologies and scale down support for carbon-intensive technologies. While the share of MDB support for renewable energy technologies (excluding hydropower) in overall commitments to power generation has grown significantly over the last decade, at a compound annual growth rate of around 13%, the share of MDB support for fossil fuels has kept pace, with a compound annual growth rate of 15.7% (Figure 7.11). At the same time, MDBs have increased financing for transmission lines, which is critical to enable energy access and reduce losses. A detailed review of energy sector pipelines for the World Bank, IFC and AsDB in 2015-16 conducted by the World Resources Institute shows that while very few projects in the near future would be considered misaligned with the goals of the Paris Agreement, the major share of projects could have either a positive or negative impact on GHG emissions, depending on how they are designed and implemented (Christianson et al., 2016). In addition to financing infrastructure, MDBs provide technical assistance and advisory services, and support capacity building in the energy sector, which helps developing countries strengthen policies and regulations critical for attracting private investment into low-emission, climate-resilient technologies (Chapter 5).

Figure 7.11. **Trends in commitments to renewable and non-renewable power generation, selected MDBs, by sub-sector, 2006-15**

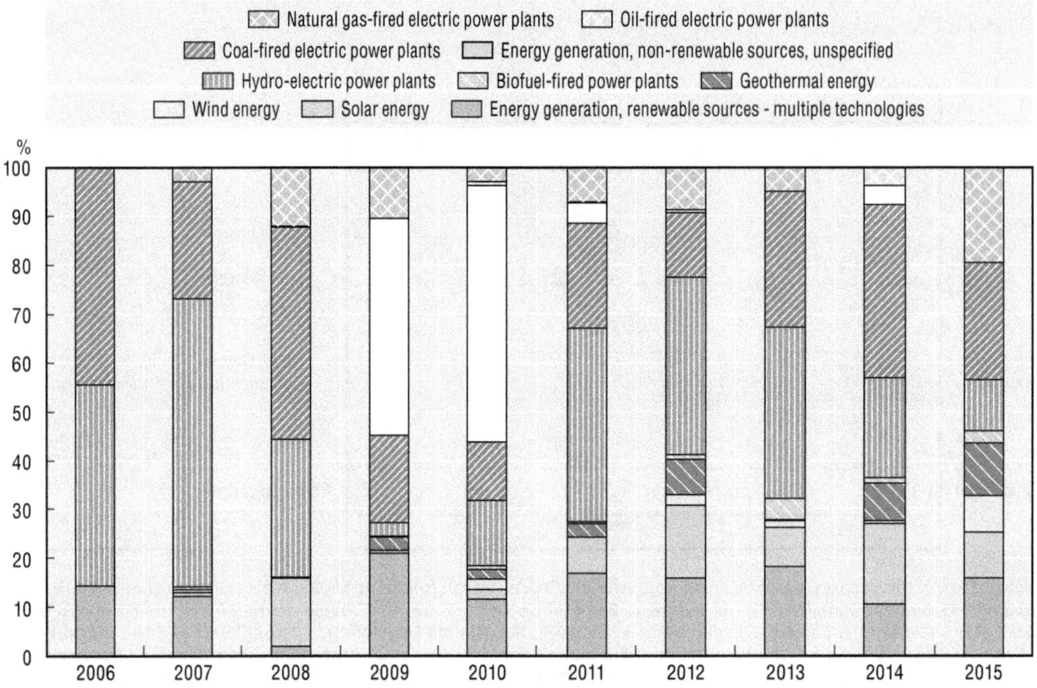

Notes: This graph is based on data reported to OECD-DAC by the following MDBs: African Development Bank (AfDB), the Asian Development Bank (AsDB), the Inter-American Development Bank (IDB), the Islamic Development Bank (IsDB) and the World Bank. Other MDBs are excluded from this analysis to avoid inconsistencies arising out of data gaps. MDB commitments include concessional and non-concessional support.

Source: OECD-DAC statistical system.

StatLink http://dx.doi.org/10.1787/888933484382

MDBs could mobilise additional investment for the transition

MDBs could scale up additional investment for low-emission, climate-resilient infrastructure by leveraging their shareholder base and mobilising more from capital markets. While MDB operations are limited by the capital allocated to them by shareholders, MDBs could make more efficient use of their existing capital, while retaining their existing credit ratings (Humphrey, 2015). In 2015, G20 governments initiated an action plan for MDBs to optimise their balance sheets so they can expand operations to meet the investment needs of delivering the SDGs.[28] The plan includes actions to improve capital efficiency, use concessional financing more innovatively (within prudential limits) and use risk mitigation more effectively to crowd-in private capital. Some banks have already taken steps in this direction. AsDB has combined its concessional and non-concessional windows, enabling greater leverage on its capital (Birdsall, Morris and Rueda-Sabater, 2015). As a result, AsDB could expand its operations by up to 50%, from USD 13 billion per year to USD 20 billion per year (AsDB, 2015). This additional financing is targeting low-income countries in the Asia Pacific region and could increasingly be targeted to low-emission, climate-resilient infrastructure. The Inter-American Development Bank (IDB) has taken similar steps, and the World Bank is introducing a new "private sector window" in its operations targeting low-income countries.[29]

MDBs could also scale up efforts to mobilise private capital by using well-designed, diversified risk mitigation instruments including guarantees, coverage of political and regulatory risks, credit enhancements, and more diversified insurance offerings (Table 7.3).

IFC, for example, has committed to a target of catalysing USD 13 billion of private sector support annually by 2020 by using de-risking and aggregation approaches in its climate change implementation plan (IFC, 2016). While private sector operations of the MDBs have focused significantly on renewable energy, as these technologies become commercially viable in many developing countries MDBs will need to focus on other areas in support of the transition.

Blended approaches – using MDB finance to mobilise private capital – can also be useful, especially to bridge viability gaps for investment, but should be governed by the standards and principles adhered to by development banks and DFIs to avoid crowding out the private sector. Such standards are in place among many institutions already; further harmonisation and co-ordination among these would be valuable. Better synergies between MDBs and the private sector would also be useful, including through co-financing facilities, insurance pools, a wider range of currency hedging tools, and investment platforms and partnerships where governments, local finance institutions and MDBs can co-invest alongside financial sponsors.

MDB support is influenced by country policy frameworks as well as access to targeted, concessional climate finance

MDBs could play a significant role in supporting countries to adopt low-carbon infrastructure choices, but this support is partly dependent on the availability of concessional climate finance. While the Paris Agreement has created momentum for countries to act on climate change, many countries have not yet centralised climate concerns within development policy and infrastructure planning, and it will take time for these enabling policy frameworks to take hold and have an impact. In the meantime, MDB support – especially efforts to mobilise private capital – is dependent on concessional support to bridge the viability gap for investments in low-carbon infrastructure and in climate-proofing infrastructure (Trabacchi et al., 2016). Concessional finance is particularly important where there are policies in place but technologies may not have enough of a track record to attract investment.

MDBs have increasingly been a channel of external concessional finance for climate change, for example through trust funds supported by bilateral donors. MDBs have also played the role of implementing entities for multilateral climate funds such as the Nordic Development Fund and Climate Investment Funds (CIFs), as well as the Global Environment Facility. Access to concessional finance has helped MDBs to gain experience and to begin climate-proofing their own investments, and could even be a pre-condition for MDBs to meet their climate targets. The World Bank's climate target, for example, is conditional in part on access to concessional finance (World Bank, 2016). Changes in the global climate finance architecture, including the scaling down of the CIFs[30] and the initiation of the newer Green Climate Fund, will have implications for how MDBs can scale up operations and meet their ambitious commitments to provide climate finance support for developing countries.

MDBs are facing demands to help countries cope with a wide range of other global challenges in addition to climate action. The World Bank, for example, identifies competition for support on social protection and budget support (for other issues) as a key risk to demand for climate support (World Bank, 2016). Through technical assistance, knowledge sharing and advice, and demonstration and piloting, MDBs can help raise awareness of stakeholders in borrowing countries, particularly in infrastructure sector ministries, and highlight the win-win cases for climate and development. These can focus on projects that have clear mutual benefits for development and for climate action.

Bilateral development banks and development finance institutions

G20 countries play a significant global role in cross-border development finance, providing 77% of concessional development finance (ODA and ODA-like flows) between 2010 and 2014.[31] While aid supports only a small share of global infrastructure investment, it plays a critical role in low-income countries, where it is difficult to mobilise domestic and other external finance. If aligned with a low-emission pathway, such support can help developing countries – both middle and low-income countries – achieve their NDCs more rapidly and "leapfrog" the emissions-intensive pathways of developed countries. G20 countries that report to OECD-DAC are mainstreaming climate considerations into their support for infrastructure (Figure 7.12).

Figure 7.12. **Share of climate-related official development finance to infrastructure, commitments, 2013-15 average, G20 members reporting to OECD-DAC**

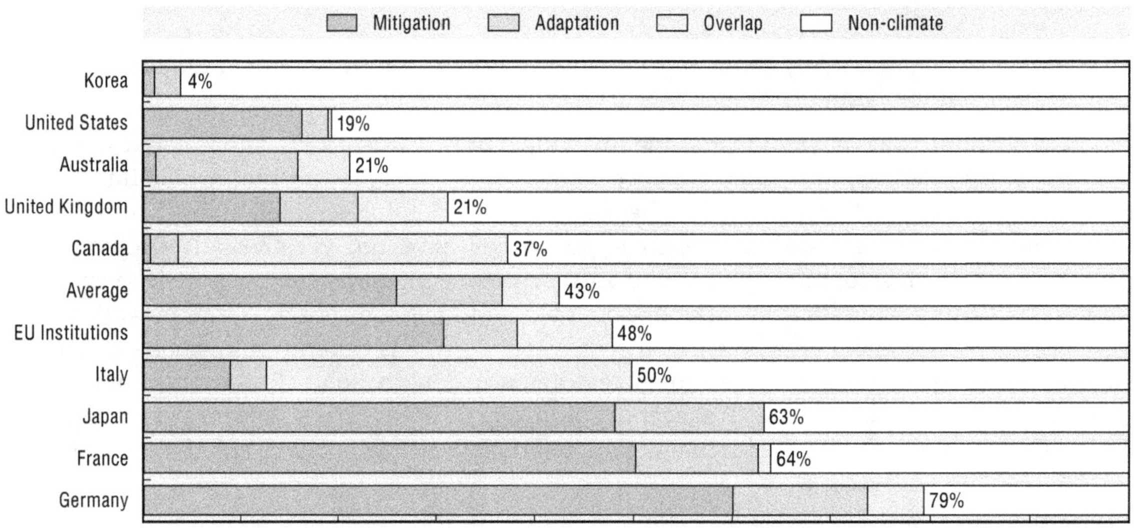

Note: This graph is based on Official Development Finance (Official Development Assistance and Other Official Finance) commitments data reported to the OECD-DAC Creditor Reporting System (CRS) database. Infrastructure sectors include transport, energy, water supply and sanitation, and communications. 'Overlap' refers to activities that simultaneously pursue mitigation and adaptation objectives.

Source: OECD-DAC Statistical System.

StatLink http://dx.doi.org/10.1787/888933484398

Bilateral support is often channelled through bilateral development banks, such as Germany's KfW and France's AFD; bilateral development finance institutions, such as CDC in the United Kingdom and the Overseas Private Investment Corporation (OPIC) in the United States; and through export-import agencies, such as China EXIM Bank, Japan Bank of International Cooperation (JBIC) and Korea EXIM Bank (KEXIM). Bilateral development banks work in a similar way to MDBs: they raise capital at attractive rates based on their governments' high credit ratings, and support both public and private sector projects in developing countries. Some bilateral development finance institutions specialise in engaging with businesses and mobilising private capital in support of development outcomes. These "private-sector focused" bilateral DFIs include those set up by various European countries (Box 7.2).

Box 7.2. European development finance institutions and support for low-emission, climate-resilient infrastructure

The European DFI Association (EDFI) brings together 15 bilateral DFIs from EU member countries. These institutions provide financing in the form of equity, loans and guarantees, on commercial terms, to international companies as well as businesses based in developing countries. Together, European DFIs represented EUR 6 billion in new commitments in 2015, equivalent to one-tenth of ODA from EU member states and the EU. The three largest European DFIs in terms of new commitments per year are FMO (Netherlands), DEG (Germany) and CDC (United Kingdom). The major focus areas of EDFI operations are infrastructure and the financial sector, which each make up roughly one-third of the overall portfolio of EDFI members in 2015. On average between 2011 and 2015, 22% of EDFI members' overall commitments were related to climate. Within the power sector, renewable energy (including hydro) made up 70% of power investments in 2015, and commitments to renewable energy have been increasing year on year. Support for adaptation is still at a nascent stage, making up 1% of climate-related finance flows from European DFIs in 2015.

Figure 7.13. **Climate-related annual commitments by EDFI members, as share of commitments and by sub-sector, 2011-15**

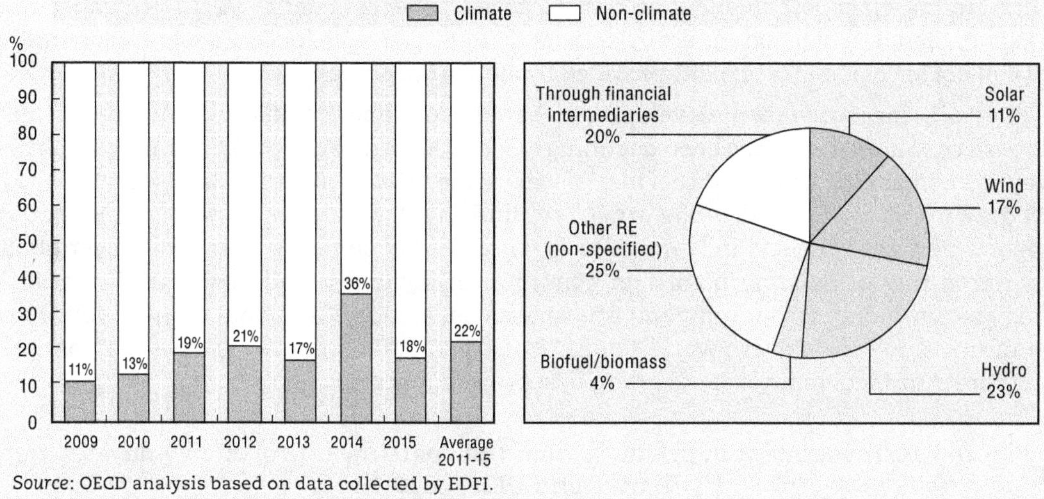

Source: OECD analysis based on data collected by EDFI.

Bilateral development banks need to scale up support for climate resilience

Climate finance targets and commitments among bilateral development banks and DFIs are driven by countries' commitments in the context of the UNFCCC negotiations, specifically the commitment to mobilise 100 billion USD per year by 2020. In recognition of this, some bilateral development banks are increasingly mainstreaming support for low-carbon infrastructure. On average between 2013 and 2015, 68% of AFD's, 58% of KfW's and 40% of JICA's financing for infrastructure targeted climate change directly.[32] Within this support, however, the major share went towards climate change mitigation, with only a minor share going towards increasing the climate resilience of infrastructure projects.

Across bilateral DFIs that target private sector operations through largely non-concessional financing, support for renewable energy has been increasing. Despite this the share of climate finance in their portfolios is lower, as is the share of climate finance going to adaptation, when compared with bilateral development banks that support both public and private sector projects.

Bilateral providers also increasingly work together to co-finance projects related to climate change. Initiatives such as the Interact Climate Change Facility and the EU Electrification Financing Initiative (ElectriFI), for example, bring together EDFI members and other bilateral and multilateral development banks to source and support renewable energy and energy efficiency projects in low and middle-income countries.

Development banks, DFIs and export credit agencies also support carbon-intensive infrastructure

As well as supporting climate-friendly projects, some bilateral DFIs and export-credit agencies continue to support more carbon-intensive forms of energy in developing countries. Investment by development banks and finance institutions (national, multilateral and bilateral) and export credit agencies in privately financed coal projects ranged from USD 682 million to USD 2.6 billion per year in the last five years (Figure 7.14). Export credits in support of power generation from G20 countries that report to the OECD overwhelmingly supported fossil fuel technologies over the last decade (Chapter 3). In the future, any export credits provided by OECD member countries for coal power will support super-critical and ultra-supercritical coal technologies, which have lower emissions than traditional coal technologies but remain high-carbon relative to other power generation options. However, this agreement – the Sector Understanding on export credits for coal-fired power projects under the OECD's Arrangement on Export-Credits, which took effect in January 2017 – is only valid for OECD member countries, and will not apply to export credits provided by all G20 countries.

Overall, investment by development banks and finance institutions in privately financed coal-fired power has been declining since 2010, and development banks are moving away from financing coal-fired thermal power. Some MDBs – such as EBRD and EIB – have withdrawn support for coal, while others are limiting their support. The World Bank only supports new coal projects in "rare circumstances". KfW supports new coal power plants and retrofitting of existing plants on condition of eligibility and sustainability criteria being met, including minimum plant efficiencies, existence of national climate policy and commitment to renewables (World Bank, 2013; KfW, 2015b). Some banks – such as EIB – incorporate the economic cost of carbon into project appraisal.

Figure 7.14. **Investment in privately financed coal power projects by development banks and DFIs, 2010-16**

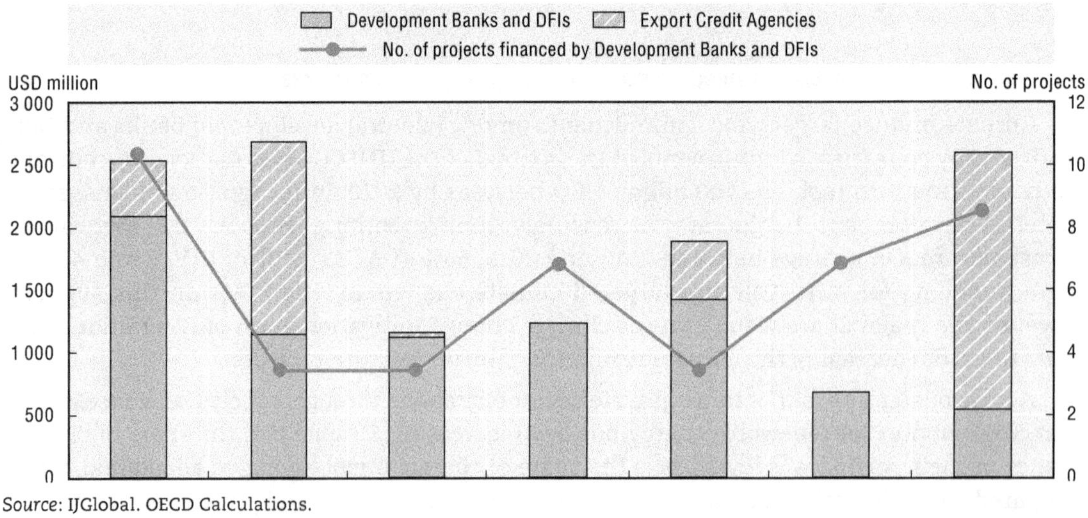

Source: IJGlobal. OECD Calculations.

StatLink http://dx.doi.org/10.1787/888933491515

Mobilising private investment for low-emission, climate-resilient infrastructure

Impediments to private investment

Projects with clear revenue streams supported by end users, such as renewable energy projects, have a strong potential for private financing. Projects face difficulties attracting capital from private investors, however, where a revenue stream is not readily available (e.g. climate proofing of existing infrastructure, or climate resilience infrastructure such as flood protection), or where revenue streams are not predictable and dependable (e.g. basic infrastructure projects in countries with weak governance and regulation). Financing is also influenced by the general risks involved in infrastructure investment, due to the project stage and other risk factors, such as being located in less creditworthy countries, or facing market risks that are difficult to quantify and mitigate. Other barriers have been identified in this report, such as the absence of a steady pipeline of infrastructure investments (Chapter 3), and the absence of many of the investment-related policies needed (Chapter 5). This section discusses returns and risks related to infrastructure, and looks at the tools available to policy makers and regulators to help de-risk investment and facilitate the engagement of investors in infrastructure projects.

Risk and return drive investor demand

Most investors expect 10-15% net ROE for onshore and offshore wind, as well as solar, according to the results of a preliminary OECD survey on renewable energy finance.[33] Some investors have high return expectations for the offshore wind sector, perhaps reflecting higher technical risk and less experience than in onshore wind and solar energy projects. Onshore wind reported a greater range of return expectations, given the maturity of markets in some regions, notably northern Europe where returns on onshore wind have been pushed down to 6%-8%. Other renewable energy sectors with greater technology risks, such as marine, have higher return expectations, but also greater uncertainty reported by investors, as expected (Figure 7.15). Further analysis of return expectations for low-emission, climate-resilient infrastructure is needed, but the general principle remains true: the greater the perceived risks of a project, the higher the returns investors will demand, and the higher the costs passed onto end users and government sources of funding.

Figure 7.15. **Expected equity return, renewable energy finance, by sector (in %)**

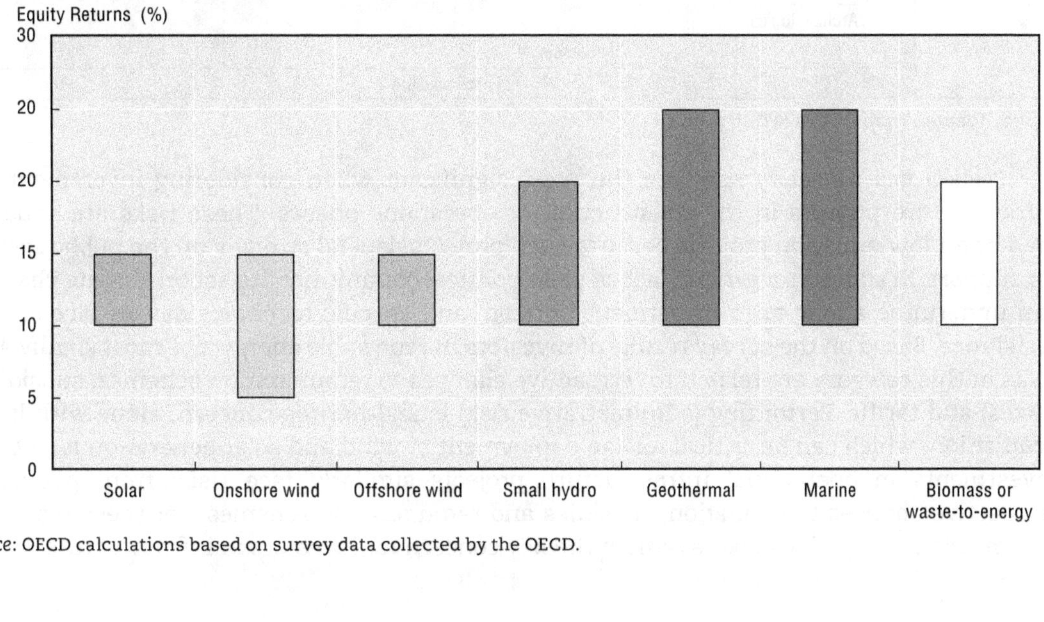

Source: OECD calculations based on survey data collected by the OECD.

Several factors determine the risks – real and perceived – associated with infrastructure investments. These investments are typically long-term, involve high upfront costs, differ from project to project and thus cannot reach economies of scale, and usually involve different stakeholders and non-standardised financing structures (OECD 2015c). As a result, infrastructure investments are commonly associated with different types of risk (Table 7.2). Some of these are further exacerbated in the case of low-emission infrastructure projects (OECD, 2012; OECD, 2015d, OECD/World Bank, 2015, Frisari, 2013).

Table 7.2. **Risks linked to infrastructure assets over the project lifecycle (shaded cells can be linked to climate change risks)**

<table>
<tr><th>Risk Categories</th><th>Development Phase</th><th>Construction Phase</th><th>Operation Phase</th><th>Termination Phase</th></tr>
<tr><td rowspan="10">Political and regulatory</td><td>Environmental review, land acquisition</td><td>Cancellation of permits</td><td rowspan="3">Change in tariff regulation</td><td>Contract duration</td></tr>
<tr><td rowspan="3">Rise in pre-construction costs (longer permitting process)</td><td rowspan="3">Contract renegotiation</td><td>Decommission</td></tr>
<tr><td>Asset transfer</td></tr>
<tr><td colspan="2">Currency convertibility</td></tr>
<tr><td colspan="4">Change in taxation</td></tr>
<tr><td colspan="4">Social acceptance</td></tr>
<tr><td colspan="4">Change in regulatory or legal environment</td></tr>
<tr><td colspan="4">Changes in climate change policy and support schemes</td></tr>
<tr><td colspan="4">Enforceability of contracts, collateral and security</td></tr>
<tr><td colspan="4"></td></tr>
<tr><td rowspan="9">Macroeconomic and business</td><td>Prefunding</td><td colspan="3">Default of counterparty</td></tr>
<tr><td colspan="2" rowspan="4">Financing availability</td><td colspan="2">Refinancing risk</td></tr>
<tr><td colspan="2">Liquidity</td></tr>
<tr><td colspan="2">Volatility of demand/market risk</td></tr>
<tr><td colspan="2">Liability risks - compensation from victims of climate change</td></tr>
<tr><td colspan="4">Inflation</td></tr>
<tr><td colspan="4">Real interest rates</td></tr>
<tr><td colspan="4">Exchange rate fluctuation</td></tr>
<tr><td colspan="4">Long pay-back period for climate change mitigation investment</td></tr>
<tr><td rowspan="6">Technical</td><td colspan="3">Governance of the project</td><td rowspan="5">Termination value different from expected / stranded assets</td></tr>
<tr><td colspan="3">Environmental</td></tr>
<tr><td>Project feasibility and inclusion in investments plan*</td><td rowspan="2">Reliability of forecasts for construction costs and delivery time</td><td rowspan="2">Qualitative deficit of the physical structure/ service</td></tr>
<tr><td>Archaeological</td></tr>
<tr><td colspan="3">Obsolescence</td></tr>
<tr><td colspan="4">Force Majeure</td></tr>
</table>

Source: Adapted from OECD (2015c), p. 48.

Political and regulatory risks are the most significant when considering investment in infrastructure projects in the construction or operations phases. These risks are usually greater for low-emission projects because such projects depend strongly on the public sector for support. In addition, a general lack of clear political commitment to act on climate change can undermine a long-term investment horizon, and specific technologies can face social resistance. Based on the survey results of investors in renewable energy, the most significant risks in this category are related to retroactive changes to remuneration schemes, subsidies (taxes) and tariffs. Permitting (administrative risk) is another top concern, along with land acquisition, which can be critical for the deployment of wind and solar generation facilities. Investments in operational infrastructure projects similarly face risks from potential retroactive changes to regulation, subsidies and remuneration schemes. For these reasons, commitments by policy makers remain the key concern for investors in infrastructure.

Macroeconomic and business risks are also of concern to investors. According to the OECD survey, the interest rate environment, state of the economy and (global) energy prices are the risks most often cited by investors in energy sector projects in the construction phase. Counterparty, sovereign default and currency convertibility risk were also mentioned. With respect to investment in the operations phase of infrastructure projects with commercial risk, a decline in prices and demand are the macroeconomic and business risks that most concern investors.

Technical risks are also relevant for investors. During construction, the most important relate to the reliability of cost and time forecasts, as well as uncertainties about the deployment of new technologies. For renewable generation projects in the operating phase, technical risks include connectivity problems (grid, offshore transmission), technical failure and reliability issues. The current value and future performance of an infrastructure asset can also be reduced if climate change increases the frequency and intensity of natural disasters. For this reason, it is vital to question the extent to which access to climate risk insurance influences the cost of capital for infrastructure projects.

Matching expected returns with acceptable levels of risk

Private sector investment in infrastructure is based on matching the expected risks and returns of infrastructure assets with the investor's own tolerances for returns and risks. The involvement of the private sector can also help to increase operational efficiency, thus providing support beyond mere financing. Projects that are clearly commercially viable are typically able to attract private sector finance. When projects face financing shortfalls, governments can use risk mitigation techniques and incentives to reduce risks to match a suitable level of return, or increase return to match a given level of risk, facilitating the pricing of investment. Lowering costs can also improve the economic viability of some projects and improve the overall efficiency of public capital for low-emission infrastructure. Reducing the cost of finance, both for debt and equity, can dramatically improve the ability of renewable energy technologies to compete with fossil fuels, especially in developing countries (UNDP, 2015).

Risk mitigation instruments and blended finance can facilitate private investment

Enabling a pipeline of bankable low-emission infrastructure projects will require efforts to mitigate the risks that investors face, and crowd in private capital. Governments, development banks and DFIs utilise a range of tools to achieve these aims, including guarantees, insurance and hedging, as well as syndication and debt subordination (Table 7.3). Each tool mitigates different types of risk. Guarantees can be useful in protecting investors from the risk of governments not honouring their obligations, for example, while currency hedging can reduce exposure to fluctuations in foreign currencies.

Many such approaches are supported by "blended finance" – the strategic use of public capital, concessional or non-concessional, to mobilise additional investment. For example, in some subordinated debt approaches, transactions can be structured so that public capital can take a "first loss" position to attract private capital to the project. While such tools are useful in engaging private capital, particularly for technologies that are yet to mature or in countries where the perceived risk of investment is high, there is a need to balance blended finance interventions – and risk mitigation tools more broadly – against issues of moral hazard[34] and market distortions.[35] MDBs are increasingly working together to develop common guidelines in the use of concessional finance in private sector operations. Similar efforts are under way among OECD-DAC members.

Public support for project preparation and development also plays an important role in promoting low-emission infrastructure. Dedicated project preparation and development facilities can support different stages of the project cycle, such as advisory support for design and conceptualisation, prefeasibility and feasibility studies, and help identifying potential investors (Nassiry et al., 2016). Such facilities also provide broader support such as technical assistance and capacity building, which are major barriers to project development in developing countries. Such approaches should be scaled up, and facilities with a multi-country or regional focus should be better co-ordinated to maximise their impact and avoid duplication or overlap.

Table 7.3. **Approaches and instruments used to mitigate risks and mobilise private capital**

Instrument / approach	Usage	Example
Guarantees	Political risk guarantees can enable investment in higher risk countries.	The Multilateral Investment Guarantee Agency (MIGA) provided a guarantee against risks related to expropriation, war and civil disobedience to a Dutch solar company (SunE Solar) to cover investment in the development of three solar power plants in Honduras.
	Credit guarantees can incentivise institutional investment.	MIGA provided a guarantee to cover the non-honouring of financial obligations issued to back a USD 361 million loan by Banco Santander S.A. of Spain to the state of São Paulo, Brazil, to improve the sustainability of São Paulo's transport system.
Currency hedging	Currency hedging mitigates the risk of currency fluctuations for foreign investors, important for infrastructure investment in developing countries.	The Currency Exchange Fund (TCX), supported by the German and Dutch governments, is a currency hedging fund. When the private fund manager ResponsAbility Investments provided local currency-indexed loans to M-Kopa, a Kenya-based solar home system company that provides decentralised solar solutions in Kenya, Tanzania and Uganda, it used a hedging contract with a specialist currency hedging provider, that either passes on the currency risks to commercial counterparties or to TCX, if the former do not exist.
Loan syndication	Loan syndication mitigates business risks for private investors and builds on due diligence capacity of development banks	MDBs usually take the role of lead arrangers, and the private sector steps in to provide additional capital. The division of the loan amount leads to risk diversification, and it builds on the due diligence capabilities and reputation of the public sector (MDB) arrangers. By doing this, the MDBs boost investor confidence and reduce transaction costs.
Debt subordination	Debt subordination diversifies risks through structuring and facilitates private investment.	The Green for Growth fund was launched by KfW and the EIB with the financial support of the European Commission, the German Federal Ministry for Economic Co-operation and Development (BMZ), and the EBRD.[36] The fund invests in energy efficiency projects as well as projects increasing renewable energy sources in Southeast Europe. The public donors are invested in the most junior tranche, while private institutional investors invest in the most senior tranches, thereby benefiting from the risk cushion provided by the public (donor) investment.
Co-investment platforms and funds	Co-investment vehicles pool investor capital to deploy directly in infrastructure projects, bypassing intermediaries.	The Marguerite/2020 European Fund for Energy, Climate Change and Infrastructure brings together several NDBs and financing institutions, including CDC, CDP, EIB, KfW and PKO Banco Polski, to co-invest in renewable energy and sustainable transport infrastructure in Europe, with a focus on greenfield projects (minimum 65% of the fund) (Della Croce and Sharma, 2014).
	Blended finance funds pool public and private capital to mobilise additional capital	Climate Investor One, launched by FMO and Phoenix InfraWorks and supported by funding from the government of the Netherlands, is a new fund of funds structure designed to invest at the development, construction and refinancing stage of large-scale solar, wind and hydro projects, with initial projects financed in Rwanda and Zambia.
Project development facilities and technical assistance	Supports the development of bankable infrastructure projects.	The Asia Pacific Project Preparation Facility, funded by Australia, Canada and Japan, supports AsDB member country governments and their public-sector agencies in preparing and structuring PPP transactions for infrastructure, prioritising infrastructure with climate change and sustainable development elements. AsDB will provide the technical assistance.[37]

Facilitating investment using diversified financial instruments and techniques

This section presents a range of financial instruments that mobilise investment through capital markets channels, and that benefit from pooling smaller projects, from partnerships, and from engaging corporate and institutional investors, while taking advantage of the expertise of established players such as utilities and banks.

These instruments build on previous recommendations endorsed by the G20 in the *G20/OECD Guidance Note on Diversification of Financial Instruments for Infrastructure and SMEs.*[38] For developing countries, the formation and deepening of local capital markets – debt and equity – is essential to attract domestic and foreign investors. The Guidance Note presents other important pre-conditions that can strengthen the financing environment for infrastructure.[39]

Diverse equity instruments for financing low-emission infrastructure

Private market instruments such as private equity-style funds have played a major role in regions where institutional investment has been taking place in infrastructure. For example, CNP Assurances (an insurance company) and Meridiam (asset manager), jointly launched the Meridiam Transition Fund, an unlisted equity fund targeting green infrastructure investment. This is just one example from a spectrum of investment strategies, levels of fees, and terms and conditions that exist. Asset management industries, combined with a competitive bidding process for assets and a project pipeline, are conducive to investment funds raising capital for deployment in low-emission infrastructure projects.

Box 7.3. **Financing sustainable transportation infrastructure**

The financing of passenger rail and metros is often constrained by higher upfront capital costs, lower returns and longer development and payback periods, compared with toll highways. In addition, direct user fares are often set too low to cover operational costs, due to social affordability concerns. Several financial tools and risk-sharing mechanisms are available to improve the relative risk-return profile of sustainable transport infrastructure projects.

Public-private partnerships (PPPs), which allow for private sector participation and risk sharing, have often been used for bus rapid transit systems, highly used and specific rail links, and shared-use vehicle and bicycle systems. However, the right institutional capacities and processes must be in place (see Chapter 5).

Land value capture tools secure revenues from the indirect and proximity benefits generated by transport infrastructure (e.g. increased real estate value) to help fund transport projects. Examples of land value capture tools include tax increment financing (TIF) districts, development charges, development rights and joint development. To date, these tools have been applied mainly to roads, metros and rail. New York City (NYC) is financing Hudson Yards subway line extension and station through the issuance of bonds by a special purpose vehicle (SPV), the Hudson Yards Infrastructure Corporation, with debt service guaranteed by innovative sources of revenues, including: tax equivalency payments, provided by NYC in anticipation of future tax revenues from land value increases; payments in lieu of taxes, which offer land tax exemptions to project developers in a specific area; and transferable development rights from the transfer of public property land and building rights. Effective land-use planning is instrumental in promoting value creation through new infrastructure (see Chapter 5).

Source: OECD (n.d.).

Since many funds invest in public-private partnerships (PPPs), a supportive project finance environment, coupled with liquidity in local debt markets, is conducive to investment, since private equity investors also seek to secure debt financing for investment. Private market funds and direct equity investment can support new, greenfield investments; however, most investors prefer the risk/return profile of operational assets that generate attractive cash yields. Countries could review the availability of unlisted equity funds in local markets, for both domestic and foreign investors, taking note of product diversity and fund activity in renewable energy markets, as well as direct equity investments by institutional investors and the secondary market conditions that can support such investment.

Listed equity instruments also have potential to channel investment for the transition, particularly through retail channels or in defined contribution pension systems. Closed-end funds, real estate investment trusts and master limited partnerships (MLPs) are designed principally as holding companies to pass through income to shareholders, with some structures offering potential tax advantages for investors. Low-emission and climate-resilient infrastructure assets could be included, where rules permit them as qualifying assets. Countries could, where appropriate, review the rules for qualifying assets for listed equity vehicles for renewable energy assets and other relevant infrastructure.

Over the past few years, yieldcos have emerged in North America as an equity-based financing model for clean energy projects such as wind and solar. Although this market has been volatile recently, this model of finance represents an innovative channel for investors to gain exposure to clean energy assets.[40] Some closed-end funds have also been launched specifically to finance renewable energy, particularly in the United Kingdom, such as Greencoat Capital.

Bond issuances such as green bonds facilitate private investment

Bond issuances have become increasingly important in recent years as a means to mobilise private finance for low-emission infrastructure projects. Project bonds can provide long-term debt capital to such projects by directly financing or refinancing specific assets through the capital markets. Such issues may be designated "green bonds", although green bonds are not specifically defined, and both officially labelled and unlabelled types exist (OECD 2017b).[41] The major issuers of green bonds are corporations and development banks (Box 7.4) (OECD, 2017b). "Climate-aligned bonds" is another label for bonds that are specifically linked to the financing of low-emission investments. As of 2016, outstanding climate-aligned bonds amounted to USD 694 billion (CBI, 2016). From 2015 to 2016, the share of climate-aligned bonds that are officially labelled rose from 11% to 17%, reflecting the growing demand for transparency in the sector (CBI, 2016).

Governments could encourage diverse channels of debt financing for relevant infrastructure projects, in particular through non-bank channels, including syndication of bank loans through capital markets, the development of a robust project finance market, revival or innovative use of infrastructure project bonds and sub-sovereign bonds, and linking to trends in green bond markets. In 2017, the French government issued its first green bond, raising EUR 7 billion. The issue is the largest and longest-dated benchmark green bond in the world, and was launched to encourage the development of a robust green bond market. In developing countries, the expansion of local currency debt markets is crucial for infrastructure projects to secure large amounts of debt financing, while minimising currency mismatches and over-reliance on external sources of debt finance (G20/OECD, 2016).

Securitisation supports the bundling of small-scale loans, tapping capital markets' liquidity

Securitisation mobilises capital markets' liquidity through the issuance of rated bonds, and diversifies risks by aggregating small-scale assets, which is particularly important for investment in energy efficiency or decentralised energy. The pooling of assets also facilitates access to infrastructure investment by institutional investors, who may have minimum investment thresholds but who may also not have the capacity or expertise to provide financing on a project-by-project basis. Some recent examples of securitisation include:

- Up to March 2016, SolarCity (a corporation that produces, installs, finances, and services solar energy installations in the United States) has raised over USD 680 million over a series of six securitisations in the past three years (Marathon, 2016).

- Loan warehouse facilities for securitisations have also taken place. In the United States, the Warehouse for Energy Efficiency Loans (WHEEL) purchases and aggregates home energy efficiency loans on a national scale, and issued its first asset-backed security in June 2015, totalling USD 12.6 million, comprising unsecured home energy efficiency loans up to USD 20 000.

Securitisation could lower the cost of finance for solar PV installations by enhancing liquidity and pooling issues across geographies, though there are challenges to overcome. As a collateral type, cash flows generated through long-term contracts on decentralised power (leases, loans, or PPAs) have a short history, and such collateral is usually unsecured in that the asset has little or low resale value, which can lead to higher losses given default. Backing energy efficiency loans with adequate collateral can sometimes be challenging. In developing countries, a lack of credit scoring, credit history and insufficient data are impediments to overcome. Standardising contracts for small-scale financing at the domestic level would be beneficial for securitisation transactions, along with mitigating certain policy risks such as net metering rules (utility buy-back of surplus generation) (Marathon, 2016), and credit guarantees. Strong growth prospects for the rooftop solar industry, driven by falling technology costs, regulatory and policy support for solar power, and acceptance of the asset class by institutional investors, are the key drivers of the sector's expansion in developed economies (Moody's, 2015).

Box 7.4. **Tapping the debt capital market by issuing green bonds**

EDF's green bond programme: The state-owned French utility EDF launched a green bond programme in 2013 with a EUR 1.4 billion bond issuance of eight-year tenure, at a 2.25% fixed rate. A second issuance of USD 1.25 billion took place in 2015, maturing in 2025, and a third issuance of EUR 1.75 billion in October 2016. EDF uses the debt financing to invest in greenfield renewable power generation projects as well as renovation and upgrade of existing hydropower facilities in France; that is, collected funds are directed to 13 projects (first issuance, as of December 2015) and 3 projects (second issuance, as of December 2015). The EDF example shows how investors can invest in tradable corporate fixed-income securities directed to low-emission, climate-resilient projects, thereby achieving an attractive risk-return relationship: the first note offered a competitive return at low risk – S&P rated it investment grade (A+).

IDB's securitised green bonds for energy efficiency: In 2014, IDB and the Clean Technology Fund (CTF) set up a USD 125 million financing project for energy efficiency projects developed by Mexican energy service companies (ESCOs). As a first step, the IDB engaged in warehousing receivables of two Mexican ESCOs of USD 50 million, which will be issued via securitisation in green bonds and sold in local debt capital markets, thereby enabling institutional investors to increase their exposure to the energy efficiency sector. This transaction will enable the ESCOs to access long-term funding to implement small-scale energy efficiency projects. The portfolio of these projects is backed by a partial credit guarantee provided by the CTF.

Using municipal and resilience bonds to finance climate resilience and adaptation

Adapting to the expected impacts of climate change, such as more frequent inland and coastal flooding, will require investment in structural protection infrastructure such as flood barriers and sea walls. However, the potential for private investment in such projects is limited because unlike toll roads or airports, structural mitigation projects do not generate revenue with which to repay investors. An emerging bond type, known as "resilience bonds" or "municipal adaptation bonds", is similar to the concept of green bonds whereby the issuer commits to use the funds for specific adaptation purposes. For example, such bonds have been issued by Nederlandse Waterschapsbank (NWB Bank) to finance loans to Dutch water authorities for water management measures, including flood protection measures. The finance of urban resilience infrastructure, of which there is great need (see Chapter 3), may be particularly suited for resilience bonds, although capacities to finance may be limited by the creditworthiness of local municipalities.

The non-life insurance companies that would face reduced claims in a better-protected municipality could have incentives to provide a portion of the financing for resilience bonds that are structured to provide a return to investors. One approach could be to monetise the expected reduction in insurance premiums that would accompany investments in resilience as a means to generate returns for investors. For example, Vajjhala and Rhodes (2015) have outlined an approach to linking investments in resilience to pre-defined rebates (insurance savings) on catastrophe bonds that could be used to fund the project costs. A significant increase in interest in catastrophe bond issuance by public agencies would be a prerequisite, as this risk transfer instrument is rarely used by public entities.

Tax equity financing models for renewable energy projects

In the United States, tax credit mechanisms initially targeted at project developers have attracted third-party tax equity providers. Project developers could not easily use the tax credits due to their small size, lack of profitability and lack of tax exposure (Sharif, et al., 2011). Third-party tax equity investors comprise banks and corporates, receiving remuneration in the form of tax credits instead of cash. Tax equity investors can also access accelerated depreciation of investments, providing further tax benefits and cost recovery for investment. Tax equity partners have expanded the amount of capital available for investment in renewable energy projects in the United States.

Although banks dominate the market, new actors are entering, including other institutional investors and corporates. In June 2011, Google invested USD 280 million in a tax equity fund with the US solar power company SolarCity to finance residential solar projects. In 2015, primary tax equity financing for renewables reached nearly USD 1 billion (IJ Global). Other sources cite much larger numbers: Renewable Energy World reported that USD 11.5 billion in tax equity financing was secured in 2015 for new wind and solar projects.[42]

Limitations to the tax equity model include scheduled phase-outs of supportive legislation that creates such incentives, political risk, and cyclical demand (tax credits only have value to investors when they have profits to offset). This model is also unique to the United States.

Emerging finance models for energy efficiency, decentralised power generation, and carbon markets

The availability of finance often determines the uptake of new technologies in energy efficiency and decentralised power generation; for example, many homeowners at first eschewed investing in solar panels due to the upfront costs and limited access to finance. As more financing models are emerging, however, there are more ways for policy makers to foster technology uptake. In developing countries, governments can enable investment in decentralised power generation by promoting innovative financing models for solar PV, small scale loans or leases for households and businesses deploying technologies, and securitisation to recycle capital into new lending. In advanced economies, governments can help to offset the costs of installing decentralised power by supporting net metering policies, with fair economics.

There is massive potential to achieve energy efficiency in buildings, infrastructure and industry (Chapter 3). It can be difficult to monetise returns in the necessary investments, however, so financing for energy efficiency has not been happening at the scale needed. To change this situation, effective investment policies are crucial (Chapter 5).

Nonetheless, new business models and financing techniques have emerged in some countries in response to specific tax policies. This includes accounting standards, such as accelerated depreciation, which can shorten the payback period on investments. Some innovative financing models include:

- With the advent of technologies such as off-grid solar coupled with mobile payment solutions, several new pay-as-you-go (PAYG) solar companies have emerged with significant potential to improve access to renewable energy in low-income countries. Companies such as M-Kopa and PEG have expanded access to solar electricity to off-grid communities and households in Africa, based on low-level weekly or month payments.
- SunFunder launched its Structured Asset Finance Instrument (SAFI) in May 2016, with a first USD 2 million deal with SolarNow in Uganda. Sunfunder has also launched a fund, initially seeded by OPIC with USD 15 million, to provide project finance loans and short-term inventory loans to solar companies deploying off-grid projects in Africa, India, Pakistan and the Philippines.
- Touting an "infrastructure-as-a-service" business model, Generate Capital, a balance-sheet corporation, raises capital through issuing shares or debt, and deploys flexible financing options for innovative solutions in energy, agriculture and water infrastructure. It focuses on sustainability and small-scale downstream opportunities with a high development impact. Generate owns the infrastructure asset, and collects usage payments.
- Using tax finance for energy efficiency, the U.S. Property Assessed Clean Energy (PACE) comprises a loan to a building owner to retrofit a building. The loan is attached to the property and reimbursed through local taxes by the occupant. The fact that payments are integrated in local taxes enhances their creditworthiness, since taxes have almost zero non-payment rates in the United States and are senior to any other debt.

Specialised loan products have been developed that focus increasingly on renewable energy and energy efficiency, such as green credit lines. In Germany, KfW provides loans and grants for energy efficiency, which are distributed through local banks. Several bilateral and multilateral providers use a similar approach to finance energy efficiency in developing countries, by working through NDBs. For example, EBRD's Turkish Sustainable Energy Financing Facility, a USD 260 million credit line facility, works with several local banks to help Turkish SMEs finance energy efficiency improvements. In the United States, energy efficient mortgages for qualifying buyers reduce borrowing costs when the mortgage takes into consideration lower utility costs, which could help support a slightly larger mortgage payment.

To scale up emerging and innovative financing, governments should implement policies that support investment in energy efficiency (such as standards and labelling for equipment, and building codes), while encouraging the private sector to develop financing tools such as long-term loans, leases, securitisation, "green mortgages", PAYG, and other products to create a competitive financing landscape. The development of standardised measurement and verification frameworks, including energy efficiency audits, has helped to create a framework for energy efficiency investment in the United States and other advanced economies:

- In April 2016, the California utility PG&E submitted plans for a residential pay-for performance pilot programme that measures results at the meter and rewards customers for savings.
- The International Renewable Energy Agency (IRENA) and Terrawatt Initiative launched the Solar Energy Standardization Initiative in June 2016 to work with 15 law firms and over 20 financial institutions worldwide to develop streamlined and standardised contract documentation for solar PV.

Other financing models are evolving that mobilise third-party investment from institutional investors and capital markets by using closer links to carbon markets. Emissions trading schemes directly engage companies and organisations, moving capital around to unlock investments in energy efficiency, renewable energy, or other low-carbon investments (Jones, 2016). Among institutional investors, carbon pricing could also affect asset allocation decisions, potentially unlocking further investment through capital markets. In this way, new instruments could compensate investors through the receipt of carbon credits, which could be used to offset their own emissions, or the credits could be sold in carbon markets. Such an innovative concept could unlock investment for carbon capture and storage assets that themselves generate carbon credits, renewable energy, energy efficiency, or carbon sequestration like forests and wetlands.

Forest bonds for the support of forest conservation

As well as being essential for the absorption of atmospheric carbon dioxide, forests have a strong mitigating effect on the extremes of climate change weather. An innovative forest bond issued in October 2016, backed by the IFC and BHP Billiton, raised money from institutional investors to combat deforestation in Kenya. This is an innovative instrument because investors can opt to be paid in carbon credits, which they can use to cover emissions or sell in the carbon credit market for cash. This model could be particularly attractive for entities that have large carbon footprints and emissions to offset and is an innovative example of attracting third-party investment through capital markets.

The role of the public sector and the need for improved business models

To improve the flow of finance to climate-friendly technologies, collaboration between the public and private sectors is needed. This could be led by the G20 or by individual country initiatives, and could focus on crowding-in investment at key points along the value chain and catalysing investment. Such initiatives are especially important in early stage financing, where there are shortages of capital for innovation, and in capital-intensive technologies or technologies with long development periods. For example, in innovation financing for climate change adaptation, where revenues are not always available, the public sector could take a stronger lead by awarding grants to promising solutions, or launching clean technology incubators. Another example is Smart City initiatives in numerous countries where local authorities work closely with innovators, financiers and the public to identify innovative solutions to urban challenges and climate change. A number of global initiatives have recently been launched with climate change as a central theme. Examples include:

- Mission Innovation, with 23 members (including Australia, Brazil, Chile, China, the European Union, India, Indonesia, Japan, Mexico, the United Kingdom and the United States) has committed to doubling public investments in clean energy research and development by 2020 while encouraging greater levels of private sector investment in transformative clean energy technologies.
- The Global Innovation Lab for Climate Finance, which brings together governments, project developers and development banks and finance institutions, supports the

identification and piloting of cutting-edge climate finance instruments that can drive investment and unlock new opportunities for renewable energy, energy efficiency, and adaptation efforts in developing countries.

- Infuse Ventures is an India-based technology fund seeded through public funds that invests in seed and early stage venture capital focused on sustainability and clean energy. The government of India has a first-loss position, while private sector investors and other public investors have contributed capital.
- The US-India Joint Clean Energy R&D Centre promotes collaborative development. This idea could be scaled up to other multilateral settings to enhance north-south collaboration, the uptake of new technologies in developing countries, and the application and commercialisation of technologies (Patwardhan, 2016).

To mobilise further investment in clean technologies, innovation in business models and financing is needed. The venture and private equity capital model could be improved by extending fund investment horizons for base technologies. Corporations could be encouraged to commercialise technologies by strategically investing in or acquiring start-ups, as is the case in the biotechnology sector. Additionally, shifting investors' mind-sets away from finding a "breakthrough" technology (which can be very difficult), to focusing on information technology solutions and applications, can better fit the VC model. Technology can be used to enhance the performance of existing infrastructure and move into new markets, providing for a better alignment with the VC model, which seeks to exploit untapped markets or unleash disruptive technologies on incumbent industries.

A responsive and resilient financial system for the transition

The Paris Agreement has sent a strong signal to financial markets. Crossing the threshold for entry into force "brings the horizon forward" for climate action (Carney, 2016). The financial system needs to evolve and adapt to this new environment, so as to capture profitable opportunities arising from the transition while remaining resilient to related risks. Together with an effective investment policy framework, including long-term policies supported by strong government commitment (Chapter 5), enhanced transparency and signalling would improve the ability of the financial system to support investment and innovation as the economy undergoes transformation.

The key functions that the financial system performs in the real economy include mobilising capital and allocating resources, managing and transferring risks, and price discovery. The availability of reliable financial and non-financial information is essential for the system's smooth functioning and integrity. Data gaps related to opportunities and risks linked to the transition could lead to mispriced risks and a poor allocation of capital, which could undermine economic growth and productivity, and create stresses that could abruptly shift asset prices and impact financial stability.

Improved information about the environment (e.g., the value of natural assets and the environmental impacts of corporate sector operations and investments, including GHG emissions) and about the low-carbon transition (e.g., corporate strategies and risk management, project pipelines, policy frameworks and carbon taxes) could support the functioning of the financial system and enhance decision-making about policy interventions. A fundamental question for investors is how to benchmark climate change risk to allow for the measuring of global exposure and for risk management, either by investing in assets with low climate change risk exposure, hedging this risk at reasonable cost, or avoiding the assets with the greatest exposure. This would also aid in identifying opportunities to invest using climate change as an investment thesis, or reason to invest. For the time being, there is no straightforward answer.

Understanding of climate change risks is increasing, yet more research is needed

In 2015, Mark Carney, Governor of the Bank of England, emphasised the risks to the financial system, and in particular to financial stability, posed by the physical risks, liability risks and transition risks associated with climate change. Infrastructure projects face varying exposures to such risks. According to a recent OECD survey, investors' top concerns are exposure to changes related to policies and regulations of GHG emissions, and changes related to government support schemes and incentives for low-carbon investing (broadly, the transition risks).[43] The materiality of climate change risks on business valuation and operations was also important. Social risks such as acceptance and securing land-use for infrastructure projects also ranked high. Physical risks (force majeure) and future liability risks, though important, did not concern investors as much as the other risks.

Regarding the materiality of climate change risks, there is increasing evidence that incorporating sustainability criteria in investments can improve financial performance, supported by a growing body of academic research. This academic research is summarised in a forthcoming OECD publication on investor governance and the integration of environment, social and governance (ESG) factors. Not all academic research points to such positive relationships between ESG factors and long-term returns, however. Several surveys – especially those published between 2000 and 2010 – find that the data is inconclusive. More empirical research on the effects of climate change risks and opportunities on corporate performance is needed.

Several organisations, from industry-led associations to subscription services, aim to help asset owners advance their ESG analysis and practices. Policy makers need to align such initiatives and collaborate where possible to reduce redundancies, eliminate frictions and achieve global standards on climate change risk, ESG practices and disclosure of risks. There is also a need to develop the tools necessary to analyse climate change risk, which can be a complex undertaking requiring large amounts of data and long-term analysis. Such tools could be used by all parts of the financial sector, making climate change stress testing and scenario analysis a part of due diligence and long-term financial analysis processes for banks and corporations. For institutional investors, the modelling of asset allocation, including climate change risks, and assumptions on carbon pricing could have important implications, underlining the need for quality information.

Efforts are under way to assess the exposure to climate change risk of the financial system (banking, pension and insurance, capital markets).[44] Individual countries or the G20 itself could prioritise such assessments, evaluating potential impacts on financial stability and the ability of markets to price climate change risk, ensuring the efficient allocation of capital, and supporting industry-led efforts to describe the significance of climate change risk. Climate change risks "can be part of a broader approach to prudential risk management and supervision".[45]

Emerging awareness of climate change within the banking system

Generally, environmental considerations are fully embedded in project appraisal processes as part of the integration of the Equator Principles,[46] yet bank lending models are evolving to further integrate sustainability factors in all types of infrastructure assets, and across their entire loan portfolios. Indeed, sustainable banking implies integrating ESG and risk management considerations into bank operations, financing, and capital raising activities, and mainstreaming practices in key bank functions such as credit and lending, savings products, and capital markets (UNEP, 2016a). Banks are a primary pillar of the financial system, yet green banking practices are at different stages of evolution across the G20, reflecting broader national financial and economic circumstances.

At the institutional level, perceptions and priorities also seem to vary, along with levels of adherence. For example, Société Générale divested from coal investments and has extended sustainable banking beyond renewables to other key infrastructure sectors such as transportation, water and telecommunications. Industrial and Commercial Bank of China has curtailed lending to certain industries that present a high risk of hazardous emissions (UNEP, 2016a). Deutsche Bank recently announced that it would no longer finance greenfield thermal coal power plants and coal mining. Regarding clean technologies, a shift is necessary in the banking sector's mind set. The dominant risk-averse posture of banks' risk management departments is limiting their ability to support innovative low-carbon technologies, which bear a higher risk but ultimately a higher potential return.

Banks are increasingly recognising the link between climate change and financial performance of assets, and are evolving credit and due diligence processes on loan origination to take into account ESG risks. For example, a tool developed by the Natural Capital Finance Alliance (NCFA)[47] and other sponsors GIZ and VfU enables users to integrate financial risk exposure to water scarcity into standard financial models used to assess credit risks to entities with high exposures to the water sector (NCFA, 2015). Uptake of such practices by banks varies, and there are no generally accepted definitions of sustainable investments or standards across the G20. In countries where universal banking is practiced, capital markets activities and investment banking services such as underwriting green public market equity issuance and green bonds are other banking activities aligned with sustainable banking strategies. Banking associations can also play an important role in implementing voluntary efforts to mainstream sustainable banking: market-led initiatives in Brazil, France, India, Mexico, the Netherlands, Singapore and Turkey all provide examples (UNEP, 2016a).

Regarding regulatory requirements and climate change risks, most G20 countries have been hesitant to require banks to incorporate environmental and social risk factors into risk management models, and G20 countries generally do not require banks to consider environmental risks as material risks for the calculation of regulatory capital requirements – although Brazil and China have formally incorporated environmental risk and governance standards into prudential bank regulation (UNEP, 2016b). Within the Basel III framework, there is no recognition that regulatory risk capital weights should incorporate the financial risks associated with environment sustainability risks, although G20 countries require banks to disclose all material risks regarding the firm's economic viability, through financial reporting, which could include climate change risks (ibid).

Box 7.5. **OECD empirical research on Basel III and bank capacities to lend to low-carbon infrastructure**

Financial stability rules and banking regulations implemented to pursue other objectives than climate objectives can have unintended consequences on the infrastructure investment required for the transition. The critically important Basel III framework for bank regulation was introduced after the 2008 financial crisis to strengthen the resilience of the banking sector and provide an international framework for measuring and monitoring liquidity risk.

The different components of Basel III banking rules have been introduced gradually, starting as early as 2011 for some countries, and are expected to be fully implemented by 2019. In particular, Basel III introduced a simple, transparent, non-risk based leverage ratio to act as a credible supplementary measure to the risk-based capital requirements. Results from a new OECD econometric study suggest that until 2014 (the last year of the study), the implementation of Basel III leverage ratio has hindered investment flows in renewable power generation across OECD and G20 countries (Ang, Röttgers and Burli, 2017).

Box 7.5. OECD empirical research on Basel III and bank capacities to lend to low-carbon infrastructure *(cont.)*

This result is in line with public comments from several financial stakeholders that Basel III may have unintentionally constrained the ability of banks to provide long-term debt financing for capital-intensive renewable infrastructure projects. To exclude the possibility that the result on Basel III is driven by banks' capitalisation levels and financial stability across countries, the econometric study has included a variable on regulatory capital to risk-weighted asset ratio. Still, there are important caveats on the interpretation of the Basel III result. Additional empirical research is needed to assess the impacts of Basel III on investment in low-carbon infrastructure.

Source: Ang, Röttgers and Burli (2017).

Integrating ESG factors into the governance of institutional investment

Regulation of institutional investment is increasingly focused on governance, as it moves away from quantitative constraints and towards risk-based controls and prudential standards (OECD forthcoming). Understanding ESG issues and the potential impact of ESG[48] factors on their investment strategy and the broader operating environment can be part of governance for institutional investors; in this context, climate change factors are increasingly being included in ESG investment practices. However, there is a wide range of definitions of ESG, so individual investors' circumstances and belief systems affect the uptake of ESG practices.

An OECD review found that regulatory frameworks in OECD and non-OECD countries rarely make explicit reference to ESG factors, although this is beginning to change.[49] Therefore it is up to institutional investors to decide whether and to what extent ESG integration is consistent with prudential standards, risk controls, legal requirements and any other obligations they may have towards their beneficiaries (OECD, 2017a). Similarly, risk-based controls generally do not explicitly refer to ESG or climate change factors. The focus is on solvency: pension funds and insurance companies are expected to identify, measure, and manage long-term risks and these are understood by both regulators and investors to be financial risks.

Box 7.6. Some countries have clarified the role of ESG in regulatory frameworks

In the United States, the Department of Labor confirmed that fiduciaries may legitimately consider ESG factors if they have a bearing on financial analysis and recognised that there has been an evolution in the data and methodologies that can be used in financial analysis. It also confirmed that fiduciaries may invest in "Economically Targeted Investments" (i.e. investments whose purpose is not purely financial) as long as the investment is otherwise appropriate for the plan and is financially and economically equivalent to competing investment choices.

In the United Kingdom, the Pensions Regulator published a new Defined Contribution Code and trustee guides in July 2016; these reflect the findings of the Law Commission's study of trustees' duties that there is no legal obstacle to taking ESG into account and they encourage trustees to take into account risks that affect the long-term sustainability of investments.

In South Africa, the 2011 Amendment to the Pension Funds Act states that "Prudent investing should give appropriate consideration to any factor which may materially affect the sustainable long-term performance of a fund's assets, including factors of an environmental, social and governance character."

Box 7.6. Some countries have clarified the role of ESG in regulatory frameworks *(cont.)*

The proposed revisions to the European Union's IORP (Institutions for Occupational Retirement Provision) Directive, which is expected to be passed by early 2019, go further: IORPII would make ESG integration a requirement of pension fund governance. The relevant articles of the proposed revisions are noteworthy for the extent to which they support the "ESG investor" interpretation of prudential standards, and for the influence of civil society in this aspect of the revision process.

Source: OECD (2017a).

Momentum is gathering to encourage institutional investors to clarify and disclose climate change risks in regard to investment portfolios, which is leading investors to seek more information on climate-related risks and opportunities, including through enhanced corporate disclosures. Disclosure also includes actions that investors are taking to mitigate climate change risk. Recent recommendations from the FSB's Task Force on Climate-Related Financial Disclosures (FSB-TCFD) provide a voluntary framework (FSB, 2016).

Taking stock of such initiatives, some countries have already put into place reporting requirements or voluntary disclosure of ESG practices by institutional investors. Australia requires pension funds, insurance companies and asset managers to disclose their ESG practices. France has introduced the most far-reaching requirements in terms of ESG reporting by institutional investors. Under Article 173-VI of the Energy Transition Act, asset managers, pension funds and insurance companies must provide information not only on how they integrate ESG factors in their investment and voting decisions but also on the climate risks they face and how their portfolio construction contributes to the transition to a low-carbon economy (OECD, 2017a). More countries had in place reporting requirements for pension funds, while insurance companies had fewer requirements.

In some countries, sub-national governments or regulatory bodies have passed laws that govern institutional investors' behaviour. The Canadian provinces of Alberta and Ontario have enacted regulations regarding the disclosure of climate change risks by pension funds. In 2015, California passed Senate Bill 185, which required the two largest state pension funds (also the largest public funds in the United States), CalPERS and CalSTRS, to divest holdings in publicly listed companies that generate more than 50% of their revenue from the mining of thermal coal. The legislation in California was designed to stem investment in industries particularly exposed to transition risks or "stranded assets". While funds worldwide are increasingly disclosing and reporting their ESG practices, few funds are measuring their stranded asset risk exposure or the overall carbon emissions (carbon footprint) of their portfolios. Policy makers could clarify that prudent investors may consider ESG criteria and climate change risk factors as part of their investment decision-making, especially investors with long-term liabilities and investment horizons.

Integrating environment and climate change considerations into operations of development banks

Approaches within development banks can be placed on a spectrum, ranging from establishing socio-environmental standards for risk management and decision-making to more holistic approaches that integrate socio-environmental criteria into performance management. Most bilateral development banks and finance institutions have adopted and/or developed ESG standards; however, banks vary widely in the ways they monitor, report and disclose climate risk and impact.

MDBs are the most advanced, having introduced and piloted climate risk screening tools to build climate resilience into the planning, design and implementation of projects. For example, AsDB projects are screened through a checklist, "at risk" projects are further screened using an online tool, and medium- and high-risk projects are subject to a more thorough Climate Risk and Vulnerability Assessment. The mid-term evaluation of AsDB's corporate strategy noted that the climate risk screening approach adopted by the bank is "well placed" but will need continued support, financial and technical, to ensure it is effective (AsDB, 2014a; AsDB, 2014b). Similarly, all country partnership frameworks and projects under the World Bank's concessional arm, the International Development Association (IDA), are subject to climate risk screening, with plans for this practice to be extended across the bank's non-concessional operations in 2017.

Nine international finance institutions – including major MDBs, bilateral DFIs and climate funds – have reached a framework agreement to harmonise their measuring and disclosing of their carbon impact at a project level, supported by sector-specific guidance.[50] While this is an important step in the right direction, it was developed to support the monitoring of mitigation projects – that is, to quantify GHG reductions. It requires that "at a minimum" banks report the emissions from "screened in" mitigation projects, but disclosure of "portfolio-wide net emissions" remains voluntary. Despite this, some banks are disclosing their carbon footprints. EBRD and IDB report on their portfolio-wide emissions as part of annual sustainability reports.

There is less harmonisation on disclosure of climate risks and impacts among NDBs. Most NDBs report environmental performance on a variety of metrics such as GHG emissions reduced and tonnes of coal consumption avoided. Besides the specialised green investment banks, few disclose metrics on the environmental impact of their portfolio or their portfolio's exposure to climate-related risks. BNDES (Brazil) stands out in this context, disclosing the environmental risk profile of its portfolio in amounts as well as number of projects (BNDES, 2016).

In December 2015, 26 development banks with total assets over USD 11 trillion, including five NDBs – TSKB (Turkey), IDBI Bank (India), KfW Group (Germany), DBSA (South Africa) and Caisse des Dépôts (France) – as well as some private sector finance institutions, adopted five voluntary mainstreaming principles to incorporate climate more holistically across their portfolios (EIB, 2015). While comprehensive and ambitious, the principles are voluntary and the extent to which they will be adopted is unclear.

Monitoring, reporting and verification of progress by development banks and DFIs to align their portfolios with efforts to keep average global warming to 2°C or below could be expected to help governments achieve their NDCs. Governments could call for more transparent disclosure of climate risks and impacts by development banks and DFIs, including efforts to monitor, disclose and report the climate impact/footprint of their overall portfolios, and in particular to what extent climate risks in the portfolio are being considered and addressed. A first step could be for organisations such as IDFC and EDFI to lead the way by helping their members to work collectively with other development banks and finance institutions to build on and tailor for their own purposes recommendations from the FSB-TCFD.

Disclosure of climate change risks of infrastructure assets and promoting infrastructure as an asset class

To improve transparency about the exposure of infrastructure assets to climate change-related risks, countries could promote or require embedding ESG criteria into reporting disclosures for infrastructure assets, including carbon emissions, potential contribution to country decarbonisation strategies, level of alignment to NDCs, potential contribution to the SDGs, energy and water use, social impacts and governance of infrastructure assets. Furthermore, countries could support initiatives to create infrastructure benchmarks that will in turn help to describe infrastructure as an asset class. This includes setting benchmarks that can measure climate change risk and the carbon intensity of assets. Promoting data collection could also include the consideration of a template for a preferred set of information to be collected and quantitative data on historical cash flows and performance at the project level.[51]

This fits more broadly with G20 work on long-term investment finance, which has repeated that there is a shortage of readily accessible, consistent and comparable data on investments (Chapter 3), and on the supply of and demand for long-term finance, on which to base policy analysis and conclusions. Promoting the development of infrastructure as an asset class and improving data and information could support more diversified and innovative financing of low-emission infrastructure. A potential outcome could be opening new channels of funds to low-emission, climate-resilient infrastructure. The findings may also support regulators in determining fair prices by appropriately including risk charges in the costs of capital. The same need to create new knowledge on the risks of long-term investment is also patent on the regulatory side. More accurate risk measures may require the adjustment of capital charges, and the more effective and efficient intermediation of long-term capital.

Box 7.7. Sustainability and ratings agencies

Increasingly, ratings agencies are recognising the importance of including climate change risks in long-term scenario analysis of rated debt instruments such as loans and bonds, in both infrastructure-related issuance and corporate issuance. Ratings agencies could take a pivotal role in recognising the materiality of climate change risks as financial risks, since firms with high perceived exposure would receive lower ratings on debt or be placed on watch lists, raising their cost of debt finance and incentivising businesses to address the effects of climate change on their business profile and profitability.

The signing and ratification of the Paris Agreement has facilitated the ability of the ratings agencies to model the risks of climate change, providing baseline scenarios and a benchmark to consider carbon transition risk in rated entities. Moody's, for instance, using a traditional credit process, focuses on the carbon regulatory impact on indicators such as business profile, leverage, liquidity, interest coverage, profitability and efficiency. Credit processes also consider the direct impact of climate change hazards on businesses, although such findings do not yet indicate a material impact on credit ratings, for the most part (Moody's, 2016).

Dagong's credit rating process regards environmental and social sustainability as fundamental to the construction and operational sustainability of infrastructure projects, which in turn determines the ability of a project to service its debt payments, and feeds in directly to ratings categories (Dagong, 2016).

Notes

1. The analysis of trends in this section draws on both project-based and mergers and acquisitions (M&A) data from commercial databases to provide evidence of finance trends in low-emission infrastructure sectors at disaggregated levels. Results should be interpreted with caution due to data-related gaps and challenges (see Annex 7.A1 for a detailed description of data used). Individual assets may not be labelled low-emission. The ability to track financing in energy efficiency is limited by a lack of data. The terms "investment" and "finance" of infrastructure are used interchangeably to reflect total capital value (stocks) of the projects and do not reflect gross fixed capital formation per se.
2. For the figures in this section, infrastructure sectors include power (including fossil fuel generation, transmission and distribution, and renewables), transport and water.
3. M&A refers to balance sheet and project finance transactions for corporate and individual asset acquisitions in all sectors.
4. As measured by the Thomson Reuters World Datastream Utilities Index.
5. "Pure equity investment" refers to capital invested in a renewable energy project where the output is not purchased or consumed by the financing entity. Some corporations have used this model to "offset" their conventional electricity use in other locations, reducing their emissions profile.
6. Data presented is from the 2015 OECD Survey of Large Pension Funds and Public Pension Reserve Funds. Funds from 36 countries were surveyed, including OECD countries, G20 countries, and beyond.
7. The composition of private pension markets – principally whether systems are predominately defined benefit or defined contribution, can influence the types of finance available within local markets. For example, defined benefit funds are able to invest in illiquid assets such as direct infrastructure, while defined contribution plans may have certain liquidity requirements that make it difficult to invest in illiquid assets.
8. Defined as early-stage equity investment in clean technologies and businesses that carry a substantial element of risk, as commercialisation and development of technologies may not be well proven.
9. For the purposes of this section, clean energy includes solar, wind, energy efficiency, green transportation, and advanced material and technologies.
10. Within the clean technology space, hardware may refer to, inter alia, electronics, solar panels and small-scale power stations; materials may refer to, inter alia, nanotechnologies, chemicals, biological materials and membranes.
11. Preqin 2017 Global Private Equity and Venture Capital Report.
12. In this chapter, development banks and finance institutions refer to publicly owned finance institutions with a development/policy-related mandate. National development banks refer to those that primarily work in a domestic context. Bilateral development banks support the development co-operation policy of a country and work in developing countries (supporting both public and private sector activities), and bilateral development finance institutions (DFIs) are agencies set up specifically to work with the private sector in developing countries.
13. MDBs, for example, maintain strong credit ratings due to the support of their shareholders, which allows them to borrow resources from private capital markets at attractive rates and on-lend these resources to developing countries with enough margin to cover administrative costs (Humphrey, 2015).
14. This includes NDBs with a specific infrastructure mandate, as well as those with broader industrial and other development mandates (which also cover some infrastructure financing).
15. Examples of sub-national development banks include NRW Bank in Germany, Banco do Nordeste and Banco de Desenvolvimento de Minas Gervais (BDMG) in Brazil. Examples of agriculture related national development banks include the Agricultural Bank of China, the National Bank for Agriculture and Rural Development in India, Financiera Rural in Mexico, Russian Agricultural Bank and the Land and Agricultural Development Bank of South Africa.
16. The Small Industries Development Bank of India (SIDBI), which targets SMEs; the Industrial Finance Corporation of India (IFCI), which caters to the long-term finance needs of the industrial sector, the India Infrastructure Finance Company Limited (IIFCL) and Infrastructure Development Finance Company, whose focus is on infrastructure, the National Housing Bank (NHB), which promotes and finances housing and the National Bank for Agriculture and Rural Development (NABARD) which focuses on agriculture.
17. OECD (2017) defines green investment banks (GIBs) as "publicly capitalised entities established specifically to facilitate private investment into domestic low-carbon, climate-resilient infrastructure and other green sectors such as water and waste management".

18. IDFC is a network of development banks. As of 2017 its membership includes 23 banks, including: national development banks: Brazil (BNDES), Chile (BE), Peru (COFIDE), Columbia (Bancoldex), Mexico (NAFIN), Morocco (CDG), South Africa (DBSA), Burundi (PTA), China (CDB), India (SIDBI), Indonesia (Eximbank), Korea (KDB), Croatia (HBOR), Germany (KfW), Russia (VEB) and Turkey (TSKB); regional development banks: Central American Bank for Economic Integration (BCIE), Development Bank of Latin America (CAF), Banque Ouest Africaine de Développement (BOAD), Black Sea Trade and Development Bank (BSTD), Islamic Corporation for the Development of the Private Sector (ICD); and bilateral development finance providers: France (AFD), Japan (JICA).
19. According to the IDFC methodology, "green finance" includes climate finance as well as finance for "other environmental objectives", such as environmental protection, remediation and biodiversity.
20. The definitions of "green" and "environmental" vary from bank to bank, making it difficult to compare banks. For instance, for TSKB "sustainability-themed" includes renewable energy, energy and resource efficiency, sustainable tourism and APEX loans.
21. KfW Group's domestic activities are broken down into SME banking and municipal development. The KfW SME banking dedicated 45% of its financing in 2014 to environmentally friendly activities such as environmental protection, renewable energy and energy efficiency. This included two lines of credits for demonstration projects in green finance: the KfW-BMUB Green Innovation Programme and the KfW-EU NER 300 funding programme. With regard to KfW's municipal banking activities, 64% of financing for housing development goes towards energy-efficient construction and refurbishment, while only 6.2% of infrastructure financing is considered environmentally friendly.
22. By the end of 2014, CDB's outstanding green credit loans – which include loans to environmental protection, energy conservation and emissions reduction ¬– were RMB 958.5 billion, whereas total loans to coal-related projects were RMB 174.5 billion (CDB, 2014).
23. Public climate finance for 2013-14 (on average) included USD 22.8 billion from bilateral sources, USD 17.9 billion from multilateral sources (of which USD 15.5 billion was from the MDBs) and USD 1.6 billion from export credits (OECD, 2015b).
24. MDB climate finance includes own resources and MDB-managed external resources. The share represents MDB Climate Finance as a percentage of total MDB operations (i.e. MDB internal resources and MDB-managed external resources) as reported in the joint MDB report (2015 Joint Report on Multilateral Development Banks).
25. MDBs' Joint Declaration of Aspirations on Actions to Support Infrastructure Investment, available at www.g20chn.org/English/Documents/Current/201608/P020160815360318908738.pdf
26. Most MDBs were originally set up to deliver infrastructure financing as a way of supporting social and economic development, and delivering poverty reduction.
27. This includes EIB, EBRD, WBG (including IFC), AsDB, AfDB, IDB and IsDB.
28. MDB Response to the G20 Action Plan for MDB Balance Sheet Optimisation July 2016, available at www.g20chn.org/English/Documents/Current/201608/P020160815361155807206.pdf.
29. The private sector window will be introduced in the funding envelope for International Development Association (IDA) which provides grants and low interest loans to low-income countries.
30. Within this context, the CIFs have played a particularly influential role in the MDBs' support for climate action. Initiated in 2008, the CIFs are a multilateral climate fund with a budget of approximately USD 8 billion. A key feature of the CIFs is that their support is programmed and implemented by MDBs, in contrast to other climate funds which are implemented by a range of different entities e.g. UN agencies, governments, local financing institutions as well as MDBs and other DFIs (Nakhooda et al., 2016). The CIFs accounted for just under half the external concessional climate finance implemented by the MDBs in 2013-14 (Trabacchi et al., 2016).
31. The analysis is based on OECD-DAC data reported by DAC members as well as several non-DAC members along with estimates of ODA-like flows from other countries, including Brazil, Chile, China, Colombia, Costa Rica, India, Indonesia, Mexico, Qatar and South Africa. For a full list of DAC and non-DAC member countries that report to the OECD, please see http://www.oecd.org/development/stats/non-dac-reporting.htm.
32. To ensure conservative estimates, this includes AFD, KfW and JICA support for infrastructure for which climate was the principal objective of the project only. If activities where climate change is a significant objective of the project are considered, the shares of infrastructure financing in 2013-15 that could be considered climate-related are much higher: 68% of AFD, 89% for KfW and 67% for JICA. Bilateral development providers report on climate-related development finance to OECD-DAC using the Rio markers approach, through which each activity is marked as to whether climate change mitigation and/or adaptation is the "principal" or "significant" objective.

33. The OECD issued a survey in 2016 on renewable energy finance, with institutional investors, asset managers, and corporations responding to questions regarding the financing environment of renewable energy.
34. Instances where certain risks are transferred from the private sector to the public sector that dis-incentivises the private sector to manage risks. For example, a public guarantee on project debts may dis-incentivise debt holders to monitor the project entity.
35. Any risk mitigation instruments, particularly examples where concessional financing is provided, should try to limit potential distortionary effects on market competition.
36. Please see www.ggf.lu/about-green-for-growth-fund/institutional-structure/.
37. See https://www.adb.org/site/funds/funds/asia-pacific-project-preparation-facility.
38. For detailed recommendations delivered to the G20 in 2016 on diversifying financial instruments for the financing of infrastructure, refer to the Guidance Note.
39. Refer also to the G20/OECD High-level Principles of Long-term Investment Financing by Institutional Investors for greater details on the enabling environment for infrastructure finance.
40. For more on yieldcos, refer to Chapter 5 of the OECD Business and Finance Outlook 2016
41. The Climate Bond Initiative (CBI) is working towards labelling green bonds, and also tracks the development of unlabelled bonds which are climate-aligned.
42. http://www.renewableenergyworld.com/articles/2016/01/wind-solar-secure-11-5-billion-in-tax-equity-deals-in-2015.html.
43. A 2016 OECD survey put questions to institutional investors, asset managers and corporations about the financing environment of renewable energy.
44. For example, a scientific advisory committee to the European Systematic Risk Board recently recommended that future stress tests of the pensions sector include climate-related risks.
45. Geoff Summerhayes, Executive Board Member of the Australian Prudential Regulation Authority, 17 February 2017, "Australia's New Horizon: Climate Change Challenges and Prudential Risk"
46. As of December 2016, 86 financial institutions have officially adopted the Equator Principles covering project finance debt, which reportedly covers 70% of international project finance debt in emerging markets
47. The NCFA consists of 30 signatories (financial institutions).
48. ESG integration is defined as: the recognition in the institutional investor's investment policy or principles that ESG factors (of which climate change may be included) may impact portfolio performance and so affect the investor's ability to meet its obligations; and using analysis of those impacts to inform asset allocation decisions and securities valuation models (or employing third parties to do so) (OECD, 2017a).
49. See OECD publication "Investment Governance and the Integration of ESG Factors" for the full study.
50. International Financial Institution Framework for a Harmonized Approach to Greenhouse Gas Accounting (2015), available at https://www.thegef.org/sites/default/files/file_attach/IFI-Harmonisation-Framework-GHG%20Accounting-2015.pdf.
51. Building on current work developed by Global infrastructure Hub (GIH), EDHEC (Ecole des Hautes Etudes Commerciales) and the OECD, and on note circulated to the G20 in 2015 on Addressing Data Gaps in Long-term Investment.

References

AsDB (2015), "Frequently Asked Questions: Enhancing ADB's Financial Capacity by Up to 50% for Reducing Poverty in Asia and the Pacific: Combining ADB's ADF OCR Resources", webpage, Asian Development Bank, Manila, www.adb.org/news/features/frequently-asked-questions-enhancing-adbs-financial-capacity-50-reducing-poverty-asia (accessed on 28 March 2017).

AsDB (2014a), "Climate Risk Management in ADB Projects", Asian Development Bank, Manila, webpage, www.adb.org/sites/default/files/publication/148796/climate-risk-management-adb-projects.pdf, (accessed on 28 March 2017).

AsDB (2014b), "Real-Time Evaluation of ADB's Initiatives to Support Access to Climate Finance", Asian Development Bank, Manila, webpage, www.adb.org/sites/default/files/evaluation-document/111028/files/in109-14.pdf, (accessed on 28 March 2017).

Ahmad, E. (2015), "Public Finance Underpinnings for Infrastructure Financing in Developing Countries", *Infrastructure Finance in the Developing World Working Paper Series, Intergovernmental Group of Twenty-Four and Global Green Growth Institute, Washington DC,* www.g24.org/wp-content/uploads/2016/05/MARGGK-WP05.pdf.

Ang, G., D. Röttgers and P. Burli (2017), "The Empirics of Enabling Investment and Innovation in Renewable Energy", OECD Environment Working Papers, OECD Publishing, Paris, http://dx.doi.org/10.1787/19970900.

Birdsall, N., S. Morris and E. Rueda-Sabater (2015), "The ADB's Bold Move and What It Can Mean for the MDBs", Center for Global Development, Washington DC, www.cgdev.org/publication/adbs-bold-move-and-what-it-can-mean-mdbs.

BNEF (2016), "Corporate Renewable Energy Procurement Monthly", December, Bloomberg New Energy Finance, https://about.bnef.com/blog/corporate-renewable-energy-procurement-monthly-december/.

BNDES (2016), "Annual Report 2015", Banco Nacional de Desenvolvimento Econômico e Social, June.

Carney, M. (2016), "Resolving the climate paradox", Arthur Burns Memorial Lecture, Berlin, 22 September 2016, http://www.fsb.org/wp-content/uploads/Resolving-the-climate-paradox.pdf.

CBI (2016), "Bonds and Climate Change: The state of the market in 2016", Climate Bond Initiative, www.climatebonds.net/resources/publications/bonds-climate-change-2016.

CDB (2014), *2014 Annual Report, China Development Bank, Beijing,* www.cdb.com.cn/English/gykh_512/ndbg_jx/2014/.

CEP (2016), Global Clean Energy Review Analytics and Graphs, Clean Energy Pipeline, 2016

Climate Home (2016), "Brazil shifts funds from coal to solar power", Climate Home, London, www.climatechangenews.com/2016/10/04/brazil-shifts-funds-from-coal-to-solar-power/.

Christianson, G. et al. (2016), "Financing the Carbon Transition: A Closer Look at How the World Bank, International Finance Corporation, and Asian Development Bank Invest in Energy Supply," World Resources Institute, Washington DC.

Dagong (2016), "Dagong Global Infrastructure Credit Rating Methodology", Dagong Global Credit Rating, Beijing.

Della Croce, R. and R. Sharma (2014), "Pooling of Institutional Investors Capital – Selected Case Studies in Unlisted Equity Infrastructure", OECD G20/OECD Task Force Working Paper, OECD Publishing, Paris, www.oecd.org/finance/OECD-Pooling-Institutional-Investors-Capital-Unlisted-Equity-Infrastructure.pdf.

DLA Piper (2016), *2016: The Year of PPAs and the Corporate Green Agenda*, DLA Piper, www.dlapiper.com/~/media/Files/Insights/Publications/2016/06/PPA_Paper_Energy_2016_Jul.pdf.

EIB (2015), "Mainstreaming climate action within financial institutions", webpage, www.eib.org/about/global-cooperation/climate/fi-climate-mainstreaming.htm (accessed 12 December 2016).

Frisari, G. et al. (2013), *Risk Gaps: A Map of Risk Mitigation Instruments for Clean Investments, Climate Policy Initiative, San Francisco,* http://climatepolicyinitiative.org/wp-content/uploads/2013/01/Risk-Gaps-A-Map-of-Risk-Mitigation-Instruments-for-Clean-Investments.pdf.

FSB (2016), "Recommendations of the Task Force on Climate-related Financial Disclosures", Financial Stability Board Task Force on Climate-related Financial Disclosures, Basel, www.fsb-tcfd.org/publications/recommendations-report/.

G20/OECD (2016), "Support Note on Diversification of Financial Instruments for Infrastructure", OECD Publishing, Paris, www.oecd.org/daf/fin/private-pensions/G20-OECD-Support-Note-Diversification-Financial-Instruments.pdf.

Gaddy, B., V. Sivaram and F. O'Sullivan (2016), "Venture and Cleantech: The Wrong Model for Clean Energy", MIT Energy Initiative, Cambridge, Massachusetts, http://energy.mit.edu/publication/venture-capital-cleantech/.

Humphrey, C. (2015), "Challenges and Opportunities for Multilateral Development Banks in 21st Century Infrastructure Finance", Global Green Growth Institute, Seoul, and Intergovernmental Group of Twenty Four on Monetary Affairs and Development (G-24), Washington DC, http://g24.org/wp-content/uploads/2016/05/MARGGK-WP08.pdf.

IDFC (2014), "IDFC Green Finance Mapping For 2014", International Development Finance Club, Frankfurt, www.idfc.org/Downloads/Publications/01_green_finance_mappings/IDFC_Green_Finance_Mapping_Report_2015.pdf.

IEA (2016), *World Energy Investment 2016*, IEA/OECD Publishing, Paris.

IFC (2016), "IFC Climate Implementation Plan", International Finance Corporation, Washington DC.

IMF (2017), "Macroeconomic Developments and Prospects in Low Income Developing Countries", International Monetary Fund, Washington DC.

KfW (2015a), Annual Report 2015, KfW, Frankfurt.

KfW (2015b), KfW Group guidelines on the financing of coal-fired power plants, KfW, Frankfurt www.kfw.de/nachhaltigkeit/PDF/Nachhaltigkeit/KfW-Guidelines_Coal_Financing_2015-03-17_EN.pdf.

Kumar, N. (2016), "National Development Banks and Sustainable Infrastructure in South Asia", Global Economic Governance Initative, Boston University, www.bu.edu/pardeeschool/files/2016/07/NDBSandSouthAsia.Final_.pdf.

Jones, A. (2016), "Bridging the Climate Finance Gap", in *Road from Paris: Ensuring Effective and Equitable Climate Action*, Observer Research Foundation, New Delhi, http://cf.orfonline.org/wp-content/uploads/2016/09/Road-from-Paris.pdf.

Marathon (2016), "The Solar Securitization Opportunity", Marathon Capital, LLC, March 2016, http://www.marathon-cap.com/resources/recent-news/news-item/2016/03/14/the-solar-securitization-opportunity.

Miyamoto, K. and E. Chiofalo (2016), "Official Development Finance for Infrastructure: With a Special Focus on Multilateral Development Banks", OECD Publishing, Paris, http://dx.doi.org/10.1787/9e93790e-en.

Moody's (2016), "Moody's to Analyse Carbon Transition Risk Based on Emissions Reduction Scenario Consistent with Paris Agreement", Moody's Investor Service, New York, www.eenews.net/assets/2016/06/29/document_cw_01.pdf.

Moody's (2015), "2016 Outlook – Solar Securitisation Market to Grow in 2016 as Rooftop Solar Costs Continue to Drop", Moody's Investor Service, New York.

Nakhooda, S. et al. (2016), "The Global Climate Finance Architecture", Overseas Development Institute, London, www.odi.org/sites/odi.org.uk/files/resource-documents/11021.pdf.

Nassiry, D. et al. (2016), "Finding the pipeline: Project preparation for sustainable infrastructure", Overseas Development Institute, London, www.odi.org/sites/odi.org.uk/files/resource-documents/11075.pdf.

NCE (2016), *The Sustainable Infrastructure Imperative: Financing for Better Growth and Development, New Climate Economy/Global Commission on the Economy and Climate*, http://newclimateeconomy.report/2016/.

NCFA (2015), "Integrating Water Stress into Corporate Bond Credit Analysis", Natural Capital Finance Alliance, Geneva, http://naturalcapitaldeclaration.org/documents/wgi/INTEGRATING%20WATER%20STRESS%20REPORT_FINAL.pdf.

OECD (2017a), *Investment Governance and the Integration of Environmental, Social and Governance Factors, OECD Publishing, Paris*, www.oecd.org/finance/Investment-Governance-Integration-ESG-Factors.pdf.

OECD (2017b), *Mobilising Bond Markets for a Low-Carbon Transition, OECD Publishing, Paris*, DOI: http://dx.doi.org/10.1787/9789264272323-en.

OECD (2016a), "Chapter 5: Fragmentation in Clean Energy Investment and Financing", *OECD Business and Finance Outlook 2016, OECD Publishing, Paris*, http://dx.doi.org/10.1787/9789264257573-en.

OECD (2016b), *Annual Survey on Large Pension Funds and Public Pension Reserve Funds: Report on Pension Funds' Long-Term Investments 2015, OECD Publishing, Paris*, www.oecd.org/daf/fin/private-pensions/2015-Large-Pension-Funds-Survey.pdf.

OECD (2016c), *2020 Projections of Climate Finance Towards the USD 100 Billion Goal: Technical Note*, OECD Publishing, Paris, www.oecd.org/environment/cc/Projecting%20Climate%20Change%202020%20WEB.pdf.

OECD (2016d), *Green Investment Banks: Innovative Public Financial Institutions Scaling up Private, Low-carbon Investment*, OECD Publishing, Paris, DOI: http://dx.doi.org/10.1787/9789264245129-en.

OECD (2015a), "Chapter 3: Shifting Risks and the Search for Yield in Financial Markets", *OECD Business and Finance Outlook 2015, OECD Publishing, Paris,* http://dx.doi.org/10.1787/9789264234291-en.

OECD (2015b), *Climate Finance in 2013-14 and the USD 100 Billion Goal: A report by the OECD in collaboration with Climate Policy Initiative, OECD Publishing, Paris,* http://dx.doi.org/10.1787/9789264249424-en.

OECD (2015c), *Infrastructure Financing Instruments and Incentives*, OECD Publishing, Paris, http://www.oecd.org/finance/private-pensions/Infrastructure-Financing-Instruments-and-Incentives.pdf.

OECD (2015d), *Mapping Channels to Mobilise Institutional Investment in Sustainable Energy, OECD Publishing, Paris,* http://dx.doi.org/10.1787/9789264224582-en.

OECD (2012), *Towards a Green Investment Policy Framework: The Case of Low-emission, Climate-resilient Infrastructure, OECD Publishing, Paris,* http://dx.doi.org/10.1787/5k8zth7s6s6d-en.

OECD (n.d.), "Mobilising Private Investment in Sustainable Transport Infrastructure", www.oecd.org/env/cc/financing-transport-brochure.pdf.

OECD/ITF (2016), *Shared Mobility: Innovation for Livable Cities, Corporate Partnership Board Report, OECD Publishing, Paris,* https://www.itf-oecd.org/sites/default/files/docs/shared-mobility-liveable-cities.pdf.

OECD/World Bank (2015), *Risk and Return Characteristics of Infrastructure Investment in Low-income Countries, OECD Publishing, Paris,* https://www.oecd.org/g20/topics/development/Report-on-Risk-and-Return-Characteristics-of-Infrastructure-Investment-in-Low-Income-Countries.pdf.

Patwardhan, A. (2016), "The Technology Challenge", in *Road from Paris: Ensuring Effective and Equitable Climate Action*, Observer Research Foundation, New Delhi, http://cf.orfonline.org/wp-content/uploads/2016/09/Road-from-Paris.pdf.

Sharif, D, A. Grace and M. Di Capua, "The return – and returns – of tax equity US renewable energy projects", Bloomberg New Energy Finance, US Clean Energy - White Paper, 21 November, https://about.bnef.com/blog/the-return-and-returns-of-tax-equity-for-us-renewable-projects/.

Smallridge, D. et al. (2013), "The Role of National Development Banks in Catalysing International Climate Finance", Inter-American Development Bank, Washington DC, https://publications.iadb.org/handle/11319/3478.

Studart, R. and K.P. Gallagher (2016), "Infrastructure for Sustainable Development: The role of National Development Banks", Global Economic Governance Initative, Boston University, www.bu.edu/pardeeschool/files/2016/08/Infrastructure.Sustainable.Final_.pdf.

Torres, E. and R. Zeidan (2016), "The life-cycle of national development banks: The experience of Brazil's BNDES", *The Quarterly Review of Economics and Finance, Vol. 62, Elsevier, Amsterdam, pp. 97-104,* https://doi.org/10.1016/j.qref.2016.07.006.

Trabacchi et al. (2016), "The role of Climate Investment Funds in meeting investment needs", Climate Policy Initiative, San Francisco, https://climatepolicyinitiative.org/wp-content/uploads/2016/06/The-role-of-the-Climate-Investment-Funds-in-meeting-investment-needs.pdf.

TSKB (2015), *TSKB Sustainability Report 2015*, Industrial Development Bank of Turkey, Istanbul.

UNDP (2015), "Derisking Renewable Energy Investment", United Nations Development Programme, New York, www.undp.org/content/undp/en/home/librarypage/environment-energy/low_emission_climateresilientdevelopment/derisking-renewable-energy-investment.html.

UNEP (2016a), "Greening the Banking System: Taking Stock of G20 Green Banking Market Practice", United Nations Environment Programme, Geneva, http://wedocs.unep.org//handle/20.500.11822/10604.

UNEP (2016b), "Greening Banking Policy", United Nations Environment Programme, Geneva, http://unepinquiry.org/wp-content/uploads/2016/09/10_Greening_Banking_Policy.pdf.

Vajjhala, S. and J. Rhodes (2015), "Leveraging Catastrophe Bonds as a Mechanism for Resilient Infrastructure Project Finance", Rebound: Insuring for Resilience, La Jolla, California, www.refocuspartners.com/reports/RE.bound-Program-Report-December-2015.pdf.

World Bank (2016), *World Bank Group Climate Change Action Plan,* http://pubdocs.worldbank.org/en/677331460056382875/WBG-Climate-Change-Action-Plan-public-version.pdf.

World Bank (2013), *Toward a Sustainable Energy Future for All: Directions for the World Bank Group's Energy Sector,* http://documents.worldbank.org/curated/en/745601468160524040/pdf/795970SST0SecM00box377380B00PUBLIC0.pdf.

Annex 7.A1

About the Database

The main source for the trends analysis is the IJGlobal transaction database. Between 2010 and 2016, 4 596 projects were included across the transport, power, renewables, oil and gas, water and telecoms sectors, for a total volume of USD 2.4 trillion. Deals in which the SPV or the sponsor is 100% state-owned are not eligible for project or corporate finance ranking. The figures presented refer to financing activity and not to capital expenditure.

The study included the use of two different databases: IJGlobal Transactions and Bloomberg New Energy Finance. IJGlobal Transactions data was used to study financing trends in the transport, power and water sectors. The transport sector refers to roads, ports, transit infrastructure, tunnels, maritime transport infrastructure, heavy rail, bridges and airports. The power sector refers to transmission and distribution, coal-fired power plants, gas-fired power plants, oil-fired power plants, independent water and power projects, hydro, carbon capture and storage, and co-generation. The water sector refers to water treatment, distribution and desalination.

For the renewable energy sector, the study of financing trends relied on the Bloomberg New Energy Finance database. The sectors included are wind, small hydro, solar, biomass, biofuels, geothermal, digital energy, energy storage, fuel cells, efficiency, carbon markets, hydrogen, and energy retail and supply.

The study only included **private sector related** transactions reaching financial closure; therefore, the dates refer to the financial close date.

Some transactions' proceeds target several regions. When it was impossible to identify the main region the financing targeted, the transaction was not included in the analysis of the regional infrastructure trends.

The study only included **transactions with a reported transaction amount**; no estimates have been performed for transactions with unknown amounts.

Only transactions with a clear capital structure description (debt vs. equity) were included; tracking and spotting trends in the transactions financing mix was among the main objectives of the analysis. The study deliberately excluded transactions with a reported amount but an unreported capital structure.

For the analysis of the infrastructure-related syndicated loans trends, the Thomson Reuters syndicated loans database was used. The study included the following sectors: alternative energy sources, power, water, waste water, and waste management, internet infrastructure and transport infrastructure.

Thomson Reuters was also used to show trends in terms of cost of financing for infrastructure-related syndicated loans. The margins over the benchmark presented for each region were weighted by transactions' sizes.

Table A7.1. **Categories included in the financing types for each database**

	IJGlobal Transactions	Bloomberg New Energy Finance
Primary Financing	Project Finance – Primary Financing	New Built
	Corporate Finance– Primary Financing	
Refinancing	Project Finance – Primary Financing	Refinancing
	Corporate Finance– Primary Financing	
M&A	Project Finance – Asset Acquisition / Corporate Acquisition	Corporate M&A
	Corporate Finance – Asset Acquisition / Corporate Acquisition	Asset Finance - Acquisition

Data include the roll-out phase on **primary financing** (financing of primary assets/ projects) and **secondary market activities** not associated with new activity, including investment projects that do not contribute directly to new assets or company financing, such as corporate mergers and acquisitions (M&As) and asset refinancing and acquisitions. Main categories used are:

- **Project Finance:** a single-purpose infrastructure asset or portfolio financed with commercial debt on a non-recourse or limited recourse basis. The transaction is secured on the project's long-term future cash flows and assets of the project or target company (SPV). Acquisitions financed with this structure and associated with new and existing infrastructure assets – such as the transfer or sale of assets or an asset-based holding company – will be included in project finance.
- **Corporate Infrastructure Finance:** transactions related to the general development of infrastructure and not classified as non-recourse and limited recourse project finance. This includes hybrid finance with recourse to corporate balance sheets and corporate loans made to companies that own and/or operate assets. Corporate Infrastructure Finance also includes mergers and acquisitions of companies that own and/or operate assets such as vertically integrated utilities with retail businesses and other companies that cannot be valued on assets alone. Mergers and acquisitions can be financed on-balance sheet or with commercial debt guaranteed by the sponsor.

ORGANISATION FOR ECONOMIC CO-OPERATION AND DEVELOPMENT

The OECD is a unique forum where governments work together to address the economic, social and environmental challenges of globalisation. The OECD is also at the forefront of efforts to understand and to help governments respond to new developments and concerns, such as corporate governance, the information economy and the challenges of an ageing population. The Organisation provides a setting where governments can compare policy experiences, seek answers to common problems, identify good practice and work to co-ordinate domestic and international policies.

The OECD member countries are: Australia, Austria, Belgium, Canada, Chile, the Czech Republic, Denmark, Estonia, Finland, France, Germany, Greece, Hungary, Iceland, Ireland, Israel, Italy, Japan, Korea, Latvia, Luxembourg, Mexico, the Netherlands, New Zealand, Norway, Poland, Portugal, the Slovak Republic, Slovenia, Spain, Sweden, Switzerland, Turkey, the United Kingdom and the United States. The European Union takes part in the work of the OECD.

OECD Publishing disseminates widely the results of the Organisation's statistics gathering and research on economic, social and environmental issues, as well as the conventions, guidelines and standards agreed by its members.

OECD PUBLISHING, 2, rue André-Pascal, 75775 PARIS CEDEX 16
(97 2017 06 1 P) ISBN 978-92-64-27351-1 – 2017

www.ingramcontent.com/pod-product-compliance
Lightning Source LLC
Chambersburg PA
CBHW061201240926
51846CB00006B/1080
* 9 7 8 9 2 6 4 2 7 3 5 1 1 *